T0191737

Cyber Deception

Sushil Jajodia • V.S. Subrahmanian
Vipin Swarup • Cliff Wang
Editors

Cyber Deception

Building the Scientific Foundation

Springer

Editors
Sushil Jajodia
Center for Secure Information Systems
George Mason University
Fairfax, VA, USA

V.S. Subrahmanian
Department of Computer Science
University of Maryland
College Park, MD, USA

Vipin Swarup
The MITRE Corporation
McLean, VA, USA

Cliff Wang
Computing & Information Science Division
Information Sciences Directorate
Triangle Park, NC, USA

ISBN 978-3-319-81349-3   ISBN 978-3-319-32699-3   (eBook)
DOI 10.1007/978-3-319-32699-3

Preface

This volume is designed to take a step toward establishing scientific foundations for cyber deception. Here we present a collection of the latest basic research results toward establishing such a foundation from several top researchers around the world. This volume includes papers that rigorously analyze many important aspects of cyber deception including the incorporation of effective cyber denial and deception for cyber defense, cyber deception tools and techniques, identification and detection of attacker cyber deception, quantification of deceptive cyber operations, deception strategies in wireless networks, positioning of honeypots, human factors, anonymity, and the attribution problem. Further, we have made an effort to not only sample different aspects of cyber deception, but also highlight a wide variety of scientific techniques that can be used to study these problems.

It is our sincere hope that this volume inspires researchers to build upon the knowledge we present to further establish scientific foundations for cyber deception and ultimately bring about a more secure and reliable Internet.

Fairfax, VA, USA   Sushil Jajodia
College Park, MD, USA   V.S. Subrahmanian
McLean, VA, USA   Vipin Swarup
Triangle Park, NC, USA   Cliff Wang

Acknowledgments

We are extremely grateful to the numerous contributors to this book. In particular, it is a pleasure to acknowledge the authors for their contributions. Special thanks go to Susan Lagerstrom-Fife, senior publishing editor at Springer for her support of this project. We also wish to thank the Army Research Office for their financial support under the grant numbers W911NF-14-1-0116, W911NF-15-1-0576, and W911NF-13-1-0421.

Contents

Integrating Cyber-D&D into Adversary Modeling for Active Cyber Defense

Frank J. Stech, Kristin E. Heckman, and Blake E. Strom

Abstract This chapter outlines a concept for integrating cyber denial and deception (cyber-D&D) tools, tactics, techniques, and procedures (TTTPs) into an adversary modeling system to support active cyber defenses (ACD) for critical enterprise networks. We describe a vision for cyber-D&D and outline a general concept of operation for the use of D&D TTTPs in ACD. We define the key elements necessary for integrating cyber-D&D into an adversary modeling system. One such recently developed system, the Adversarial Tactics, Techniques and Common Knowledge (ATT&CK™) Adversary Model is being enhanced by adding cyber-D&D TTTPs that defenders might use to detect and mitigate attacker tactics, techniques, and procedures (TTPs). We describe general D&D types and tactics, and relate these to a relatively new concept, the cyber-deception chain. We describe how defenders might build and tailor a cyber-deception chain to mitigate an attacker's actions within the cyber attack lifecycle. While we stress that this chapter describes a concept and not an operational system, we are currently engineering components of this concept for ACD and enabling defenders to apply such a system.

Traditional approaches to cyber defense increasingly have been found to be inadequate to defend critical cyber enterprises. Massive exploitations of enterprises,

---

The original version of this chapter was revised. An erratum to this chapter can be found at DOI 10.1007/978-3-319-32699-3_13

Authors: Frank J. Stech, Kristin E. Heckman, and Blake E. Strom, the MITRE Corporation (stech@mitre.org, kheckman@mitre.org, and bstrom@mitre.org). Approved for Public Release; Distribution Unlimited. Case Number 15-2851. The authors' affiliation with The MITRE Corporation is provided for identification purposes only, and is not intended to convey or imply MITRE's concurrence with, or support for, the positions, opinions or viewpoints expressed by the authors. Some material in this chapter appeared in Kristin E. Heckman, Frank J. Stech, Ben S. Schmoker, Roshan K. Thomas (2015) "Denial and Deception in Cyber Defense," *Computer*, vol. 48, no. 4, pp. 36–44, Apr. 2015. http://doi.ieeecomputersociety.org/10.1109/MC.2015.104

F.J. Stech (✉) • K.E. Heckman • B.E. Strom
MITRE Corporation, Mclean, VA, USA
e-mail: stech@mitre.org; kheckman@mitre.org; bstrom@mitre.org

© Springer International Publishing Switzerland 2016
S. Jajodia et al. (eds.), *Cyber Deception*, DOI 10.1007/978-3-319-32699-3_1

1

commercial (e.g., Target[1]) and government (e.g., OMB[2]), demonstrate that the cyber defenses typically deployed over the last decade (e.g., boundary controllers and filters such as firewalls and guards, malware scanners, and intrusion detection and prevention technologies) can be and have been bypassed by sophisticated attackers, especially the advanced persistent threats (APTs[3]). Sophisticated adversaries, using software exploits, social engineering or other means of gaining access, infiltrate these defended enterprises, establish a persistent presence, install malware and backdoors, and exfiltrate vital data such as credit card records, intellectual property and personnel security information. We must assume, then, that an adversary will breach border defenses and establish footholds within the defender's network. We must also assume that a sophisticated adversary will learn from and attempt to evade technology-based defenses, so we need new ways to engage the adversary on the defender's turf, and to influence the adversary's moves to the defender's advantage. One such means of influence is deception, and we argue a key component in the new paradigm of active cyber defense[4] is cyber denial and deception (cyber-D&D).

---

[1] Jim Walter (2014) "Analyzing the Target Point-of-Sale Malware," McAfee Labs, Jan 16, 2014. https://blogs.mcafee.com/mcafee-labs/analyzing-the-target-point-of-sale-malware/ and Fahmida Y. Rashid (2014) "How Cybercriminals Attacked Target: Analysis," *Security Week*, January 20, 2014. http://www.securityweek.com/how-cybercriminals-attacked-target-analysis

[2] Jim Sciutto (2015) OPM government data breach impacted 21.5 million," *CNN*, July 10, 2015. http://www.cnn.com/2015/07/09/politics/office-of-personnel-management-data-breach-20-million/ Jason Devaney (2015) "Report: Feds Hit by Record-High 70,000 Cyberattacks in 2014," *NewsMax*, 04 Mar 2015. http://www.newsmax.com/Newsfront/cyberattacks-Homeland-Security-Tom-Carper-OMB/2015/03/04/id/628279/

[3] Advanced persistent threats (APTs) have been defined as "a set of stealthy and continuous computer hacking processes, often orchestrated by human(s) targeting a specific entity. APT usually targets organizations and/or nations for business or political motives. APT processes require a high degree of covertness over a long period of time. The "advanced" process signifies sophisticated techniques using malware to exploit vulnerabilities in systems. The "persistent" process suggests that an external command and control system is continuously monitoring and extracting data from a specific target. The "threat" process indicates human involvement in orchestrating the attack." https://en.wikipedia.org/wiki/Advanced_persistent_threat A useful simple introduction and overview is Symantec, "Advanced Persistent Threats: A Symantec Perspective—Preparing the Right Defense for the New Threat Landscape," no date. http://www.symantec.com/content/en/us/enterprise/white_papers/b-advanced_persistent_threats_WP_21215957.en-us.pdf A detailed description of an APT is Mandiant (2013) *APT1: Exposing One of China's Cyber Espionage Units*, www.mandiant.com, 18 February 2013. http://intelreport.mandiant.com/Mandiant_APT1_Report.pdf

[4] The U.S. Department of Defense (DOD) defined *active cyber defense* (ACD) in 2011: "As malicious cyber activity continues to grow, DoD has employed active cyber defense to prevent intrusions and defeat adversary activities on DoD networks and systems. Active cyber defense is DoD's synchronized, real-time capability to discover, detect, analyze, and mitigate threats and vulnerabilities.... using sensors, software, and intelligence to detect and stop malicious activity before it can affect DoD networks and systems. As intrusions may not always be stopped at the network boundary, DoD will continue to operate and improve upon its advanced sensors to detect, discover, map, and mitigate malicious activity on DoD networks." Department of Defense (2011) *Strategy for Operating in Cyberspace*, July 2011, p. 7.

The goal of D&D is to influence the adversary to behave in a way that gives the defender (i.e., the deceiver) an advantage, creating a causal relationship between the psychological state created by the influence on the adversary and the adversary's behavior. Denial TTTPs[5] actively conceal facts and fictions and create perceptual ambiguity to prevent the adversary from accurately perceiving information and stimuli (Table 1, right column). Deception TTTPs (Table 1, middle column) reveal facts and fictions to provide misleading information and stimuli to the adversary to actively create and reinforce the adversary's perceptions, cognitions, and beliefs in the defender's deception cover story.

Both denial and deception methods affect the adversary's situational awareness systems (e.g., for orientation and observation) and operational systems (e.g., for decision and action), that is, D&D influences the adversary's "OODA loop."[6] D&D aims to either: (1) generate a mistaken certainty about what is and is not real, making the adversary erroneously confident and ready to act, or (2) create sufficient uncertainty about what is real such that the adversary wastes time and/or resources

---

Cyber researcher Dorothy Denning differentiated active and passive cyber defense: "*Active Cyber Defense* is direct defensive action taken to destroy, nullify, or reduce the effectiveness of cyber threats against friendly forces and assets. *Passive Cyber Defense* is all measures, other than active cyber defense, taken to minimize the effectiveness of cyber threats against friendly forces and assets. Whereas active defenses are direct actions taken against specific threats, passive defenses focus more on making cyber assets more resilient to attack." Dorothy E. Denning (2014) "Framework and Principles for Active Cyber Defense," *Computers & Security*, 40 (2014) 108–113. http://www.sciencedirect.com/science/article/pii/S0167404813001661/pdfft? md5=68fecd71b93cc108c015cac1ddb0d430&pid=1-s2.0-S0167404813001661-main.pdf
In both the DOD's and Denning's definitions, actions are defensive. Thus ACD is NOT the same as hacking back, offensive cyber operations, or preemption. However, ACD options are active and can involve actions outside of one's own network or enterprise, for example, collecting information on attackers and sharing the information with other defenders.

[5]"Tactics, techniques, and procedures (TTPs)" is a common military expression and acronym for a standardized method or process to accomplish a function or task. We added 'tools' because cyber adversaries use a variety of different tools in their tactics, techniques, and procedures. On the analysis of TTPs, see Richard Topolski, Bruce C. Leibrecht, Timothy Porter, Chris Green, and R. Bruce Haverty, Brian T. Crabb (2010) *Soldiers' Toolbox for Developing Tactics, Techniques, and Procedures (TTP)*, U.S. Army Research Institute for the Behavioral and Social Sciences Research Report 1919, February 2010.http://www.dtic.mil/dtic/tr/fulltext/u2/a517635.pdf

[6]Jeffrey Rule quotes its creator, John R. Boyd, describing the OODA loop: "orientation shapes observation, shapes decision, shapes action, and in turn is shaped by the feedback and other phenomena coming into our sensing or observing window. ...the entire "loop" (not just orientation) is an ongoing many-sided implicit cross-referencing process of projection, empathy, correlation, and rejection." Jeffrey N. Rule (2013) "A Symbiotic Relationship: The OODA Loop, Intuition, and Strategic Thought," Carlisle PA: U.S. Army War College. http://oai.dtic.mil/oai/oai? verb=getRecord&metadataPrefix=html&identifier=ADA590672 Rule and other exponents of the OODA loop concept see the utility of D&D to isolate the adversary inside their own OODA loop and thus separate the adversary from reality. Osinga sees Boyd's OODA theory as affirming the use of deception against the adversary's OODA loop: "Employ a variety of measures that interweave menace, uncertainty and mistrust with tangles of ambiguity, deception and novelty as the basis to sever an adversary's moral ties and disorient or twist his mental images and thus mask, distort and magnify our presence and activities." Frans P. B. Osinga (2007) *Science, Strategy and War: The strategic theory of John Boyd*, Oxford UK: Routledge, p. 173.

Table 1 D&D methods matrix

Deception objects	Deception: Mislead (M)-type methods	Denial: Ambiguity (A)-type methods
	Revealing	Concealing
Facts	Reveal facts: Nonessential elements of friendly information Reveal true information to the target Reveal true physical entitles, events, or processes to the target	Conceal facts (dissimulation): Essential elements of friendly information Conceal true information from the target Conceal true physical entitles, events, or processes from the target
Fiction	Reveal fiction (simulation): Essential elements of deception information Reveal false information to the target Reveal false physical entitles, events, or processes to the target	Conceal fiction: Nondisclosable deception information Conceal false information from the target Conceal false physical entitles, events, or processes from the target

Kristin E. Heckman, Frank J. Stech, Ben S. Schmoker, Roshan K. Thomas (2015) "Denial and Deception in Cyber Defense", *Computer*, vol. 48, no. 4, pp. 36–44, Apr. 2015. http://doi.ieeecomputersociety.org/10.1109/MC.2015.104 and Kristin E. Heckman and Frank J. Stech (2015) "Cyber Counterdeception: How to Detect Denial & Deception (D&D)," in Sushil Jajodia, Paulo Shakarian, V.S. Subrahmanian, Vipin Swarup, & Cliff Wang eds. (2015) *Cyber Warfare; Building the Scientific Foundation*. Switzerland: Springer

via poor decisions or attempts to increase their certainty. We describe in this chapter how defenders can integrate cyber-D&D into a system of adversary modeling for ACD.

1   Vision for Cyber-D&D in Active Cyber Defense

Inclusion of Cyber-Denial & Deception (D&D) as Standard Operating Procedure (SOP) As adversaries' attack techniques evolve, defenders' cyber systems must also evolve to provide the best active and continuous defense. We envision cyber-D&D as a key part of the standard operating procedures (SOPs) of cyber defensive and security operations, along with cyber threat intelligence[7] and cyber

---

[7]"Cyber threat intelligence" is still an evolving concept. See, for example, Cyber Intelligence Task Force (2013) *Operational Levels of Cyber Intelligence,* Intelligence and National Security Alliance, September 2013. http://www.insaonline.org/i/d/a/Resources/Cyber_

operations security (OPSEC).[8] Engineering cyber systems to better detect adversarial tactics and to actively apply D&D against APTs will force adversaries to move more slowly, expend more resources, and take greater risks. In doing so, defenders may possibly avoid, or at least better fight through, cyber attacks.

Assist Defender Use of Cyber-D&D Tools, Tactics, Techniques, and Procedures (TTTPs) in Conjunction with Other Defensive Mitigation TTPs Cyber-D&D is a relatively new concept and cyber defenders have relatively little experience in engineering defensive cyber-D&D operations. The cyber-D&D concept envisions specific aids to provide cyber defenders with information on when and how to use cyber-D&D TTTPs against specific attack TTPs.

Table 2 shows the two-dimensional D&D framework (from Table 1) adapted to cyber-D&D operations. The cyber-D&D defender uses denial to prevent the detection of the *essential elements of friendly information* (EEFI) by hiding what's real, and uses deception to induce misperception by using the *essential elements of deception information* (EEDI) to show what's false. The deceiver also has to hide the false information—that is, the *nondisclosable deception information* (NDDI)—to protect the D&D plan, and additionally show the real information—the *nonessential elements of friendly information* (NEFI)—to enhance the D&D cover story. Deception is a very dynamic process, and deception planners will benefit from the interplay of techniques from more than one quadrant in Table 2 for conducting a defensive deception operation.

Defend Against APT Threats During the Cyber Attack Lifecycle (Both Pre- and Post-Exploit) Ultimately, we envision cyber-D&D TTTPs being engineered to counter APT threats before, during, and after APT exploitation of the defended cyber enterprise vulnerabilities, that is, throughout the lifecycle of the cyber attack, or

---

Intelligence.aspx; Cyber Intelligence Task Force (2014) *Operational Cyber Intelligence,* Intelligence and National Security Alliance, October 2014. http://www.insaonline.org/i/d/a/Resources/OCI_wp.aspx; Cyber Intelligence Task Force (2014) *Strategic Cyber Intelligence,* March 2014. http://www.insaonline.org/CMDownload.aspx?ContentKey=71a12684-6c6a-4b05-8df8-a5d864ac8c17&ContentItemKey=197cb61d-267c-4f23-9d6b-2e182bf7892e; David Chismon and Martyn Ruks (2015) *Threat Intelligence: Collecting, Analysing, Evaluating.* mwrinfosecurity.com, CPNI.gov.uk, cert.gov.uk. https://www.mwrinfosecurity.com/system/assets/909/original/Threat_Intelligence_Whitepaper.pdf

[8]The concept of "cyber operations security (OPSEC)" has had little systematic development or disciplined application in cyber security. One analyst wrote, "Social media, the internet, and the increased connectivity of modern life have transformed cyber space into an OPSEC nightmare." Devin C. Streeter (2013) "The Effect of Human Error on Modern Security Breaches," *Strategic Informer: Student Publication of the Strategic Intelligence Society:* Vol. 1: Iss. 3, Article 2. http://digitalcommons.liberty.edu/si/vol1/iss3/2 See also Mark Fabro, Vincent Maio (2007) *Using Operational Security (OPSEC) to Support a Cyber Security Culture in Control Systems Environments, Version 1.0 Draft,* Idaho Falls, Idaho: Idaho National Laboratory, INL Critical Infrastructure Protection Center, February 2007. http://energy.gov/sites/prod/files/oeprod/DocumentsandMedia/OpSec_Recommended_Practice.pdf

Table 2 D&D methods matrix with cyber-D&D techniques

Deception objects	Deception: M-type methods	Denial: A-type methods
	Revealing	Concealing
Facts	Reveal facts: Nonessential elements of friendly information Publish true network information Allow disclosure of real files Reveal technical deception capabilities Reveal misleading, compromising details Selectively remediate intrusion	Conceal facts (dissimulation): Essential elements of friendly information Deny access to system resource Hide software using stealth methods Reroute network traffic Silently intercept network traffic
Fiction	Reveal fiction (simulation): Essential elements of deception information Misrepresent intent of software Modify network traffic Expose fictional systems Allow disclosure of fictional information	Conceal fiction: Nondisclosable deception information Hide simulated information on honey pots Keep deceptive security operations a secret Allow partial enumeration of fictional files

what Lockheed Martin termed "the cyber kill chain."[9] Our concept applies similarly to the "left-of-exploit" APT TTPs. In this chapter, however, we focus on APT post-exploit TTPs, or "right-of-exploit."

An Integrated Approach that Facilitates Communication and Coordination Between Analysts, Operators, and D&D Planners Deception operations require careful planning, preparation, and execution; communications among cyber operations, cyber threat intelligence, and cyber OPSEC functions; close coordination of ongoing denial and deception activities and other defensive and mitigation operations; feedback from cyber threat intelligence to deception operations on the success or failure of deception activities; and close monitoring of the APT adversary's actions and

---

[9]E.M. Hutchins, M.J. Cloppert, and R.M. Amin, "Intelligence-Driven Computer Network Defense Informed by Analysis of Adversary Campaigns and Cyber Kill Chains," presented at the 6th Ann. Int'l Conf. Information Warfare and Security, 2011; www.lockheedmartin.com/content/dam/lockheed/data/corporate/documents/LWhite-Paper-Intel-Driven-Defense.pdf

reactions to maintain OPSEC for the defensive cyber deception operations.[10] Our concept for cyber-D&D envisions a shared vocabulary, comprehensive framework, and a systems approach to cyber-D&D operations to facilitate communication and coordination between cyber analysts, cyber defensive operators, and cyber-D&D operators.

2   Key Elements of Integrating Cyber-D&D into Adversary Modeling

The cyber-D&D defensive concept described in this chapter envisions several key elements that must be integrated to communicate and coordinate deception with other defensive operations to influence sophisticated adversaries to the benefit of the enterprise defenders.

Cyber Threat Intelligence The cyber-D&D defensive system depends on threat intelligence on the attacker's TTPs. Defenders must be able to obtain, process, and use both public and private threat information; collect observable systems and network data; correlate sensor data to indicators of possible attacker TTPs; specify the characteristics of the attacker TTPs; monitor and track the frequencies of attacker use of various TTPs; and specify known and possible attack patterns of TTPs used by APTs.[11]

Systems and Network Sensors Enterprise networks require host, server and network sensors, as well as associated storage and processing infrastructure, to observe and characterize system behavior and attacker TTPs. They also require sensors to monitor cyber-D&D influence attempts and the effectiveness of defensive mitigations in influencing attacker behaviors. Sensors enable defenders to determine if mitigations and deceptions are working or not, when to reinforce such efforts, or when to switch to back-up plans for defenses.

Intrusion Detection and Adversary Behavior Analysis Defenders will require an analytic platform for analysis of data collected from systems and network sensors for the purpose of intrusion detection and situational awareness. The platform should allow for correlation of data to detect adversary presence and scope of intrusion. When sufficient data is unavailable, it should also allow defenders to

---

[10]Bodner et al. argue "Working in unison is the only way we can reverse the enemy's deception," and offer an overview on applying integrated cyber-D&D to defense operations and implementing and validating cyber-D&D operations against the adversary's OODA loop; Sean Bodmer, Max Kilger, Gregory Carpenter, and Jade Jones (2012) *Reverse Deception: Organized Cyber Threat Counter-Exploitation,* New York: McGraw-Hill, p. 354.

[11]See, for example, David Chismon and Martyn Ruks (2015) Threat Intelligence: Collecting, Analysing, Evaluating. mwrinfosecurity.com, CPNI.gov.uk, cert.gov.uk. https://www.mwrinfosecurity.com/system/assets/909/original/Threat_Intelligence_Whitepaper.pdf

estimate, with some confidence level, that an attacker may be within the network and enable further investigation to increase confidence. Ideally, this analytic platform should not be located in or directly accessible from the enterprise network, which may allow for the attacker to gain awareness of intrusion detection and cyber-D&D defensive capabilities.

Defender Mitigations As defenders characterize attacker TTPs, they need specific mitigations and cyber-D&D TTTPs to use against the attacker TTPs that defenders detect are being used against their enterprise. Defender mitigations and specific cyber-D&D TTTPs should be crafted for specific attacker TTPs.

RED—BLUE Experimentation To some extent, effectiveness of attacker detection, TTP mitigations, and cyber-D&D TTTPs can be estimated and measured technically. That is, some attacker TTPs can be completely defeated or deceived by technical means. However, mitigating, denying, or deceiving some attacker TTPs depends upon successfully influencing the attacker. To measure the success of mitigation recommendations and cyber-D&D TTTPs, experiments and exercises (i.e., RED teams emulating attacker behavior pitted against BLUE teams using the analytic platform and employing cyber-D&D) may be necessary to gain confidence that the mitigations and cyber-D&D methods will influence the adversary as needed.

3 Adversarial Tactics, Techniques, and Common Knowledge (ATT&CK™)

The recently developed MITRE Adversarial Tactics, Techniques, and Common Knowledge (ATT&CK™) model[12] provides a taxonomy and framework describing APT actions and behavior in an enterprise network during the post-exploit phases of the cyber attack lifecycle. The model incorporates technical descriptions of TTPs, details specific sensor data observables and indicators for each TTP, describes detection analytics for the TTPs, and specifies potential mitigations. ATT&CK describes techniques that can be used against Microsoft Windows-based enterprise network environments, but the concept and methodology for deconstructing TTPs with an adversarial mindset can be extended into other technologies and operating systems.

While there is significant research on initial exploitation and use of perimeter-focused cyber defenses, there is a gap in public knowledge of adversary process after initial compromise and access to an enterprise network. ATT&CK incorporates information on adversary TTPs gathered from various sources, including MITRE research, public threat intelligence, penetration testing, vulnerability research, and from RED versus BLUE team exercises and experiments. ATT&CK collects knowledge characterizing the post-exploit activities of cyber adversaries. ATT&CK

---

[12]See https://attack.mitre.org/wiki/Main_Page for details on the ATT&CK model and framework.

Table 3 MITRE ATT&CK matrix™—overview of tactics and techniques described in the ATT&CK model

Persistence	Privilege Escalation	Defense Evasion	Credential Access	Host Enumeration	Lateral Movement	Execution	C2	Exfiltration
Legitimate Credentials			Credential Dumping	Account enumeration	Application deployment software	Command Line	Commonly used port	Automated or scripted exfiltration
Accessibility Features		Binary Padding					Comm through removable media	
AddMonitor		DLL Side-Loading	Credentials in Files	File system enumeration	Exploitation of Vulnerability	File Access		Data compressed
DLL Search Order Hijack		Disabling Security Tools	Network Sniffing	Group permission enumeration	Logon scripts	PowerShell	Custom application layer protocol	Data encrypted
Edit Default File Handlers						Process Hollowing		Data size limits
New Service		File System Logical Offsets	User Interaction	Local network connection enumeration	Pass the hash	Registry		
Path Interception					Pass the ticket	Rundll32	Custom encryption cipher	Data staged
Scheduled Task		Process Hollowing			Peer connections	Scheduled Task		Exfil over C2 channel
Service File Permission Weakness				Local networking enumeration	Remote Desktop Protocol	Service	Data obfuscation Fallback channels	Exfil over alternate channel to C2 network
Shortcut Modification						Manipulation		
BIOS	Bypass UAC			Operating system enumeration	Windows management instrumentation	Third Party Software	Multiband comm	Exfil over other network medium
Hypervisor Rootkit	DLL Injection	Indicator blocking on host			Windows remote management	Windows management instrumentation	Multilayer encryption	
Logon Scripts	Exploitation of Vulnerability	Indicator removal from host		Owner/User enumeration	Remote Services	Windows remote management	Peer connections	Exfil over physical medium
Master Boot Record		Indicator removal from tools		Process enumeration	Replication through removable media		Standard app layer protocol	From local system
Mod. Exist'g Service		Masquerad-ing		Security software enumeration			Standard non-app layer protocol	From network resource
Registry Run Keys		NTFS Extended Attributes		Service enumeration	Shared webroot	Remote Services through removable media	Standard encryption cipher	From removable media
Serv. Reg. Perm. Weakness					Taint shared content			
Windows Mgmt Instr. Event Subsc.		Obfuscated Payload		Window enumeration	Windows admin shares		Uncommonly used port	Scheduled transfer
Winlogon Helper DLL		Rootkit						
		Rundll32						
		Scripting						
		Software Packing						

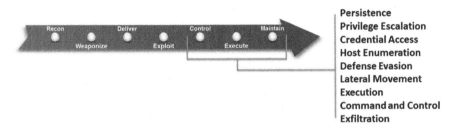

Persistence
Privilege Escalation
Credential Access
Host Enumeration
Defense Evasion
Lateral Movement
Execution
Command and Control
Exfiltration

Fig. 1 ATT&CK™ adversary model. *Source*: https://attack.mitre.org/wiki/File:9_tactics.png

includes the TTPs adversaries use to make decisions, expand access, and execute their objectives. ATT&CK describes an adversary's steps at a high enough level to be applied widely across different platforms, while maintaining enough details to be technically applicable to cyber defenses and research.

The nine tactic categories for ATT&CK (Table 3, top row), ranging from Persistence to Exfiltration, were derived from the later stages (Control, Maintain, and Execute) of the seven stage cyber attack lifecycle (Fig. 1).

Focusing on these post-exploit tactic categories provides a deeper level of granularity in describing what can occur during an intrusion, after an adversary has acquired access to the enterprise network. Each of the nine tactic categories (e.g., Persistence) lists specific techniques an adversary could use to perform that tactic (e.g., Legitimate Credentials). Note that several techniques may serve more than one post-exploit tactic. For example, the technique of using "Legitimate Credentials" can serve three tactic categories: Persistence, Privilege Escalation, and Defense Evasion.

We are currently developing and adding cyber-D&D TTTPs to the ATT&CK Matrix cells. These cyber-D&D TTTPs will complement or substitute for the defensive mitigations in the matrix. To understand the efficacy of cyber-D&D TTTPs for defense, we are currently exploring experimental validation. We have an experimental design with control and test conditions which balances scientific rigor with ecological validity. In this design, participants will be using a number of steps from the ATT&CK Matrix to attack a network, which will be defended by a cyber-D&D tool in one of the test conditions, and in the other test condition, the network will be "defended" by participants' belief that the network uses cyber-D&D defenses. The cyber-D&D tool captures command line commands, silently fails to execute a set of specified commands, but reports execution success.

Much like using ATT&CK to identify gaps in detection coverage of adversary behavior, applying empirical adversary threat intelligence to the ATT&CK model helps focus cyber-D&D on the commonly used techniques across current threat activity. This forms the basis of knowledge necessary to construct cyber-D&D methods against techniques most likely to be used by threats to a particular network or organization.

4   D&D Types and Tactics

Just as the D&D methods matrix includes both facts and fictions, and concealing (i.e., ambiguity-producing denial methods) as well as revealing (i.e., misleading-type deception methods), each of the four cells in the D&D methods matrix (Tables 1 and 2) can contain a variety of D&D tactics, as shown in Fig. 2.[13]

In turn, these D&D tactics can be engineered into cyber-D&D TTTPs. For example, Fig. 3 shows a number of attacker (top) and defender (bottom) D&D TTPs mapped to the D&D methods matrix. Conversely, the D&D methods matrix and associated D&D tactics are being used to develop defensive cyber-D&D TTTPs for the specific post-exploit adversary TTPs shown in the ATT&CK matrix.

In other words, defenders using cyber-D&D need to have cyber-D&D TTTPs available and ready to employ when attacker TTPs are detected (e.g., through the ATT&CK model techniques). As adversaries develop new attack TTPs, and adjust and adapt old ones to counter cyber defenses, the D&D matrix can help

---

[13]Definitions of the D&D tactics shown in Fig. 2 are provided in Kristin E. Heckman, Frank J. Stech, Roshan K. Thomas, Ben Schmoker, and Alexander W. Tsow *Cyber Denial, Deception& Counterdeception: A Framework for Supporting Active Cyber Defense.* Switzerland: Springer, ch. 2.

defenders adapt old, or develop new, cyber-D&D TTTPs. To be fully effective, cyber-D&D defensive operations can be planned, prepared, and executed using the cyber deception chain.

5 Cyber-Deception Chain

The deception chain is a high-level meta-model for cyber-D&D operations management from a lifecycle perspective (Fig. 4). Analogous to Lockheed Martin's "cyber kill chain" model,[14] the deception chain is adapted from Barton Whaley's ten-step process for planning, preparing, and executing deception operations.[15] The deception chain facilitates the integration of three systems—cyber-D&D, cyber threat intelligence, and cyber operations security (OPSEC)—into the enterprise's larger active defense system to plan, prepare, and execute deception operations. Deception operations are conducted by a triad of equal partners working those three systems interactively: cyber-D&D planners, cyber threat intelligence analysts, and cyber-OPSEC specialists. This triad (planners, analysts, and specialists) is essential for a threat-based active cyber defense. Just as computer network defense (CND) is not any one tool but a system that deploys new technologies and procedures as they

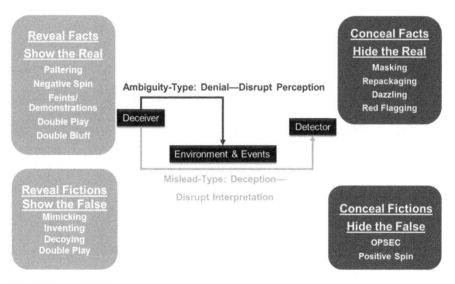

Fig. 2 Types of D&D tactics

---

[14]E.M. Hutchins, M.J. Cloppert, and R.M. Amin, "Intelligence-Driven Computer Network Defense Informed by Analysis of Adversary Campaigns and Cyber Kill Chains," Op cit.

[15]Barton Whaley, "Toward a General Theory of Deception," J. Gooch and A. Perlmuter, eds., *Military Deception and Strategic Surprise*, Routledge, 2007, pp. 188–190.

Attacker TTPs

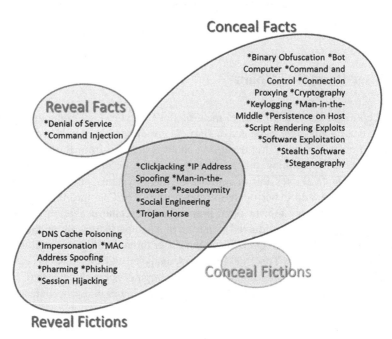

Defender TTPs

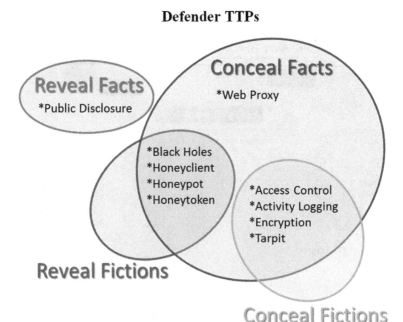

Fig. 3 Cyber D&D TTPs

become available, cyber-D&D must be thought of as an active defensive operational campaign, employing evolving tools, tactics, techniques, and procedures (TTTPs). We believe the deception chain is a flexible framework for embedding advanced TTTPs into operational campaigns while focusing on an organization's mission objectives. There are eight phases in the deception chain (Fig. 4).

Purpose This initial phase helps enterprise managers define the strategic, operational, or tactical goal for the deception operations—in other words, the purpose of the deception—and the criteria that would indicate the deception's success. Since deception operations fundamentally aim to influence the adversary's behavior, the purpose of cyber-deception operations should be defined in terms of the desired influence on the behaviors of the adversary. That is, the deception operation's goal is to influence the adversary to take action or inaction to the benefit of the defender.

Collect Intelligence In the next phase of the cyber-deception chain, D&D planners define how the adversary is expected to behave in response to the deception operation. Defining expected behaviors is done in part through the planners' partnership with cyber threat intelligence, to determine what the adversary will observe, how the adversary might interpret those observations, how the adversary might react (or not) to those observations, and how defenders will monitor the adversary's behavior. This threat intelligence will help planners during the last two phases (monitor and reinforce) to determine whether the deception is succeeding. Cyber threat intelligence can inform D&D planners on what the adversary already knows, believes, and potentially their expectations.

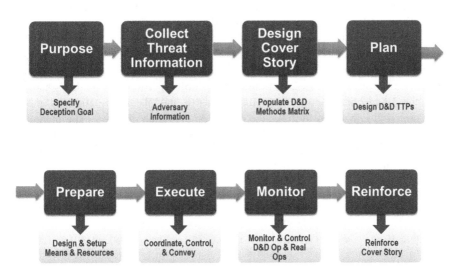

Fig. 4 The cyber-deception chain

One internal source of cyber intelligence is intrusion campaign analysis.[16] Broadly speaking, an intrusion campaign is a framework for grouping related intrusion events and artifacts into knowledge about particular threats to an organization. Analytic methodologies such as the Diamond Model of Intrusion Analysis are also useful in grouping activities in a consistent way.[17] Threat-sharing partnerships are another source of cyber threat intelligence and might involve government, private industry, or non-profit organizations. Information may be shared through various means. Two examples of efforts created for scalable and secure sharing of threat information are the Structured Threat Information eXpression (STIX; http://stix.mitre.org) and Trusted Automated eXchange of Indicator Information (TAXII; http://taxii.mitre.org) systems, sponsored by the Office of Cybersecurity and Communications at the US Department of Homeland Security. STIX and TAXII provide structured formats for defenders to share threat indicators in a manner that reflects the trust relationships inherent in such transfers. STIX is a community-driven language used to represent structured cyber threat information. It contains a structured format for the cyber-deception chain. TAXII enables the sharing of information across organization and product boundaries to detect and mitigate cyber threats. A threat seen by one partner today might be the threat facing another partner in the near future. All of these sources of cyber threat intelligence can aid D&D planners in assessing an adversary's cyber attack capability maturity, which in turn supports the development of an appropriately customized cyber-D&D operation.

Design Cover Story The *cover story* is what the cyber-D&D planner wants the adversary to perceive and believe. The D&D planner will consider the critical components of the D&D operation, assess the adversary's observation and analysis capabilities, and develop a convincing story that "explains" the operation's components observable to the adversary, but misleads the adversary as to the meaning and significance of those observations. The D&D planner will decide what information must be hidden (the EEFI and the NDDI, Table 2) and what information must be created and revealed (the EEDI and the NEFI, Table 2). The D&D methods matrix in Tables 1 and 2 aid planners by capturing the true and false information that must be revealed or concealed to make the deception operation effective. The planners and cybersecurity operators must decide what information "belongs" in the four cells of the matrix and get buy-in from enterprise managers for the deception goals and cover story.

Plan In this phase, cyber-D&D planners analyze the characteristics of the real events and activities that must be hidden to support the deception cover story, identify the corresponding signatures that would be observed by the adversary, and plan to use denial tactics (such as masking, repackaging, dazzling, or red

---

[16]MITRE, Threat- Based Defense: A New Cyber Defense Playbook, 2012; www.mitre.org/sites/default/files/pdf/cyber_defense_playbook.pdf

[17]Sergio Caltagirone, Andrew Pendergast, and Christopher Betz, "Diamond Model of Intrusion Analysis," Center for Cyber Threat Intelligence and Threat Research, Hanover, MD, Technical Report ADA586960, 05 July 2013.

flagging, Fig. 2) to hide the signatures from the adversary. Planners also analyze the characteristics of the notional events and activities that must be portrayed and observed to support the cover story, identify corresponding signatures the adversary would observe, and plan to use deception tactics (such as mimic, invent, decoy, or double play or double bluff,[18] Fig. 3) to mislead the adversary. In short, D&D planners turn the matrix cell information into operational activities that reveal or conceal the key information conveying the cover story. These steps must be coordinated with cyber OPSEC activities so that the D&D steps are as realistic and natural as possible, and the deception should allow the adversary to observe real operational events that support the cover story.

Prepare In this phase, D&D planners design the desired perceptual and cognitive effects on the adversary of the deception operation and explore the available means and resources to create these effects. This entails coordination with OPSEC specialists on the timing for developing the notional and real equipment, staffing, training, and other preparations to support the deception cover story.

Execute As the deception preparations and real operational preparations are synchronized and supported, the D&D planners and OPSEC specialists must coordinate and control all relevant ongoing operations so they can consistently, credibly, and effectively support and execute the deception cover story, without hindering or compromising the real operations.

---

[18]The tactics "mimic, invent, decoy" are all commonly used terms describing actions to mislead. The tactic "double play," or "double bluff," requires some explanation. A double play or a double bluff is a ruse to mislead an adversary (the deception "target") in which "true plans are revealed to a target that has been conditioned to expect deception with the expectation that the target will reject the truth as another deception," Michael Bennett and Edward Waltz (2007) *Counterdeception: Principles and Applications for National Security.* Boston: Artech House, p. 37. A poker player might bluff repeatedly (i.e., bet heavily on an obviously weak hand) to create a bluffing pattern, and then sustain the bluff-like betting pattern when dealt an extremely strong hand to mislead the other players into staying in the betting to call what they believe is a weak hand. The double play formed the basis of a famous Cold War espionage novel, John le Carré's, *The Spy Who Came in from the Cold.* An intelligence service causes a defector to reveal to an adversary the real identity of a double agent mole spying on the adversary. The defector is then completely discredited as a plant to deceive the adversary, causing the adversary to doubt the truth, that the mole could actually be a spy. The Soviet Union *may* have used a version of the double play to manipulate perceptions of a defector (Yuriy Nosenko), i.e., "too good to be true," to mislead the CIA to doubt the bona fides of Nosenko and other Soviet defectors (e.g., Anatoliy Golitsyn). See Richards J. Heuer, Jr. (1987) "Nosenko: Five Paths to Judgment," *Studies in Intelligence,* vol. 31, no. 3 (Fall 1987), pp. 71–101. Declassified, originally classified "Secret." http://intellit.muskingum.edu/alpha_folder/H_folder/ Heuer_on_NosenkoV1.pdf However, "blowing" an actual mole with a double play is atypical of what CIA counterintelligence agent, Tennent Bagley, called "Hiding a Mole, KGB-Style." He quotes KGB colonel Victor Cherkashin as admitting the KGB dangled "Alexander Zhomov, an SCD [Second Chief Directorate] officer," in an "elaborate double-agent operation in Moscow in the late 1980s to protect [not expose] Ames [the KGB's mole in the CIA]." Tennent H. Bagley (2007) *Spy Wars: Moles, Mysteries, and Deadly Games.* New Haven: Yale University Press, p. 226. http:// cdn.preterhuman.net/texts/government_information/intelligence_and_espionage/Spy.Wars.pdf

Monitor D&D planners work with cyber threat intelligence analysts and OPSEC specialists to monitor and control the deception and real operations. This entails monitoring both friendly and adversary operational preparations, carefully watching the observation channels and sources selected to convey the deception to the adversary, and monitoring the adversary's reaction to the "performance," that is, the cover story execution. These targeted channels must remain open to the adversary, convey the planned deception, and be observed by the adversary to convey the cover story. Most importantly, cyber-D&D operators must monitor the adversary to determine if deception operations are having the desired effect on adversary behavior.

Reinforce If cyber intelligence on the adversary indicates that the deception operation does not seem to be "selling" the cover story to the adversary and creating the desired effects on the attacker's behavior, the D&D planners may need to reinforce the cover story through additional deceptions, or to convey the deception operation to the adversary through other channels or sources. The planners may have to revisit the first phase of the deception chain, execute a back-up deception, or plan another operation.

6   The Deception Chain and the Cyber Kill Chain

Adversaries follow a common pattern of behavior to compromise valuable information in a target network. Adversaries generally employ a cyberattack strategy, divided into the six phases of the cyber kill chain or attack lifecycle. Like the cyber kill chain, the deception chain is not always linear. Progression through the phases can be recursive or disjoint. One run of the kill chain models a single intrusion, but a campaign spanning multiple engagements builds on previous results and omits phases as necessary (Fig. 5). Similarly, D&D planners and cyber defense operators will selectively run through the deception chain to achieve their goals. The deception chain is also applicable at each phase of the cyber kill chain, and deception operational goals may be associated with each kill chain phase, as suggested by the following hypothetical D&D tactics:

Recon: If defenders are aware of adversarial reconnaissance efforts, provide the adversary with a set of personae and a Web footprint for defensive targeting efforts in the delivery phase. Note that deception operations can be used to influence adversary actions in future kill chain phases.

Weaponize: Making the adversary (wrongly) feel certain about an organization's vulnerabilities, defense posture, or capabilities could enable the organization to recognize or defend against the adversary's weaponized payload. If the recon phase was successful, the adversary will attempt to deliver the weaponized payload to one or more of the false personae.

Exploit: Recognizing exploitation attempts, defenders may redirect the adversary to a honeypot environment, which appears to be part of a network that contains

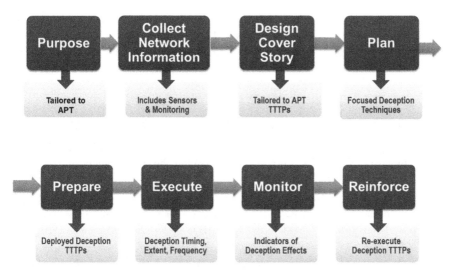

Fig. 5 Building a cyber-deception chain defense

valuable information but is actually isolated and monitored by defenders. The goal is to conceal all honeypot "tells" or indicators to delay the adversary.

Control: When the adversary has "hands on keyboard" access, provide the adversary with a high interaction honeypot with a rich variety of information, designed with the D&D planners, to help identify the adversary's motives, intentions, and capability maturity.

Execute: Slow the adversary down by simulating system interrupts to collect cyber intelligence.

Maintain: Keep up the appearance of realism in a high-interaction honeypot by adding or retiring false personae, as well as maintaining existing personae and their "pocket litter," such as files, email, password change history, login history, and browser history.

These examples also show that there may be a need for more than one deception operation during a single intrusion.

The example below shows a cyber-deception chain built to deflect the attacker TTP of using Legitimate Credentials. The ATT&CK Matrix defines this TTP as:

Adversaries may steal the credentials of a specific user or service account using Credential Access techniques. Compromised credentials may be used to bypass access controls placed on various resources on hosts and within the network and may even be used for persistent access to remote systems. Compromised credentials may also grant an adversary increased privilege to specific systems or access to restricted areas of the network. The adversary may choose not to use malware or tools in conjunction with the legitimate access those credentials provide to make it harder to detect their presence.[19]

---

[19]https://attack.mitre.org/wiki/Legitimate_Credentials

The overlap of credentials and permissions across a network is of concern because the adversary may be able to use Legitimate Credentials to pivot across accounts and bypass any access controls. The Legitimate Credentials TTP serves the attack tactic categories of Persistence, Privilege Escalation, and Defense Evasion.

6.1 Purpose: Legitimate Versus Compromised Credentials

The defensive cyber-D&D operators begin to plan the cyber-deception chain by establishing a *Deception Goal:*

> To influence the adversary to utilize compromised credentials and enable them to further their operations via known false accounts whose usage can be tracked and whose pocket litter can be prepared to further enable the adversary to operate in a means controlled by the defense.

The defensive cyber-D&D operators specify *Effectiveness and Influence Measures* to monitor and measure the effectiveness of their deception operations:

> Detection delay; operational delay; adversary confusion; self-doubt; extra actions; exposed malware; wasted time; dwell time; and dwell ratio.

6.2 Collect Information: Legitimate Credentials: Tactics and Technical Description

Development of the cyber-deception chain continues by *Collecting Network Information* such as sensor data to detect the specific attacker technique, as shown in Table 4.

6.3 Design Cover Story: D&D Methods Matrix

The cyber-D&D defenders then design the deception operation *Cover Story* using the denial and deception (D&D) methods matrix (Tables 1 and 2) to specify the facts and fictions that must be revealed or concealed (Table 5) and the deception operations needed.

6.4 Plan: Legitimate Credentials: Detection and Mitigation

The building of the cyber-deception chain continues with the *Planning* of cyber-D&D TTTPs based on an understanding of enterprise network attacker detection

Table 4 Legitimate Credentials attacker technique and detection methods

Attacker technique: Legitimate Credentials	Detection
Permissions required: user, administrator Effective permissions: user, administrator Data sources: authentication logs, process monitoring Defense bypassed: antivirus, firewall, host intrusion prevention systems, network intrusion detection system, process whitelisting, system access controls	Suspicious account behavior across systems that share accounts, either user, admin, or service accounts. Examples: one account logged into multiple systems simultaneously; multiple accounts logged into the same machine simultaneously; accounts logged in at odd times or outside of business hours. Activity may be from interactive login sessions or process ownership from accounts being used to execute binaries on a remote system as a particular account. Correlate other security systems with login information; e.g. a user has an active login session but has not entered the building or does not have VPN access

https://attack.mitre.org/wiki/Legitimate_Credentials

capabilities, and the set of mitigations defined for the Legitimate Credentials ATT&CK TTP (Table 6).

Planning concludes with the identification of a set of parallel defensive cyber-D&D TTTPs that can complement or substitute for the defender mitigations against the attacker TTP Legitimate Credentials (Table 7).

The defensive cyber-D&D operators then *Prepare, Execute, Monitor*, and *Reinforce* the cyber-deception chain defenses, coordinating with the ongoing enterprise operations, cyber threat intelligence, and cyber OPSEC.

7 Summary

Paradigms for cyber defense are evolving, from static and passive perimeter defenses to concepts for outward-looking active defenses that study, engage, and influence the cyber adversary. This evolution opens the door at tactical, operational, and strategic levels for cyber-D&D to enhance defenses for cyber enterprises,

Table 5 Designing the cover story using the D&D methods matrix

	Reveal-deception	Hide-denial
Facts	• Deception bait accounts with less guarded credentials • Drop clues: "Baited deception accounts are being used to trap attackers" • Some deception bait accounts are harder to access than other deception accounts	• Deception bait accounts facilitate lateral movement to honey traps and other defender controlled systems • Make all deception accounts easier to access than real accounts • Attacker is under observation (and possibly control)
Fictions	• Defender actions: "All legitimate credential account accesses are treated equally"	• Defender actions: Hacked real accounts receive extra deception defenses to make them appear to be bait accounts

Table 6 Preparing deployed deception TTTPs

Detection	Mitigation
Suspicious account behavior across systems that share accounts, either user, admin, or service accounts. Examples: one account logged into multiple systems simultaneously; multiple accounts logged into the same machine simultaneously; accounts logged in at odd times or outside of business hours. Activity may be from interactive login sessions or process ownership from accounts being used to execute binaries on a remote system as a particular account. Correlate other security systems with login information; e.g. a user has an active login session but has entered the building or does not have VPN access	1. Take measures to detect or prevent credential dumping or installation of keyloggers. 2. Limit credential overlap across systems to prevent access if passwords and hashes are obtained. 3. Ensure local administrator accounts have complex, unique passwords across all systems on the network. 4. Do not put user or admin domain accounts in the local administrator groups across systems unless they are tightly controlled as this is often equivalent to having a local administrator account with the same password on all systems.

such as outlined in the concepts described in this chapter. Cyber-D&D is an emerging interdisciplinary approach to cybersecurity, combining human research, experiments, and exercises into cyber-D&D operations with the engineering and testing of appropriate cyber-D&D TTTPs.

Table 7 Planning deception TTTPs to be deployed

Attacker technique: Legitimate Credentials	
ATT&CK mitigations	Parallel D&D techniques
1. Take measures to detect or prevent credential dumping or installation of keyloggers.	Set up deception accounts with less guarded credentials as bait for attackers. Detect and monitor credential dumping and installation of keyloggers to determine if attackers find deception bait accounts.
2. Limit credential overlap across systems to prevent access if passwords and hashes are obtained.	Set up deception bait accounts with less guarded credentials, with credentials overlap across systems to facilitate attacker access if passwords and hashes are obtained for bait accounts.
3. Ensure local administrator accounts have complex, unique passwords across all systems on the network.	Make some deception bait accounts harder to access than other deception accounts, but make all deception accounts easier to access than real accounts. Link bait accounts to honey traps, etc.
4. Do not put user or admin domain accounts in the local administrator groups across systems unless they are tightly controlled as this is often equivalent to having a local administrator account with the same password on all systems.	Put deception bait accounts in the local administrator groups across systems with looser controls. Prepare deception materials to support cover story, in the event real accounts are compromised, that deception accounts are being used.

There is currently no coherent intellectual "center of gravity" for developing an integrated concept for cyber-D&D systems for active cyber defenses such as the concept we have described. Such a center of gravity would conduct and coordinate innovative research, standards development, cyber-D&D systems engineering, shared repositories, and training curriculum creation for cyber-D&D. Integration is lacking for developing policies and programs in cyber-D&D, and for coordinating the development and use of cyber-D&D defenses. The concept of a defensive cyber-D&D center of gravity would involve three action areas:

- Standards, methodologies, and shared repositories;
- Research and operational coordination; and
- Active defense cyber-D&D enterprise organization.

The first area focuses on best practices and standards for defensive cyber-D&D:

- Cataloging ongoing offensive and defensive cyber-D&D techniques;
- Mapping ongoing cyber threats to appropriate D&D techniques to support cyber threat intelligence on adversaries and intrusion campaign analysis;
- Conducting outcome analysis of operational cyber-D&D defensive techniques, impacts, and effectiveness; and
- Enabling the sharing of standards and methodologies through repositories of tools and practices to counter cyber threats with defensive D&D.

The second area focuses on facilitating four types of information exchange:

- Strategic, to formulate cyber-D&D policy, programs, and sponsorship for participants and stakeholders;
- Research management, to formulate a strategic cyber-D&D research roadmap with cyber researchers and operational community participation;
- Research, to share technical cyber-D&D research; and
- Transformation, to formulate an operational roadmap that incorporates cyber-D&D research results into cyber defense operations.

Government-sponsored research needs to lead the effort and commercial technology needs to make substantial investments.

Success in the third area requires an organization to serve as a trusted intermediary to broker cyber-D&D operational sharing, collaboration, and networking to manage cyber-D&D defenses in the threat landscape, while safeguarding such techniques from compromise to cyber adversaries.

This organization would also organize sharing of cyber-D&D information exchange at highly detailed technical levels via technical exchange meetings, shared repositories, and standards. The organization would support the identification of near-term and long-term research needs and threat-based defense gaps, and help government, industry, and academia to meet those needs with cyber-D&D defenses. Finally, the defensive cyber-D&D center of gravity organization would foster the development of cyber-D&D training and educational curricula.

Cyber Security Deception

Mohammed H. Almeshekah and Eugene H. Spafford

Abstract Our physical and digital worlds are converging at a rapid pace, putting a lot of our valuable information in digital formats. Currently, most computer systems' predictable responses provide attackers with valuable information on how to infiltrate them. In this chapter, we discuss how the use of deception can play a prominent role in enhancing the security of current computer systems. We show how deceptive techniques have been used in many successful computer breaches. Phishing, social engineering, and drive-by-downloads are some prime examples. We discuss why deception has only been used haphazardly in computer security. Additionally, we discuss some of the unique advantages deception-based security mechanisms bring to computer security. Finally, we present a framework where deception can be planned and integrated into computer defenses.

1 Introduction

Most data is digitized and stored in organizations' servers, making them a valuable target. Advanced persistent threats (APT), corporate espionage, and other forms of attacks are continuously increasing. Companies reported 142 million unsuccessful attacks in the first half of 2013, as reported by Fortinet [1]. In addition, a recent Verizon Data Breach Investigation Report (DBIR) points out that currently deployed protection mechanisms are not adequate to address current threats [1]. The report states that 66 % of the breaches took months or years to discover, rising from 56 % in 2012. Furthermore, 84 % of these attacks only took hours or less to infiltrate computer systems [1]. Moreover, the report states that only 5 % of these breaches were detected using traditional intrusion detection systems (IDSs) while 69 % were detected by external parties [1].

These numbers are only discussing attacks that were discovered. Because only 5 % of the attacks are discovered using traditional security tools, it is likely that the

M.H. Almeshekah (✉)
King Saud University, Riyadh, Saudi Arabia
e-mail: meshekah@ksu.edu.sa

E.H. Spafford
Purdue University, West Lafayette, IN, USA
e-mail: spaf@purdue.edu

© Springer International Publishing Switzerland 2016
S. Jajodia et al. (eds.), *Cyber Deception*, DOI 10.1007/978-3-319-32699-3_2

reality is significantly worse as there are unreported and undiscovered attacks. These findings show that the status quo of organizations' security posture is not enough to address current threats.

Within computer systems, software and protocols have been written for decades with an intent of providing useful feedback to every interaction. The original design of these systems is structured to ease the process of error detection and correction by informing the user about the exact reason why an interaction failed. This behavior enhances the efforts of malfeasors by giving them information that helps them to understand why their attack was not successful, refine their attacks and tools, and then re-attack. As a result, these systems are helpful to attackers and guide them throughout their attack. Meanwhile, targeted systems learn nothing about these attempts, other than a panic in the security team. In fact, in many cases multiple attempts that originate from the same entity are not successfully correlated.

Deception-based techniques provide significant advantages over traditional security controls. Currently, most security tools are responsive measures to attackers' probes to previously known vulnerabilities. Whenever an attack surfaces, it is hit hard with all preventative mechanisms at the defender's disposal. Eventually, persistent attackers find a vulnerability that leads to a successful infiltration by evading the way tools detect probes or by finding new unknown vulnerabilities. This security posture is partially driven by the assumption that "hacking-back" is unethical, while there is a difference between the act of "attacking back" and the act of deceiving attackers.

There is a fundamental difference in how deception-based mechanisms work in contrast to traditional security controls. The latter usually focuses on attackers' actions—detecting or preventing them—while the former focuses on attackers' perceptions—manipulating them and therefore inducing adversaries to take actions/inactions in ways that are advantageous to targeted systems; traditional security controls position themselves in response to attackers' actions while deception-based tools are positioned in prospect of such actions.

1.1 Definition

One of the most widely accepted definitions of computer-security deception is the one by Yuill [2]; Computer Deception is "Planned actions taken to mislead attackers and to thereby cause them to take (or not take) specific actions that aid computer-security defenses." We adapt this definition and add "confusion" as one of goals of using deceit (the expression of things that are not true) in computer system protection. Therefore, the definition of defensive computer deception we will use throughout this chapter is

Definition 1. Deception is "Planned actions taken to mislead and/or confuse attackers and to thereby cause them to take (or not take) specific actions that aid computer-security defenses."

2 A Brief History

Throughout history, deception has evolved to find its natural place in our societies and eventually our technical systems. Deception and decoy-based mechanisms have been used in security for more than two decades in mechanisms such as honeypots and honeytokens. An early example of how deception was used to attribute and study attackers can be seen in the work of Cheswick in his well-known paper "An Evening with Berferd" [3]. He discusses how he interacted with an attacker in real time providing him with fabricated responses. Two of the earliest documented uses of deceptive techniques for computer security are in the work of Cliff Stoll in his book "The Cuckoo's Egg" [4] and the work of Spafford in his own lab [5]. The Deception Toolkit (DTK),[1] developed by Fred Cohen in 1997 was one of the first publicly available tools to use deception for the purpose of computer defenses.

In late 1990s, "honeypots"—"a component that provides its value by being attacked by an adversary" i.e. deceiving the attacker to interact with them—have been used in computer security. In 2003, Spitzner published his book on "Honeypots" discussing how they can be used to enhance computer defenses [6]. Following on the idea of honeypots, a proliferation of "honey-*" prefixed tools have been proposed. Additionally, with the release of Tripwire, Kim and Spafford suggested the use of planted files that should not be accessed by normal users, with interesting names and/or locations and serving as bait that will trigger an alarm if they are accessed by intruders [7].

2.1 Honey-Based Tools

2.1.1 Honeypots

Honeypots have been used in multiple security applications such as detecting and stopping spam[2] and analyzing malware [8]. In addition, honeypots have been used to secure databases [9]. They are starting to find their way into mobile environments [10] where some interesting results have been reported [11].

Honeypots in the literature come in two different types: server honeypot and client honeypot. The server honeypot is a computer system that contains no valuable information and is designed to appear vulnerable for the goal of enticing attackers to access them. Client honeypots are more active. These are vulnerable user agents that troll many servers actively trying to get compromised [12]. When such incidents happen, the client honeypots report the servers that are infecting users' clients. Honeypots have been used in computing in four main areas as we discuss in the following paragraphs.

---

[1]http://www.all.net/dtk/.

[2]http://www.projecthoneypot.org.

Detection

Honeypots provide an additional advantage over traditional detection mechanisms such as *Intrusion Detection Systems (IDS)* and anomaly detection. First, they generate less logging data as they are not intended to be used as part of normal operations and thus any interaction with them is illicit. Second, the rate of false positive is low as no one should interact with them for normal operations. Angnostakis et al. proposed an advanced honeypot-based detection architecture in the use of *shadow honeypots* [13]. In their scheme they position *Anomaly Detection Sensors (ADSs)* in front of the real system where a decision is made as whether to send the request to a *shadow* machine or to the normal machine. The scheme attempts to integrate honeypots with real systems by seamlessly diverting suspicious traffic to the shadow system for further investigation. Finally, honeypots are also helpful in detecting industry-wide attacks and outbreaks, e.g. the case of the Slammer worm as discussed in [14].

Prevention

Honeypots are used in prevention where they assist in slowing down the attackers and/or deterring them. *Sticky honeypots* are one example of machines that utilize unused IP address space and interact with attackers probing the network to slow them down [15]. In addition, Cohen argues that by using his Deception ToolKit (DTK) we can deter attackers confusing them and introducing risk on their side [16]. However, we are not aware of any studies that investigated those claims.

Beyond the notion of enticement and traps used in honeypots, deception has been studied from other perspectives. For example, Rowe et al. present a novel way of using honeypots for deterrence [17]. They protect systems by making them look like a honeypot and therefore deter attackers from accessing them. Their observation stemmed from the developments of anti-honeypots techniques that employ advanced methods to detect if the current system is a honeypot [18].

Response

One of the advantages of using honeypots is that they are totally independent systems that can be disconnected and analyzed after a successful attack on them without hindering the functionality of the production systems. This simplifies the task of forensic analysts as they can preserve the *attacked* state of the system and extensively analyze what went wrong.

Research

Honeypots are heavily used in analyzing and researching new families of malware. The honeynet project[3] is an "international non-profit security research organization, dedicated to investigating the latest attacks and developing open source security tools to improve Internet security." For example, the HoneyComb system uses honeypots to create unique attack signatures [19]. Other more specific tools, such as *dionaea*,[4] are designed to capture a copy of computer malware for further study. Furthermore, honeypots help in inferring and understanding some widespread attacks such as Distributed Denial of Service (DDoS) [20].

2.1.2 Other Honey Prefixed Tools

The prefix "honey-*" has been used to refer to a wide range of techniques that incorporate the act of deceit in them. The basic idea behind the use of the prefix word "honey" in these techniques is that they need to entice attackers to interact with them, i.e. fall for the bait—the "honey." When such an interaction occurs the value of these methods is realized.

The term honeytokens has been proposed by Spitzner [21] to refer to honeypots but at a smaller granularity. Stoll used a number of files with enticing names and distributed them in the targeted computer systems, acting as a beaconing mechanism when they are accessed, to track down Markus Hess [4]. Yuill et al. coined the term *honeyfiles* to refer to these files [22]. HoneyGen was also used to refer to tools that are used to generate honeytokens [23].

Most recently, a scheme named *Honeywords* was proposed by Jules and Rivest to confuse attackers when they crack a stolen hashed password file [24] by hiding the real password among a list of "fake" ones. Their scheme augmenting password databases with an additional $(N - 1)$ fake credentials [24]. If the DB is stolen and cracked, attackers are faced with N different passwords to choose from where only one of them is the correct one. However, if they use any of the fake ones the system triggers an alarm alerting system administrators that the DB has been cracked.

2.2 Limitations of Isolated Use of Deception

Honeypot-based tools are a valuable technique used for the detection, prevention, and response to cyber attacks as we discuss in this chapter. Nevertheless, those techniques suffer from the following major limitations:

---

[3]www.honeynet.org.
[4]http://dionaea.carnivore.it/.

- As the prefix *honey-** indicates, for such techniques to become useful, the adversary needs to interact with them. Attackers and malware are increasingly becoming sophisticated and their ability to avoid honeypots is increasing [25].
- Assuming we manage to lure the attacker into our honeypot, we need to be able to *continuously* deceive them that they are in the real system. Chen et al. study such a challenge and show that some malware, such as polymorphic malware, not only detects honeypots, but also changes its behavior to deceive the honeypot itself [25]. In this situation, attackers are in a position where they have the ability to conduct counter-deception activities by behaving in a manner that is different than how would they do in a real environment.
- To learn about attackers' objectives and attribute them, we need them to interact with the honeypot systems. However, with a high-interaction honeypot there is a risk that attackers might exploit the honeypot itself and use it as a pivot point to compromise other, more sensitive, parts of the organization's internal systems. Of course, with correct separation and DMZs we can alleviate the damage, but many organizations consider the risk intolerable and simply avoid using such tools.
- As honeypots are totally "fake systems" many tools currently exist to identify whether the current system is a honeypot or not [18, 25]. This fundamental limitation is intrinsic in their design.

3   Deception as a Security Technique

Achieving security cannot be done with single, silver-bullet solutions; instead, good security involves a collection of mechanisms that work together to balance the cost of securing our systems with the possible damage caused by security compromises, and drive the success rate of attackers to the lowest possible level. In Fig. 1, we present a taxonomy of protection mechanisms commonly used in systems. The diagram shows four major categories of protection mechanisms and illustrates how they intersect achieving multiple goals.

The rationale behind having these intersecting categories is that a single layer of security is not adequate to protect organizations so multi-level security controls are needed [26]. In this model, the first goal is to deny unauthorized access and isolate our information systems from untrusted agents. However, if adversaries succeed in penetrating these security controls, we should have degradation and obfuscation mechanisms in place that slow the lateral movement of attackers in penetrating our internal systems. At the same time, this makes the extraction of information from penetrated systems more challenging.

Even if we slow the attackers down and obfuscate our information, advanced adversaries may explore our systems undetected. This motivates the need for a third level of security controls that involves using means of deceit and negative information. These techniques are designed to lead attackers astray and augment our systems with decoys to detect stealthy adversaries. Furthermore, this deceitful information will waste the time of the attackers and/or add risk during their

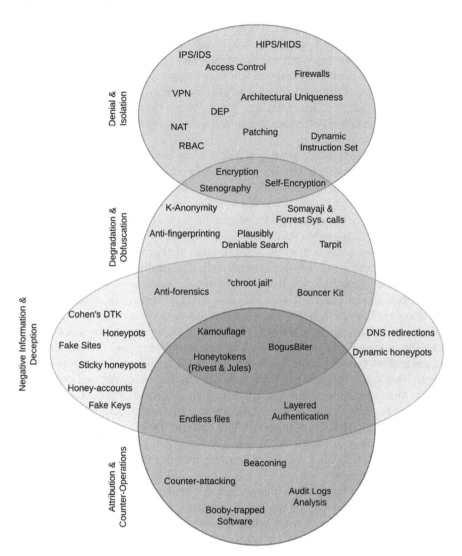

Fig. 1 Taxonomy of information protection mechanisms

infiltration. The final group of mechanisms in our taxonomy is designed to attribute attackers and give us the ability to have counter-operations. Booby-trapped software is one example of counter-operations that can be employed.

Securing a system is an economic activity and organizations have to strike the right balance between cost and benefits. Our taxonomy provides a holistic overview of security controls, with an understanding of the goals of each group and how can they interact with each other. This empowers decision makers on what and which security controls they should deploy.

Despite all the efforts organizations have in place, attackers might infiltrate information systems, and operate without being detected or slowed. In addition, persistent adversaries might infiltrat the system and passively observe for a while to avoid being detected and/or slowed when moving on to their targets. As a result, a deceptive layer of defense is needed to augment our systems with negative and deceiving information to lead attackers astray. We may also significantly enhance organizational intrusion detection capabilities by deploying detection methods using multiple, additional facets.

Deception techniques are an integral part of human nature that is used around us all the time. As an example of a deception widely used in sports: teams attempt to deceive the other team into believing they are following a particular plan so as to influence their course of action. Use of cosmetics may also be viewed as a form of mild deception. We use white lies in conversation to hide mild lapses in etiquette. In cybersecurity, deception and decoy-based mechanisms haven been used in security for more than two decades in technologies such as honeypots and honeytokens.

When attackers infiltrate the system and successfully overcome traditional detection and degradation mechanisms we would like to have the ability to not only obfuscate our data, but also lead the attackers astray by deceiving them and drawing their attention to other data that are false or intestinally misleading. Furthermore, exhausting the attacker and causing frustration is also a successful defensive outcome. This can be achieved by planting fake keys and/or using schemes such as endless files [5]. These files look small on the organization servers but when downloaded to be exfiltrated will exhaust the adversaries' bandwidth and raise some alarms. Moreover, with carefully designed deceiving information we can even cause damage at the adversaries' servers. A traditional, successful, deception technique can be learned from the well-known story of Farewell Dossier during the cold war where the CIA provided modified items to a Soviet spy ring. When the Soviets used these designs thinking they are legitimate, it resulted in a major disaster affecting a trans-Siberian pipeline.

When we inject false information we cause some confusion for the adversaries even if they have already obtained some sensitive information; the injection of negative information can degrade and/or devalue the correct information obtained by adversaries. Heckman and his team, from Lockheed Martin, conducted an experiment between a red and a blue team using some deception techniques, where they found some interesting results [27]. Even after the red team successfully attacked and infiltrate the blue system and obtained sensitive information, the blue team injected some false information in their system that led the red team to devalue the information they had obtained, believing that the new values were correct.

Another relationship can be observed between the last group of protection techniques, namely attribution, and deception techniques. Deception-based mechanisms are an effective way to lure attackers to expose themselves and their objectives when we detect them accessing things and conducting unusual activities. Other tools, such as anomaly-based IDS, have similar goals, but the advantage deception-based tools have is that there is a clear line between normal user activities and abnormal ones. This is because legitimate users are clearly not supposed to access this information.

This difference significantly enhances the effectiveness of deception-based security controls and reduces the number of false-positives, as well as the size of the system's log file.

3.1 Advantages of Using Deception in Computer Defenses

Reginald Jones, the British scientific military intelligence scholar, concisely articulated the relationship between security and deception. He referred to security as a "negative activity, in that you are trying to stop the flow of clues to an opponent" and it needs its other counterpart, namely deception, to have a competitive advantage in a conflict [28]. He refers to deception as the "positive counterpart to security" that provides false clues to be fed to opponents.

By intelligently using deceptive techniques, system defenders can mislead and/or confuse attackers, thus enhancing their defensive capabilities over time. By exploiting attackers' unquestioned *trust* of computer system responses, system defenders can gain an edge and position themselves a step ahead of compromise attempts. In general, deception-based security defenses bring the following unique advantages to computer systems [29]

1. *Increases the entropy of leaked information about targeted systems during compromise attempts.*

   When a computer system is targeted, the focus is usually only on protecting and defending it. With deception, extra defensive measures can be taken by feeding attackers false information that will, in addition to defending the targeted system, cause intruders to make wrong actions/inactions and draw incorrect conclusions. With the increased spread of APT attacks and government/corporate espionage threats such techniques can be effective.

   When we inject false information we cause some confusion for the adversaries even if they have already obtained some sensitive information; the injection of negative information can degrade and devalue the correct information obtained by adversaries. Heckman and her team, developed a tool, referred to as "Blackjack," that dynamically copies an internal state of a production server—after removing sensitive information and injecting deceit—and then directs adversaries to that instance [27]. Even after the red team successfully attacked and infiltrated the blue systems and obtained sensitive information, the blue team injected some false information in their system that led the red team to devalue the information they had obtained, believing that the new values were correct.

2. *Increases the information obtained from compromise attempts.*

Many security controls are designed to create a boundary around computer systems automatically stopping any illicit access attempts. This is becoming increasingly challenging as such boundaries are increasingly blurring, partly as a result of recent trends such as "consumerization"[5] [30]. Moreover, because of the low cost on the adversaries' side, and the existence of many automated exploitation tools, attackers can continuously probe computer systems until they find a vulnerability to infiltrate undetected. During this process, systems' defenders learn nothing about the intruders' targets. Ironically, this makes the task of defending a computer system harder after every unsuccessful attack. We conjecture that incorporating deception-based techniques can enhance our understanding of compromise attempts using the illicit probing activity as opportunity to enhance our understanding of the threats and, therefore, better protect our systems over time.

3. *Give defenders an edge in the OODA loop.*

The OODA loop (for Observe, Orient, Decide, and Act) is a cyclic process model, proposed by John Boyd, by which an entity reacts to an event [31]. The victory in any tactical conflict requires executing this loop in a manner that is faster than the opponent. The act of defending a computer system against persistent attacks can be viewed as an OODA loop race between the attacker and the defender. The winner of this conflict is the entity that executes this loop faster. One critical advantage of deception-based defenses is that they give defenders an edge in such a race as they actively feed adversaries deceptive information that affects their OODA loop, more specifically the "observe" and "orient" stages of the loop. Furthermore, slowing the adversary's process gives defenders more time to decide and act. This is especially crucial in the situation of surprise, which is a common theme in digital attacks.

4. *Increases the risk of attacking computer systems from the adversaries' side.*

Many current security controls focus on preventing the actions associated with illicit attempts to access computer systems. As a result, intruders are using this accurate negative feedback as an indication that their attempts have been detected. Subsequently, they withdraw and use other, more stealthy, methods of infiltration. Incorporating deceit in the design of computer systems introduces a new possibility that adversaries need to account for; namely that they have been detected and currently deceived. This new possibility can deter attackers who are not willing to take the risk of being deceived, and further analyzed. In addition, such technique gives systems' defenders the ability to use intruders' infiltration attempts to their advantage by actively feeding them false information.

---

[5]This term is widely used to refer to enterprises' employees bringing their own digital devises and using them to access the companies' resources.

3.2   Deception in the Cyber Kill-Chain

The cyber kill-chain introduced by Lockheed Martin researchers advocates an intelligence-driven security model [32]. The main premise behind this model is that for attackers to be successful they need to go through all these steps in the chain in sequence. Breaking the chain at any step will break the attack and the earlier that we break it the better we prevent the attackers from attacking our systems.

The cyber kill-chain model is a good framework to demonstrate the effectiveness of incorporating deception at multiple levels in the chain. With the same underlying principle of the kill-chain—early detection of adversaries—we argue that the earlier we detect adversaries, the better we are at deceiving them and learning more about their methods and techniques. We postulate that full intelligence cannot be gathered without using some means of deception techniques.

Also, the better we know our enemies the better we can defend against them. By using means of deception we can continuously learn about attackers at different levels of the kill-chain and enhance our capabilities of detecting them and reducing their abilities to attack us. This negative correlation is an interesting relationship between our ability to detect attackers and their ability to probe our resources.

There is a consensus that we would like to be at least one step ahead of adversaries when they attack our systems. We argue that by intelligently incorporating deception methods in our security models we can start achieving that. This is because the further we enhance our abilities to detect adversaries the further ahead of them we position ourselves. If we take an example of external network probing, if we simply detect an attack and identify a set of IP address and domain names as "bad," we do not achieve much: these can be easily changed and adversaries will become more careful not to raise an alarm the next time they probe our systems. However, if we go one more step to attribute them by factors that are more difficult to change it can cause greater difficulty for future attacks. For example, if we are able to deceive attackers in manners that allow us to gather more information that allows us to distinguish them based on fixed artifacts (such as distinctive protocol headers, known tools and/or behavior and traits) we have a better position for defense. The attackers will now have a less clear idea of how we are able to detect them, and when they know, it should be more difficult for them to change these attributes.

The deployment of the cyber kill-chain was seen as fruitful for Lockheed when they were able to detect an intruder who successfully logged into their system using the SecurID attack [33]. We adopt this model with slight modification to better reflect our additions.

Many deception techniques, such as honeypots, work in isolation and independently of other parts of current information systems. This design decision has been partly driven by the security risks associated with honeypots. We argue that intelligently augmenting our systems with interacting deception-based techniques can significantly enhance our security and gives us the ability to achieve deception in depth. If we examine Table 1, we can see that we can apply deception at every stage of the cyber kill-chain, allowing us to break the chain and possibly attribute

Table 1 Mapping deception to the kill-chain model

Cyber kill-chain phase	Deception
Reconnaissance	Artificial ports, fake sites
Weaponization and delivery	Create artificial bouncing back, sticky honeypots
Exploitation and installation	Create artificial exploitation response
Command and control (operation)	Honeypot
Lateral movement and persistence	HoneyAccounts, honeyFiles
Staging and exfiltration	Honeytokens, endless files, fake keys

attackers. At the reconnaissance stage we can lure adversaries by creating a site and have honey-activities that mimic a real-world organization. As an example, an organization can subscribe with a number of cloud service providers and have honey activities in place while monitoring any activities that signal external interest. Another example is to address the problem of spear-phishing by creating a number of fake persons and disseminating their information into the Internet while at the same monitoring their contact details to detect any probing activities; some commercial security firms currently do this.

3.3 Deception and Obscurity

Deception always involves two basic steps, hiding the real and showing the false. This, at first glance, contradicts the widely believed misinterpretation of Kerckhoff's principle; "no security through obscurity." A more correct English translation of Kerckhoff's principle is the one provided by Petitcolas in [34];

> The system must not require secrecy and can be stolen by the enemy without causing trouble.

The misinterpretation leads some security practitioners to believe that any "obscurity" is ineffective, while this is not the case. Hiding a system from an attacker or having a secret password does increase the work factor for the attacker—until the deception is detected and defeated. So long as the security does not materially depend on the obscurity, the addition of misdirection and deceit provides an advantage. It is therefore valuable for a designer to include such mechanisms in a comprehensive defense, with the knowledge that such mechanisms should not be viewed as primary defenses.

In any system design there are three levels of viewing a system's behavior and responses to service requests [29]:

- *Truthful.* In such systems, the processes will always respond to any input with full "honesty." In other words, the system's responses are always "trusted" and accurately represent the internal state of the machine. For example, when the user asks for a particular network port, a truthful system responds with either a real port number or denies the request giving the specific reason of such denial.

- *Naively Deceptive.* In such systems, the processes attempt to deceive the interacting user by crafting an artificial response. However, if the user knows the deceptive behavior, e.g. by analyzing the previous deceptive response used by the system, the deception act becomes useless and will only alert the user that the system is trying to deceive her. For example, the system can designate a specific port that is used for deceptive purposes. When the attacker asks for a port, without carrying the appropriate permissions, this deceptive port is sent back.

- *Intelligently Deceptive.* In this case, the systems "deceptive behavior" is indistinguishable from the normal behavior even if the user has previously interacted with the system. For example, an intelligently-deceptive system responds to unauthorized port listening requests the same as a normal allowed request. However, extra actions are taken to monitor the port, alert the system administrators, and/or sandbox the listening process to limit the damage if the process downloads malicious content.

3.4 Offensive Deception

Offensively, many current, common attacks use deceptive techniques as a cornerstone of their success. For example, phishing attacks often use two-level deceptive techniques; they deceive users into clicking on links that appear to be coming from legitimate sources, which take them to the second level of deception where they will be presented with legitimate-looking websites luring them to give their credentials. The "Nigerian 419" scams are another example of how users are deceived into providing sensitive information with the hope of receiving a fortune later.

In many of these cases, attackers focus on deceiving users as they are usually the most vulnerable component. Kevin Mitnick showed a number of examples in his book, "The Art of Deception" [35], of how he used social engineering, i.e., deceptive skills to gain access to many computer systems. Trojan horses, which are more than 30 years old, are a prime example of how deception has been used to infiltrate systems.

Phishing, Cross-site Scripting (XSS) [36], and Cross-site Request Forgery (XSRF) [37] are some examples of using deception. Despite more than a decade of research by both the academic and private sectors, these problems are causing more damage every year. XSS and XSRF have remained on the OWASP's top ten list since the first time they were added in 2007 [38]. The effectiveness of offensive deception techniques should motivate security researchers to think of positive applications for deception in security defenses.

4 A Framework to Integrate Deception in Computer Defenses

We presented a framework that can be used to plan and integrate deception in computer security defenses [39]. Many computer defenses that use deception were ad-hoc attempts to incorporate deceptive elements in their design. We show how our framework can be used to incorporate deception in many parts of a computer system and discuss how we can use such techniques effectively. A successful deception should present plausible alternative(s) to the truth and these should be designed to exploit specific adversaries' biases, as we will discuss later.

The framework discussed in this chapter is based on the general deception model discussed by Bell and Whaley in [40]. There are three general phases of any deceptive component; namely planning, implementing and integrating, and finally monitoring and evaluating. In the following sections we discuss each one of those phases in more detail. The framework is depicted in Fig. 3.

4.1 The Role of Biases

In cognitive psychology a bias refers to

> An inclination to judge others or interpret situations based on a personal and oftentimes unreasonable point of view [41]

Biases are a cornerstone component to the success of any deception-based mechanism. The target of the deception needs to be presented with a plausible "deceit" to successfully deceive and/or confuse him. If the target perceives this deceit to be non-plausible she is more inclined to reject it instead of believing it, or at least raise her suspicions about the possibility of currently being deceived. A successful deception should *exploit* a bias in the attackers' perception and provide them with one or more plausible alternative information other than the truth.

Thompson et al. discuss four major groups of biases any analysts need to be aware of: personal biases, cultural biases, organizational biases, and cognitive biases [42]. It can be seen in Fig. 2 that the more specific the bias being exploited in a deceptive security tool is, the less such a tool can be generalized, For example, exploiting a number of personal biases, specific to an attacker, might not be easily generalized to other adversaries who attack your system. However, the more specific the choice of bias enhances the effectiveness of the deceptive component. This is true partly because cognitive biases are well-known and adversaries might intentionally guard themselves with an additional layer of explicit reasoning to minimize their effects in manipulating their perceptions. In the following paragraphs we discuss each one of these classes of biases.

Fig. 2 Deception target
biases

4.1.1 Personal Biases

Personal biases are those biases that originate from either first-hand experiences or personal traits, as discussed by Jervis in [43]. These biases can be helpful in designing deceptive components/operation; however, they are (1) harder to obtain as they require specific knowledge of potential attackers and (2) they make deceptive components less applicable to a wider range of attackers while becoming more powerful against specific attackers. Personal biases have been exploited in traditional deception operations in war, such as exploiting the arrogance of Hitler's administration in World War II as part of Operation Fortitude [41].

4.1.2 Cultural Biases

Hofstede refers to cultural biases as the "software of the mind" [44]. They represent the mental and cognitive ways of thinking, perception, and action by humans belonging to these cultures. In a study conducted by Guss and Dorner, they found that cultures influenced the subjects' perception, strategy development and decision choices, even though all those subjects were presented with the same data [45]. Hofstede discusses six main dimensions of cultures and assigns quantitative values to those dimensions for each culture in his website (geerte-hofstede.com). Also, he associates different behavior that correlates with his measurements. Theses dimensions are:

1. **Power Distance Index (PDI)**—PDI is a measure of the expectation and acceptance that "power is distributed unequally." Hofstede found that cultures with high PDI tend to have a sense of loyalty, show of strength, and preference to in-group-person. This feature can be exploited by a deception planner focusing on the attacker's sense of pride to reveal himself, knowing that the attack is originating from a high PDI culture with a show-of-strength property.
2. **Individualism versus Collectivism (IVC)**—A collectivist society values the "betterment of a group" at the expense of the individual. Hofstede found that most cultures are collectivist, i.e. with low IVC index.
3. **Masculine versus Feminine (MVF)**—A masculine culture is a culture where "emotional gender roles are clearly distinct." For example, an attacker coming

from a masculine culture is more likely to discredit information and warnings written by or addressed to a female. In this case, this bias can be exploited to influence attackers' behaviors.

4. **Uncertainty Avoidance Cultures (UAI)**—This measures the cultural response to the unknown or the unexpected. High UAI means that this culture has a fairly structured response to uncertainty making the attackers' anticipation of deception and confusion a much easier task.

5. **Long-Term Orientation Versus Short-Term Orientation (LTO vs. STO)**— STO cultures usually seek immediate gratification. For example, the defender may sacrifice information of lesser importance to deceive an attacker into thinking that such information is of importance, in support of an over-arching goal of protecting the most important information.

6. **Indulgence versus Restraint (IVR)**—This dimension characterizes cultures on their norms of how they choose activities for leisure time and happiness.

Wirtz and Godson summarize the importance of accounting for cultures while designing deception in the following quote; "To be successful the deceiver must recognize the target's perceptual context to know what (false) pictures of the world will appear plausible" [46].

4.1.3 Organizational Biases

Organizational biases are of importance when designing deception for an target within a heavily structured environment [41]. In such organizations there are many keepers who have the job of analyzing information and deciding what is to be passed to higher levels of analysts. This is one example of how organizational biases can be used. These biases can be exploited causing important information to be marked as less important while causing deceit to be passed to higher levels. One example of organizational biases is uneven distribution of information led to uneven perception and failure to anticipate the Pearl Harbor attack in 1941 by the United States [41].

4.1.4 Cognitive Biases

Cognitive biases are common among all humans across all cultures, personalities, and organizations. They represent the "innate ways human beings perceive, recall, and process information" [41]. These biases have long been studied by many researchers around the world in many disciplines (particularly in cognitive psychology); they are of importance to deception design as well as computing.

Tversky and Kahneman proposed three general heuristics our minds seem to use to reduce a complex task to a simpler judgment decision—especially under conditions of uncertainty—thus leading to some predictable biases [47]. These are: representativeness, availability, and anchoring and adjustment. They defined the representativeness heuristic as a "heuristic to evaluate the probability of an event by the degree to which it is (i) similar in essential properties to its parent population; and (ii) reflects the salient features of the process by which it is generated" [47]. The availability heuristic is another bias that assess the likelihood of an uncertain event by the *ease* with which someone can bring it to mind. Finally, the anchoring/adjustment heuristic is a bias that causes us to make estimations closer to the initial values we have been provided with than is otherwise warranted.

Solman presented a discussion of two reasoning systems postulated to be common in humans: associative (system 1) and rule-based (system 2) [48]. System 1 is usually automatic and heuristic-based, and is usually governed by habits. System 2 is usually more logical with rules and principles. Both systems are theorized to work simultaneously in the human brain; deception targets System 1 to achieve more desirable reactions.

In 1994, Tversky and Koehler argued that people do not subjectively attach probability judgments to events; instead they attach probabilities to the description of these events [49]. That is, two different descriptions of the same event often lead people to assign different probabilities to their likelihood. Moreover, the authors postulate that the more *explicit* and detailed the description of the event is, the higher the probability people assign to it. In addition, they found that unpacking the description of the event into several disjoint components increases the probability people attach to it. Their work provides an explanation for the errors often found in probability assessments associated with the "conjunction fallacy" [50]. Tversky and Kahneman found that people usually would give a higher probability to the conjunction of two events, e.g. $P(X$ and $Y)$, than a single event, e.g. $P(X)$ or $P(Y)$. They showed that humans are usually more inclined to *believe* a detailed story with explicit details over a short compact one.

4.2   Planning Deception

There are six essential steps to planning a successful deception-based defensive component. The first, and often neglected, step is specifying exactly the *strategic goals* the defender wants to achieve. Simply augmenting a computer system with honey-like components, such as honeypots and honeyfiles, gives us a false sense that we are using deception to lie to adversaries. It is essential to detail exactly what are the goals of using any deception-based mechanisms. As an example, it is significantly different to set up a honeypot for the purpose of simply capturing malware than having a honeypot to closely monitor APT-like attacks.

After specifying the strategic goals of the deception process, we need to specify—in the second step of the framework—how the target (attacker) should react to the deception. This determination is critical to the long-term success of any deceptive process. For example the work of Zhao and Mannan in [51] deceive attackers launching online guessing attacks into believing that they have found a correct username and password. The strategic goal of this deception process is to direct an attacker to a "fake" account thus wasting their resources and monitoring their activities to learn about their objectives. It is crucial to analyze how the target should react after the successful "fake" login. The obvious reaction is that the attacker would continue to laterally move in the target system, attempting further compromise. However, an alternative response is that the attacker ceases the guessing attack and reports to its command and control that a successful username/password pair has been found. In consideration of the second alternative we might need to maintain the username/password pair of the fake account and keep that account information consistent for future targeting.

Moreover, part of this second step is to specify how we desire an attacker to react such that we may try to influence his perception and thus lead him to the desired reaction. Continuing with the example in the previous paragraph, if we want the attacker to login again so we have more time to monitor and setup a fake account, we might cause an artificial network disconnection that will cause the target to login again.

4.2.1 Adversaries' Biases

Deception-based defenses are useful tools that have been shown to be effective in many human conflicts. Their effectiveness relies on the fact that they are designed to exploit specific biases in how people think, making them appear to be plausible but false alternatives to the hidden truth, as discussed above. These mechanisms give defenders the ability to learn more about their attackers, reduce indirect information leakages in their systems, and provide an advantage with regard to their defenses.

Step 3 of planning deception is to understand the attackers' biases. As discussed earlier, biases are a cornerstone component to the success of any deception-based mechanisms. The deceiver needs to present a plausible deceit to successfully deceive and/or confuse an adversary. If attackers decide that such information is not plausible they are more inclined to reject it, or at least raise their suspicions about the possibility of currently being deceived. When the defender determines the strategic goal of the deception and the desired reactions by the target, he needs to investigate the attacker's biases to decide how best to influence the attacker's perception to achieve the desired reactions.

One example of using biases in developing some deceptive computer defenses is using the "confirmation bias" to lead adversaries astray and waste their time and resources. Confirmation bias is defined as "the seeking or interpreting of evidence in ways that are partial to existing beliefs, expectations, or a hypothesis in hand" [52]. A computer defender can use this bias in responding to a known adversarial probing of the system's perimeter. Traditional security defenses are intended to detect and prevent such activity, by simply dropping such requests or actively responding with an explicit denial. Taking this a step further by exploiting some pre-existing expectation, i.e. the confirmation bias, we might provide a response that the system is being taken down for some regular maintenance or as a result of some unexpected failure. With such a response, the defender manages to prevent illicit activity, provide a pause to consider next steps for the defender, and perhaps waste the adversary's time as they wait or investigate other alternatives to continue their attacks.

Cultural biases play an important role in designing deceptive responses, as discussed in Sect. 4.1.2. For example, some studies found relationships between the type of computer attacks and the culture/country from which the attack originated [53].

In computing, the conjunction fallacy bias, discussed in Sect. 4.1.4, can be exploited by presenting the deception story as a conjunction of multiple detailed components. For example, if deceivers want to misinform an attacker probing their system by creating an artificial network failure, instead of simply blocking these attempts, it is better to give a longer story. A message that says "Sorry the network is down for some scheduled network maintenance. Please come back in three hours" is more plausible than simply saying "The network is down" and thus more likely to be believed.

4.2.2 Creating the Deception Story

After analyzing attackers' biases the deceiver needs to decide exactly what components to simulate/dissimulate; namely step 4 of the framework in Fig. 3.

In Fig. 4 we provide an overview of the different system components where deception can be applied, exploiting the attacker's biases to achieve the desired reaction. Overall, deceit can be injected into the functionality and/or state of our systems. We give a discussion of each one of these categories below and present some examples.

System's Decisions

We can apply deception to the different decisions any computer system makes. As an example, Zhao and Mannan work in [51] apply deception at the system's authentication decision where they deceive adversaries by giving them access to "fake" accounts in the cases of online guessing attacks. Another system's decision we can use concerns firewalls. Traditionally, we add firewall rules that prevent specific IP addresses from interacting with our systems after detecting that they are sources of some attacks. We consider this another form of data leakage in accordance with the discussion of Zhao and Mannan in [51]. Therefore, we can augment firewalls by applying deception to their decisions by presenting adversaries with plausible responses other than simply denying access.

System's Software and Services

Reconnaissance is the first stage of any attack on any computing system, as identified in the kill-chain model [32]. Providing fake systems and services has been the main focus of honeypot-based mechanisms. Honeypots discussed earlier in this chapter are intended to provide attackers with a number of fake systems running

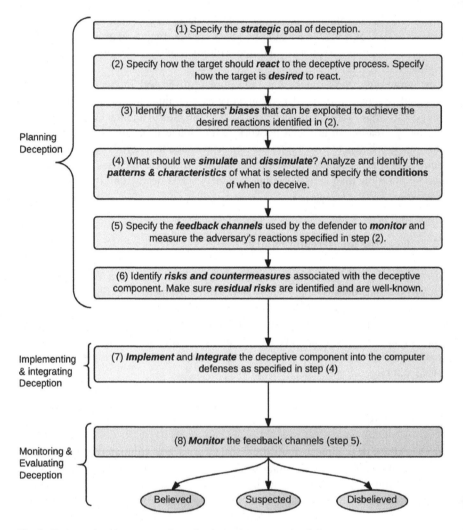

Fig. 3 Framework to incorporate deception in computer security defenses

fake services. Moreover, we can use deception to mask the identities of our current existing software/services. The work of Murphy et al. in [54] recommended the use of operating system obfuscation tools for Air Force computer defenses.

System's Internal and Public Data

A honeyfile, discussed above, is an example of injecting deceit into the system's internal data. It can be applied to the raw data in computer systems, e.g., files and directories, or to the administrative data that are used to make decisions and/or

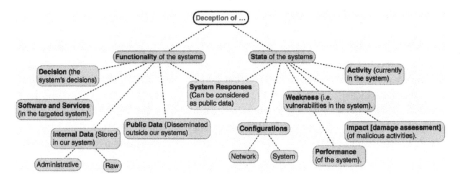

Fig. 4 Computer systems components where deception can be integrated with

monitor the system's activities. An example applying deception to the administrative data can be seen in the *honeywords* proposal [24]. Deceit can also be injected into the public data about our systems. Wang et al. made the case of disseminating public data about some "fake" personnel for the purpose of catching attacks such as spear phishing [55]. Cliff Stoll did this during the story of his book [4]. In addition, we note that this category also includes offline stored data such as back-ups that can be used as a focus of deception.

System's Activity

Different activities within a system are considered as one source of information leakage. For example, traffic flow analysis has long been studied as a means for attackers to deduce information [56]. Additionally, a system's activity has been used as a means of distinguishing between a "fake" and a real system [25]. We can intelligently inject some data about activities into our system to influence attackers' perception and, therefore, their reactions.

System's Weaknesses

Adversaries probe computer systems trying to discover and then exploit any weakness (vulnerability). Often, these adversaries come prepared with a list of possible vulnerabilities and then try to use them until they discover something that works. Traditional security mechanisms aid adversaries by quickly and promptly responding back to any attempt to exploit fixed, i.e. patched, vulnerabilities with a denial response. This response leaks information that these vulnerabilities are known and fixed. When we inject deceit into this aspect of our systems we can misinform adversaries by confusing them—by not giving them a definitive answer whether the exploit has succeeded—or by deceiving them by making it appear as if the vulnerability has been exploited.

System's Damage Assessment

This relates to the previous component; however, the focus here is to make the attacker perceive that the damage caused is more or less than the real damage. We may want the adversary to believe that he has caused more damage than what has happened so as to either stop the attack or cause the attacker to become less aggressive. This is especially important in the context of the OODA loop discussed earlier in this chapter. We might want the adversary to believe that he has caused less damage if we want to learn more about the attacker by prompting a more aggressive attack.

System's Performance

Influencing the attacker's perception of system's performance may put the deceiver at an advantageous position. This has been seen in the use of *sticky honeypots* and tarpits discussed at the beginning of this chapter that are intended to slow the adversary's probing activity. Also, tarpits have been used to throttle the spread of network malware. In a related fashion, Somayaji et al. proposed a method to deal with intrusions by slowing the operating system response to a series of anomalous system calls [57].

System's Configurations

Knowledge of the configuration of the defender's systems and networks is often of great importance to the success of the adversary's attack. In the lateral movement phase of the kill-chain adversarial model, attackers need to know how and where to move to act on their targets. In the red-teaming experiment by Cohen and Koike they deceived adversaries to attack the targeted system in a particular sequence from a networking perspective [58].

After deciding which components to simulate/dissimulate, we can apply one of Bell and Whaley's techniques discussed in [29]. We give an example of how each one of these techniques can be used in the following paragraphs.

- *Using Masking*—This has been used offensively where attackers hide potentially damaging scripts in the background of the page by matching the text color with the background color. When we apply hiding to software and services, we can hide the fact that we are running some specific services when we detect a probing activity. For example, when we receive an SSH connection request from a known bad IP address we can mask our SSHd demon and respond as if the service is not working or as if it is encountering an error.
- *Using Repackaging*—In several cases it might be easier to "repackage" data as something else. In computing, repackaging has long been used to attack computer users. The infamous cross-site scripting (XSS) attack uses this technique where

an attacker masks a dangerous post as harmless to steal the user's cookies when they view such post. Another example can be seen in the cross-site request forgery (XSRF) attacks where an adversary deceives a user into visiting some innocuous looking web pages that silently instruct the user's browser to engage in some unwanted activities. In addition, repackaging techniques are used by botnet Trojans that repackage themselves as anti-virus software to deceive users into installing them so an attacker can take control of their machines. From the defensive standpoint, a repackaging act can be seen in HoneyFiles, discussed above, that repackage themselves as normal files while acting internally as silent alarms to system administrators when accessed.

- *Using Dazzling*—This is considered to be the weakest form of dissimulation, where we confuse the targeted objects with others. An example of using dazzling can be seen in the "honeywords" proposal [24]. The scheme confuses each user's hashed password with an extra $(N - 1)$ hashes of other, similar, passwords dazzling an attacker who obtains the credentials database.
- *Using Mimicking*—In computing, phishing attacks are a traditional example of an unwanted deceiving login page mimicking a real website login. An attacker takes advantage of users by deceiving them into giving up their credentials by appearing as the real site. From a defensive perspective, we can apply mimicking to software and services by making our system mimic the responses of a different system, e.g., respond as if we are running a version of Windows XP while we are running Windows 7. This will waste attackers' resources in trying to exploit our Windows 7 machine thinking it is Windows XP, as well as increase the opportunity for discovery. This is seen in the work of Murphy et al. in operating system obfuscation [54].
- *Using Inventing*—Mimicking requires the results to look like something else; when this is not easy to achieve invention can be used instead. This technique has seen the most research in the application of deception to computer security defenses. Honeypots are one prominent example of inventing a number of nodes in an organizations with the goal of deceiving an attacker that they are real systems.
- *Using Decoying*—This technique is used to attract adversaries' attention away from the most valuable parts of a computer system. Honeypots are used, in some cases, to deceive attackers by showing that these systems are more vulnerable than other parts of the organization and therefore capture attackers' attention. This can be seen in the work of Carroll and Grosu [59].

After deciding which deceptive technique to use we need to analyze the patterns attackers perceive and then apply one or more of those techniques to achieve the desired reactions.

Deceit is an active manipulation of reality. We argue that reality can be manipulated in one of three general ways, as depicted in Fig. 5a. We can *manufacture* reality, *alter* reality, and/or *hide* reality. This can be applied to any one of the components we discussed above.

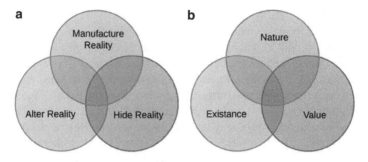

Fig. 5 Creating deceit. (**a**) Manipulation of reality. (**b**) Deception can be applied to the nature, existence and/or value of data

In addition, reality manipulation is not only to be applied to the existence of the data in our systems—it can be applied to two other features of the data. As represented in Fig. 5b, we can manipulate the reality with respect to the *existence* of data, *nature* of the data, and/or *value* of the data. The existence of the data can be manipulated not only for the present but also when the data has been created. This can be achieved for example with the manipulation of time stamps. With regard to the nature of the data, we can manipulate the size of the data, such as in the example of endless files, when and why the data has been created. The value of the data can also be manipulated. For example, log files are usually considered important data that adversaries try to delete to cover their tracks. Making a file appear as a log file will increase its value from the adversary's perspective.

At this step, it is crucial to specify exactly when the deception process should be activated. It is usually important that legitimate users' activity should not be hindered by the deceptive components. Optimally, the deception should only be activated in the case of malicious interactions. However, we recognize that this may not always be possible as the lines between legitimate and malicious activities might be blurry. We argue that there are many defensive measures that can apply some deceptive techniques in place of the traditional denial-based defenses that can make these tradeoffs.

4.2.3 Feedback Channels and Risks

Deception-based defenses are not a single one-time defensive measure, as is the case with many advanced computer defenses. It is essential to monitor these defenses, and more importantly measure the impact they have on attackers' perceptions and actions. This is step 5 in the deception framework. We recognize that if an attacker detects that he is being deceived, he can use this to his advantage to make a counter-deception reaction. To successfully monitor such activities we need to clearly identity the deception channels that can and should be used to monitor and measure any adversary's perceptions and actions.

In the sixth and final step before implementation and integration, we need to consider that deception may introduce some new risks for which organizations need to account. For example, the fact that adversaries can launch a counter-deception operation is a new risk that needs to be analyzed. In addition, an analysis needs to done on the effects of deception on normal users' activities. The defender needs to accurately identify potential risks associated with the use of such deceptive components and ensure that residual risks are accepted and well identified.

4.3   Implementing and Integrating Deception

Many deception-based mechanisms are implemented as a separate disjoint component from real production systems, as in the honeypot example. With the advancement of many detection techniques used by adversaries and malware, attackers can detect whether they are in real system or a "fake" system [25], and then change behavior accordingly, as we discussed earlier in this chapter. A successful deception operation needs to be integrated with the real operation. The honeywords proposal [24] is an example of this tight integration as there is no obvious way to distinguish between a real and a "fake" password.

4.4   Monitoring and Evaluating the Use of Deception

Identifying and monitoring the feedback channels is critical to the success of any deception operation/component. Hesketh discussed three general categories of signals that can be used to know whether a deception was successful or not [60]:

1. The target acts in the wrong time and/or place.
2. The target acts in a way that is wasteful of his resources.
3. The target delays acting or stop acting at all.

Defenders need to monitor all the feedback channels identified in step 5 of the framework. We note that there are usually three general outputs from the use of any deceptive components. The adversary might (1) *believe* it, where the defender usually sees one of the three signs of a successful deception highlighted above, (2) *suspect* it or (3) *disbelieve* it. When an attacker suspects that a deceptive component is being used, we should make the decision whether to increase the level of deception or stop the deceptive component to avoid exposure. Often deception can be enhanced by presenting more (and perhaps, true) information that makes the deception story more plausible. This can be included as a feedback loop in the framework. This observation should be analyzed by the defender to review his analysis of the attacker's biases, (i.e., step 3), and the methodology used to create the deceit (i.e., step 4). Furthermore, the deceiver might employ multiple levels of deception based on the interaction with the attacker during the attack.

When an attacker disbelieves the presented deceit we need to have an active monitoring and a detailed plan of action. This should be part the sixth step of planning in our framework where risks are assessed. In addition, during our discussions with security practitioners many have indicated that some attackers often act aggressively when they realize that they have been deceived. This can be one of the signals that is used during the monitoring stage to measure attackers' reaction of the deceptive component. In addition, this behavior can be used as one of the biases to be exploited by other deceptive mechanisms that may focus on deceiving the attacker about the *system's damage assessment*.

Acknowledgements The material in the chapter is derived from [29]. Portions of this work were supported by National Science Foundation Grant EAGER-1548114, by Northrop Grumman Corporation (NGCRC), and by sponsors of the Center for Education and Research in Information Assurance and Security (CERIAS).

References

1. Verizon, "Threats on the Horizon – The Rise of the Advanced Persistent Threat." http://www.verizonenterprise.com/DBIR/.
2. J. J. Yuill, *Defensive Computer-Security Deception Operations: Processes, Principles and Techniques*. PhD Dissertation, North Carolina State University, 2006.
3. B. Cheswick, "An Evening with Berferd in Which a Cracker is Lured, Endured, and Studied," in *Proceedings of Winter USENIX Conference*, (San Francisco), 1992.
4. C. P. Stoll, *The Cuckoo's Egg: Tracing a Spy Through the Maze of Computer Espionage*. Doubleday, 1989.
5. E. H. Spafford, "More than Passive Defense." http://goo.gl/5lwZup, 2011.
6. L. Spitzner, *Honeypots: Tracking Hackers*. Addison-Wesley Reading, 2003.
7. G. H. Kim and E. H. Spafford, "Experiences with Tripwire: Using Integrity Checkers for Intrusion Detection," tech. rep., Department of Computer, Purdue University, West Lafayette, IN, 1994.
8. D. Dagon, X. Qin, G. Gu, W. Lee, J. Grizzard, J. Levine, and H. Owen, "Honeystat: Local Worm Detection Using Honeypots," in *Recent Advances in Intrusion Detection*, pp. 39–58, Springer, 2004.
9. C. Fiedler, "Secure Your Database by Building HoneyPot Architecture Using a SQL Database Firewall." http://goo.gl/yr55Cp.
10. C. Mulliner, S. Liebergeld, and M. Lange, "Poster: Honeydroid-Creating a Smartphone Honeypot," in *IEEE Symposium on Security and Privacy*, 2011.
11. M. Wählisch, A. Vorbach, C. Keil, J. Schönfelder, T. C. Schmidt, and J. H. Schiller, "Design, Implementation, and Operation of a Mobile Honeypot," tech. rep., Cornell University Library, 2013.
12. C. Seifert, I. Welch, and P. Komisarczuk, "Honeyc: The Low Interaction Client Honeypot," *Proceedings of the 2007 NZCSRCS*, 2007.
13. K. G. Anagnostakis, S. Sidiroglou, P. Akritidis, K. Xinidis, E. Markatos, and A. D. Keromytis, "Detecting Targeted Attacks Using Shadow Honeypots," in *Proceedings of the 14th USENIX Security Symposium*, 2005.
14. D. Moore, V. Paxson, S. Savage, C. Shannon, S. Staniford, and N. Weaver, "Inside the Slammer Worm," *IEEE Security & Privacy*, vol. 1, no. 4, pp. 33–39, 2003.
15. T. Liston, "LaBrea: "Sticky" Honeypot and IDS." http://labrea.sourceforge.net/labrea-info.html, 2009.

16. F. Cohen, "The Deception Toolkit." http://www.all.net/dtk/, 1998.
17. N. Rowe, E. J. Custy, and B. T. Duong, "Defending Cyberspace with Fake Honeypots," *Journal of Computers*, vol. 2, no. 2, pp. 25–36, 2007.
18. T. Holz and F. Raynal, "Detecting Honeypots and Other Suspicious Environments," in *Information Assurance Workshop*, pp. 29–36, IEEE, 2005.
19. C. Kreibich and J. Crowcroft, "Honeycomb: Creating Intrusion Detection Signatures Using Honeypots," *ACM SIGCOMM Computer Communication Review*, vol. 34, no. 1, pp. 51–56, 2004.
20. D. Moore, C. Shannon, D. J. Brown, G. M. Voelker, and S. Savage, "Inferring Internet Denial-of-Service Activity," *ACM Transactions on Computer Systems (TOCS)*, vol. 24, no. 2, pp.115–139, 2006.
21. L. Spitzner, "Honeytokens: The Other Honeypot." http://www.symantec.com/connect/articles/honeytokens-other-honeypot, 2003.
22. J. J. Yuill, M. Zappe, D. Denning, and F. Feer, "Honeyfiles: Deceptive Files for Intrusion Detection," in *Information Assurance Workshop*, pp. 116–122, IEEE, 2004.
23. M. Bercovitch, M. Renford, L. Hasson, A. Shabtai, L. Rokach, and Y. Elovici, "HoneyGen: An Automated Honeytokens Generator," in *IEEE International Conference on Intelligence and Security Informatics (ISI'11)*, pp. 131–136, IEEE, 2011.
24. A. Juels and R. L. Rivest, "Honeywords: Making Password-Cracking Detectable," in *Proceedings of the 2013 ACM SIGSAC Conference on Computer & Communications Security*, pp. 145–160, ACM, 2013.
25. X. Chen, J. Andersen, Z. M. Mao, M. Bailey, and J. Nazario, "Towards an Understanding of Anti-Virtualization and Anti-Debugging Behavior in Modern Malware," in *IEEE International Conference on Dependable Systems and Networks*, pp. 177–186, IEEE, 2008.
26. M. Sourour, B. Adel, and A. Tarek, "Ensuring Security-In-Depth Based on Heterogeneous Network Security Technologies," *International Journal of Information Security*, vol. 8, no. 4, pp. 233–246, 2009.
27. K. Heckman, "Active Cyber Network Defense with Denial and Deception." http://goo.gl/Typwi4, Mar. 2013.
28. R. V. Jones, *Reflections on Intelligence*. London: William Heinemann Ltd, 1989.
29. M. H. Almeshekah, *Using Deception to Enhance Security: A Taxonomy, Model and Novel Uses*. PhD thesis, Purdue University, 2015.
30. M. Harkins, "A New Security Architecture to Improve Business Agility," in *Managing Risk and Information Security*, pp. 87–102, Springer, 2013.
31. J. Boyd, "The Essence of Winning and Losing." http://www.danford.net/boyd/essence.htm, 1995.
32. E. M. Hutchins, M. J. Cloppert, and R. M. Amin, "Intelligence-Driven Computer Network Defense Informed by Analysis of Adversary Campaigns and Intrusion Kill Chains," *Leading Issues in Information Warfare & Security Research*, vol. 1, p. 80, 2011.
33. K. J. Higgins, "How Lockheed Martin's 'Kill Chain' Stopped SecurID Attack." http://goo.gl/r9ctmG, 2013.
34. F. Petitcolas, "La Cryptographie Militaire." http://goo.gl/e5IOj1.
35. K. D. Mitnick and W. L. Simon, *The Art of Deception: Controlling the Human Element of Security*. Wiley, 2003.
36. P. Vogt, F. Nentwich, N. Jovanovic, E. Kirda, C. Kruegel, and G. Vigna, "Cross-Site Scripting Prevention with Dynamic Data Tainting and Static Analysis," in *The 2007 Network and Distributed System Security Symposium (NDSS'07)*, 2007.
37. A. Barth, C. Jackson, and J. C. Mitchell, "Robust Defenses for Cross-Site Request Forgery," *Proceedings of the 15th ACM Conference on Computer and Communications Security (CCS'08)*, 2008.
38. O. W. A. S. P. (OWASP), "OWASP Top 10." http://owasptop10.googlecode.com/files/OWASPTop10-2013.pdf, 2013.

39. M. H. Almeshekah and E. H. Spafford, "Planning and Integrating Deception into Computer Security Defenses," in *New Security Paradigms Workshop (NSPW'14)*, (Victoria, BC, Canada), 2014.
40. J. B. Bell and B. Whaley, *Cheating and Deception*. Transaction Publishers New Brunswick, 1991.
41. M. Bennett and E. Waltz, *Counterdeception Principles and Applications for National Security*. Artech House, 2007.
42. J. R. Thompson, R. Hopf-Wichel, and R. E. Geiselman, "The Cognitive Bases of Intelligence Analysis," tech. rep., US Army Research Institute for the Behavioral and Social Sciences, 1984.
43. R. Jervis, *Deception and Misperception in International Politics*. Princeton University Press, 1976.
44. G. Hofstede, G. Hofstede, and M. Minkov, *Cultures and Organizations*. McGraw-Hill, 3rd ed., 2010.
45. D. Gus and D. Dorner, "Cultural Difference in Dynamic Decision-Making Strategies in a Non-lines, Time-delayed Task," *Cognitive Systems Research*, vol. 12, no. 3–4, pp. 365–376, 2011.
46. R. Godson and J. Wirtz, *Strategic Denial and Deception*. Transaction Publishers, 2002.
47. A. Tversky and D. Kahneman, "Judgment under Uncertainty: Heuristics and Biases.," *Science*, vol. 185, pp. 1124–31, Sept. 1974.
48. S. A. Sloman, "The Empirical Case for Two Systems of Reasoning," *Psychological Bulletin*, vol. 119, no. 1, pp. 3–22, 1996.
49. A. Tversky and D. Koehler, "Support Theory: A Nonextensional Representation of Subjective Probability.," *Psychological Review*, vol. 101, no. 4, p. 547, 1994.
50. A. Tversky and D. Kahneman, "Extensional Versus Intuitive Reasoning: The Conjunction Fallacy in Probability Judgment," *Psychological review*, vol. 90, no. 4, pp. 293–315, 1983.
51. L. Zhao and M. Mannan, "Explicit Authentication Response Considered Harmful," in *New Security Paradigms Workshop (NSPW '13)*, (New York, New York, USA), pp. 77–86, ACM Press, 2013.
52. R. S. Nickerson, "Confirmation Bias: A Ubiquitous Phenomenon in Many Guises," *Review of General Psychology*, vol. 2, pp. 175–220, June 1998.
53. C. Sample, "Applicability of Cultural Markers in Computer Network Attacks," in *12th European Conference on Information Warfare and Security*, (University of Jyvaskyla, Finland), pp. 361–369, 2013.
54. S. B. Murphy, J. T. McDonald, and R. F. Mills, "An Application of Deception in Cyberspace: Operating System Obfuscation," in *Proceedings of the 5th International Conference on Information Warfare and Security (ICIW 2010)*, pp. 241–249, 2010.
55. W. Wang, J. Bickford, I. Murynets, R. Subbaraman, A. G. Forte, and G. Singaraju, "Detecting Targeted Attacks by Multilayer Deception," *Journal of Cyber Security and Mobility*, vol. 2, no. 2, pp. 175–199, 2013.
56. X. Fu, *On Traffic Analysis Attacks and Countermeasures*. PhD Dissertation, Texas A & M University, 2005.
57. S. A. Hofmeyr, S. Forrest, and A. Somayaji, "Intrusion Detection Using Sequences of System Calls," *Journal of Computer Security*, vol. 6, no. 3, pp. 151–180, 1998.
58. F. Cohen and D. Koike, "Misleading Attackers with Deception," in *Proceedings from the 5th annual IEEE SMC Information Assurance Workshop*, pp. 30–37, IEEE, 2004.
59. T. E. Carroll and D. Grosu, "A Game Theoretic Investigation of Deception in Network Security," *Security and Communication Networks*, vol. 4, no. 10, pp. 1162–1172, 2011.
60. R. Hesketh, *Fortitude: The D-Day Deception Campaign*. Woodstock, NY: Overlook Hardcover, 2000.

Quantifying Covertness in Deceptive Cyber Operations

George Cybenko, Gabriel Stocco, and Patrick Sweeney

Abstract A deception is often enabled by cloaking or disguising the true intent and corresponding actions of the perpetrating actor. In cyber deception, the degree to which actions are disguised or cloaked is typically called "covertness." In this chapter, we describe a novel approach to quantifying *cyber covertness*, a specific attribute of malware relative to specific alert logic that the defender uses. We propose that the covertness of an offensive cyber operation in an adversarial environment is derived from the probability that the operation is detected by the defender. We show that this quantitative concept can be computed using Covertness Block Diagrams that are related to classical reliability block diagrams used for years in the reliability engineering community. This requires methods for modeling the malware and target network defenses that allow us to calculate a quantitative measure of covertness which is interpreted as the probability of detection. Called the Covertness Score, this measure can be used by attackers to design a stealthier method of completing their mission as well as by defenders to understand the detection limitations of their defenses before they are exploited.

1 Introduction

In order to dominate the cyber battlespace, cyber mission planners and operators need to understand both their defensive postures and their cyber munitions. In kinetic warfare, military planners have successfully formalized the process of assessing munitions and have developed Joint Munitions Effectiveness Manuals (JMEMs) [1] that quantify their various properties.

G. Cybenko (✉)
Thayer School of Engineering, Dartmouth College, Hanover, NH 03755, USA
e-mail: gvc@dartmouth.edu

G. Stocco
Microsoft, Redmond, WA, USA
e-mail: gstocco@gmail.com

P. Sweeney
Air Force Research Laboratory, Dayton, OH, USA
e-mail: sweenepj@gmail.com

© Springer International Publishing Switzerland 2016   51
S. Jajodia et al. (eds.), *Cyber Deception*, DOI 10.1007/978-3-319-32699-3_3

JMEMs provide detailed quantitative information on the expected effectiveness of different types of weapons and have become critical for performing munition/target assignment, mission planning, mission execution and battle damage assessment. Proposals to extend JMEM concepts to electronic warfare (EW) and stealth have been made [2, 3].

Similar assessments for cyber munitions and defenses would be invaluable to cyber mission planners. Some high-level attributes of cyber munitions include:

- Covertness;
- Precision;
- Reusability;
- Potential for Collateral Damage;
- Persistence;
- Degree of Attribution;
- Time to Deploy on a Novel Target;
- Ease of Battle Damage Assessment;
- Reliability.

These and other possible attributes of cyber munitions have been described and documented in general terms over the past few years [4]. There is general agreement that such attributes are relevant and useful and, in fact, the limited ability to quantify cyber operations is often seen as an impediment to their more widespread use today. For example, David Sanger reported in February 2014 [5]:

> American military planners concluded after putting together [cyber] options for Mr. Obama over the past two and a half years that any meaningful attack on Syria's facilities would have to be both long enough to make a difference and targeted enough to keep from making an already suffering population even worse off.
>
> For those and other reasons, there are doubters throughout the military and intelligence establishment. 'It would be of limited utility, frankly,' one senior administration official said.

The 2015 DoD Cyber Strategy outlines five goals, one of which is to "Build and Maintain Viable Cyber Options..." Notably, and in keeping with the above sentiments, this goal includes not only using cyberspace to achieve mission objectives, but to do so "with precision, and to minimize loss of life and destruction of property" [6]. Without a well-developed Cyber JMEM we run the risk of employing cyber munitions without fully understanding or anticipating the effects it may have on our adversaries, not to mention nth order effects on ourselves.

While the need for a Cyber JMEM has been gaining more traction in recent years [7], there has been relatively little work done on the actual operational quantification, measurement and exploitation of cyber munition attributes (although specific ongoing DOD efforts require such quantification and measurement) [8]. At the Cyber Risk Standards Summit in October 2015, J. Michael Gilmore, director of operational test and evaluation at the Office of the Secretary of Defense, announced a plan to stand up a program office to address these challenges but it is a nascent science at best [9].

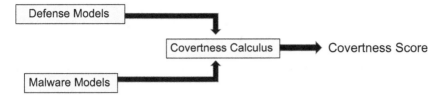

Fig. 1 Our Covertness Score quantification takes models of malware and the malware's targeted defenses to produce a Covertness Score using the Covertness Calculus

These considerations directly support the related challenge of creating the science and supporting technologies to realize effective "Cyber Quantification." Just as we need to know the physical properties, radar cross sections and maneuverabilities of aircraft and land vehicles to have meaningful *physical* battlespace awareness, we need to know corresponding properties of cyber entities for any useful "Cyber Situational Awareness" of the future. In other words, the quantification of cyber munitions informs us about the properties and dynamics of critical objects that populate the cyber battlespace.

In this chapter, we propose a computable, operationally useful quantification of covertness in the context of specific cyber operation environments. We believe that these results can assist both defensive and offensive operations. Moreover, it is possible that the techniques developed here can apply to other cyber munitions attributes as well, opening the door to formalizing such assessments and joint planning activities more broadly. By advancing the basic science and supporting technologies for cyber attribute quantification, we hope to further the pursuit of a cyber-equivalent to the JMEMs for future cyber operations.

Our approach involves modeling malware, modeling the defenses of a target environment for that malware and then combining those through a Covertness Calculus to produce a Covertness Score that is an estimate of the probability that the defenses will detect the malware. Figure 1 depicts these ingredients and their relationships.

Since our goal is to quantify covertness, our malware model consists of quantitative host and network observables that the malware produces. Such observables include measurable quantities such as CPU cycles, memory footprint, system calls, network traffic types and network traffic volume for example. We can in principle include any measurement of an observable used by standard as well as novel and exotic host and network intrusion detection systems and malware analysis tools [10–12].

Our approach is open-ended and agnostic with respect to which measurements, behaviors or signatures (alone or in combination) are being used to detect malware. Just as aircraft stealth must be assessed with respect to specific sensor phenomenologies (RF wavelengths used, bistatic vs. multistatic radars, etc.), we argue that malware covertness is dependent on the cyber sensor phenomenology employed by a defender.

Models of the defensive infrastructure are expressed in terms of a "Covertness Block Diagram" or CBD which is an analogue of Reliability Block Diagrams (RBD) used in Reliability Theory [13, 14]. A CBD is a graphical depiction of the alert logic a defender uses to notify system administrators of possible malware activity. Such alert logic is an integral part of modern Security Information Management (SIM), Security Event Management (SEM), and Security Information and Event Management (SIEM) systems such as ArcSight and Splunk [15, 16]. These systems, which are employed by essentially all enterprise security operations in both the private and government sectors, generate alerts based on rules that trigger when sensors and logs report a specified combination of events. The total collection of such rules is therefore a disjunction of conjunctions or a disjunctive normal form expression of sensor and log reports. We depict such rules as either Boolean expressions or as CBD's when appropriate for pedagogical purposes.

As a result of this generality, our methodology for estimating covertness scores can be based on any cyber defense sensor suite and alert logic.

The remainder of the chapter is organized as follows. Section 2 describes our approach for quantitative modeling of target defenses. Section 3 details our approach for modeling malware and Sect. 4 contains a description of the Covertness Calculus with a concrete numerical example. Section 5 summarizes this chapter with a discussion of current and future work in this area.

2   Defense Models

In spite of the vast amount of research on improved malware detection techniques, most government and private organizations use standard, commercially available malware detection products such as network traffic analysis platforms [17], intrusion detection systems based on packet analysis [18], traditional virus scanners [19] and file integrity checkers [20]. In additional, various log files from firewalls, web servers, database servers and operating systems are maintained and aggregated centrally for archival, analysis and fusion purposes in support of the organization's cyber situational awareness and risk assessments.

In our general model, defenders are able to place such network, host and applications sensors wherever they desire on their network. We do assume that cyber sensors report what is actually observed and that those observations cannot be tampered by an attacker. Cyber sensors can be placed on end-nodes or network resources. The sensors' observations are pooled together with any log files in an enterprise SIM, SEM or SIEM such as ArcSight or Splunk [15, 16]. Such a product "analyzes and correlates every event" across an IT infrastructure [21]. The analysis and fusion performed by these products is typically summarized by rules that trigger alerts. The rules that the defender constructs are based on type and frequency of events and are combined into logical conjunctions and disjunctions.

For example, if an administrator wishes to detect a domain fluxing attack [22], they could create a rule such as:

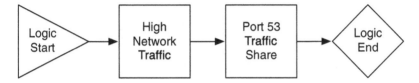

Fig. 2 An example of Basic Defender Logic with two logic tests conjoined, depicted as a Covertness Block Diagram (CBD)

If a system generates a large amount of network traffic and that system has an abnormally high percentage of traffic on port 53 (DNS) then trigger an alert.

This alert logic can be expressed as a Boolean expression or represented as a "Covertness Block Diagram" (CBD) as in Fig. 2. Unlike a classical Reliability Block Diagram [13, 14] that illustrates which components are necessary for the system to operate properly, a CBD indicates which conditions are necessary for generating an alert. The alert is triggered if there is a path from left node (the start state) to the right node (the end state).

A more complicated example is depicted in Fig. 3. Any left-to-right path from start to end nodes is a single rule which represents a conjunction of the sensors along the path being triggered. Multiple paths represent the disjunction of such conjunctions. This example is discussed in more detail later in this chapter but to help the reader understand the semantics of these CBD's, note that the logical expression in Fig. 3 is:

$$(Process\ I/O) \wedge (((Inter - DNS\ Timing) \wedge (DNS\ Packet\ Size))$$

$$\vee (DNS : Traffic\ Ratio)) \vee (Signature\ Detection)$$

Defenders can be expected to know the ground truth about the defenses used in their systems. Attackers do not typically have access to the same information initially, but aim to discover it such that the defenses can be subverted. Therefore, in order to use such a construct for developing an attacker's situational awareness, some discovery of the targeted defenses is necessary. At the same time, this process of discovery should not reveal the attacker's intentions. For example, the actual mission payload must be held in reserve until the attacker can confidently learn the defensive landscape and then maximize the covertness of their malware.

While this is an area of ongoing research, we do have some preliminary approaches. We have named one possible technique of discovery "Digital Virus Vectors." To utilize this technique the attacker measures, before deployment, the observables that their malware will generate in the targeted network. Using a baseline estimate of the normal observables on the defender's network and systems, they can then isolate those observables which are most likely to raise suspicion. Specific instrumentation and examples are provided in the next section.

The attacker then generates a vector of these suspicious observables. Each observable is weighted according to the criticality of the action that is generating

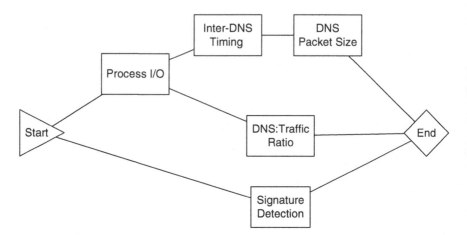

Fig. 3 A more complex example of Basic Defender Logic with five individual logic tests, depicted as a Covertness Block Diagram (CBD). A single left-to-right path represents a conjunction with parallel paths being disjunctions of the conjunctions. While this logic and the associated CBD have binary sensors now, the extension to stochastic sensors, such as those using thresholds, are developed shortly. As this example illustrates, a CBD can be more effective for understanding alert logic than the actual logical rule expressed as a Boolean expression

the observable. For example, consider an attacking malware whose mission is to scan files on the target system. In this case, reads from the logical disk are observable indicators that are critical to mission accomplishment and would be highly weighted. Using the information on observables and criticality, the attacker generates probes—specially crafted malware instances—that generate observables, but that do not contain the payload of the actual malware they wish to deploy.

The attacker will then deploy these probes onto the defender's network. If a probe is detected, the detection itself leaks information about the defender's logic to the attacker, who uses the feedback to tailor a new probe. Once a probe has gone undetected, the attacker now has the information required to deploy a covert payload, as well as information about the defender's logic for designing future covert malware against the same adversary.

For example, consider an attacker designing malware in a simple universe of observables $\{A, B, C, D\}$. In order to accomplish the mission (i.e. gain utility), the attacker must exhibit observables $((A \vee B) \wedge (C \vee D))$. The capabilities that generate each observable or set of observables results in utility for the attacker according to their payoff function, shown in Fig. 4. Prior to attempting to accomplish the mission, the attacker must determine the defender's logic so as to maximize his covertness.

The defender's logic, initially unknown by the attacker, is $(A \wedge B) \vee C$, and observables satisfying this logic equation will generate an alert on the defender's system. Figure 5 shows the defender logic laid out as a covertness block diagram.

Using a naïve greedy strategy, the attacker designs a probe that exhibits the observables in the logical combination $A \wedge B \wedge C \wedge D$, as this generates the highest

Fig. 4 The utilities when the attacker is able to generate combinations of observables

Observables	Utility
$A \wedge C$	10
$A \wedge D$	20
$B \wedge C$	5
$B \wedge D$	15

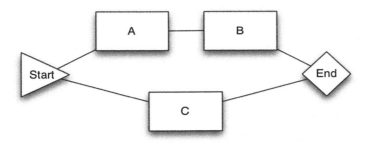

Fig. 5 A simple 3 node defender logic

Fig. 6 The attacker malware design iterations, with the resulting undetected design $A \wedge D$

Probe Iteration	Observables	Potential Utility	Detected
1	$A \wedge B \wedge C \wedge D$	50	Yes
2	$A \wedge B \wedge D$	35	Yes
3	$A \wedge C \wedge D$	30	Yes
4	$A \wedge D$	20	No

payoff of any combination for the attacker. The utility for any combination is the maximum possible sum of utilities of disjoint subsets of the combination. If the combination is not listed, for example, $C \wedge D$, it is not worth any utility.

Following this same greedy strategy, Fig. 6 shows the iterative design steps that may be used in order of decreasing utility. Ultimately, the attacker determines that he can safely design a piece of malware that exhibits observables $A \wedge D$. As each probe costs time and effort as well as exposes the attacker to additional scrutiny, he actually realizes less net utility according to the number of probes sent. This diminishing return informs the attacker when determining a probe strategy.

This example summarizes the approach that an attacker can use to gain understanding and awareness of the defender's logic for future exploitation.

3 Malware Models

To exercise our approach in a realistic scenario, we analyzed actual malware and sought to identify some measurable attributes of running programs that are indicators of malicious behavior. In particular, we are interested in programs that use network resources as well as local resources, as these are representative of most modern malware that not only performs a local mission but participates in

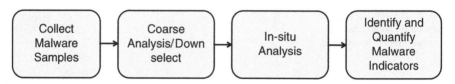

Fig. 7 Steps in the malware characterization process

some form of command and control such as in a botnet. Unlike static signature analysis and end-point anomaly detection, we combine both end-point and network observations to provide a more complete description of what a program is doing. By comparing these indicators to observations of benign activity, we can draw some conclusions about the covertness of the malware with respect to individual sensors and thresholds those sensors might have.

This is a multi-step process involving sample collection, coarse analysis/downselect, in-situ analysis, and indicator quantification as in Fig. 7. The steps are further detailed in the following subsections.

3.1   Collect Malware Samples

Several repositories of malware samples are maintained on the Internet, such as Open Malware and VirusShare [23, 24]. These sites provide categorized, named, and verified samples. For example, a search of "conficker" in the Open Malware archive brings up numerous links to infected executables.

In addition to these categorized samples, there are vast archives of uncategorized samples to draw from. VirusWatch archives, in particular, are continuously maintained lists of malicious links and executables that are live on the Internet [25]. Using a script, flagged executables from the VirusWatch archive are gathered into a local repository.

3.2   Coarse Analysis/Downselect

When the samples are collected, they are next run through the Cuckoo analysis engine [26]. Cuckoo is an automated malware analysis program that runs executables inside a virtual machine sandbox and reports on their behavior. Report items include hosts contacted, DNS inquiries, system calls executed, dropped files, registry keys accessed, etc. Since we are concerned primarily with the covertness of malware that exhibits network activity, we first narrow down, or downselect, the pool of interesting malware to those samples that attempt to communicate with external hosts. In some cases, Cuckoo will flag specific IP addresses as external hosts contacted, however that is indicative of a hardcoded "callback" address within

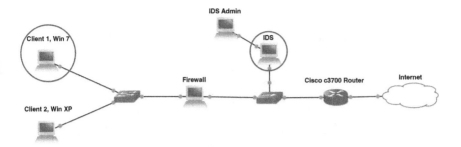

Fig. 8 The topology of the network test bed mimics a typical network enclave. Measurement data was collected on the circled nodes

the executable. Hardcoded IPs are not commonly used in modern malware due to the ease of blacklisting or firewalling the IP as a trivial defense.

More commonly, DNS inquiries are made for file sharing servers,[1] that may indicate the malware is attempting to download a secondary payload. In other malware samples, DNS inquiries are made to seemingly random domain names, which may be indicative of DNS fluxing. DNS fluxing (i.e. using domain generation algorithms) is a technique wherein the malware generates domain names according to a specific algorithm, searching for one that is linked to a command and control (C2) server. The malware owner uses the same algorithm to generate and register the domains, with the intent that the malware will eventually link up with a C2 server through a conjunction of generated names.

3.3 In-Situ Analysis

After executables of interest are identified, they are re-run through a test bed that is representative of a typical network environment. Figure 8 shows the layout of our test bed. Each system is virtualized through the use of Oracle VM VirtualBox and GNS3. The client systems run the malicious or control (benign) software and take local measurements, while the Intrusion Detection System (IDS) collects network measurements. Although the IDS system is also configured to use Snort, that data is ignored under the assumption that a covert piece of malware must, at a minimum, evade detection by commonly-used security products.

On client systems, Perfmon (the Windows Performance Monitor) is used to collect data relevant to the executable's behavior. Perfmon provides a large array of possible measurements, and our list is tailored to suit the type of malware activity we're interested in. Table 1 shows a list of these measurements. Full descriptions of

---

[1] such as www.4shared.com.

Table 1 Summary of Perfmon data collectors used

Perfmon measure	Description
Cache	Counters that monitor the file system cache as an indicator of application I/O operations
Process	Counters that monitor running application program and system processes
Processor	Counters that measure aspects of processor activity
System	Counters that apply to more than one instance of a component processors on the computer
Thread	Consists of counters that measure aspects of thread behavior
Memory	Counters that describe the behavior of physical and virtual memory on the computer
TCPv4	Counters that measure the rates at which TCP Segments are sent and received by using the TCP protocol. It includes counters that monitor the number of TCP connections in each TCP connection state
Browser	Consists of counters that measure the rates of announcements, enumerations, and other Browser transmissions
Logical disk	Counters that monitor logical partitions of a hard or fixed disk drives

the measurements, abbreviated here, are available within Perfmon as well as on the Microsoft Developer Network (MSDN) Library website [27].

The IDS collects network data via the raw packet capture utility, tcpdump. Unlike Perfmon, there is no need to parry down the collected data; all network traffic is captured for analysis.

3.4   Identify and Quantify Malware Indicators

The key to this step is identifying indicators that strongly differentiate malicious activity from benign activity. Benign activity ranges from no activity at all, to typical web-browsing and running executables.

Two data collection traces are executed on the testbed to set a baseline. A *no activity* baseline is measured where, as the name suggests, no activity is performed on the clients. A second trace, *benign activity*, is also measured wherein a client PC is used to perform various benign activities such as start up a web browser to look at articles on CNN.com and shop on Amazon.com.

With the baseline set, a malware sample of interest is executed and the trace data collected. The measured data is compared to the baselines to identify some possible indicators of malicious activity. As an example, one sample of malware exhibited very regular DNS queries as it performed DNS fluxing. An excerpt of these queries are shown in Fig. 9.

```
128 181.502684 192.168.2.10      129.170.6.4        DNS    91 Standard query 0x7252  A debizdaiheaeeqaehcofhqcbqqs.com
129 181.545510 129.170.6.4       192.168.2.10       DNS   164 Standard query response 0x7252 No such name
130 183.034275 192.168.2.10      129.170.6.4        DNS    84 Standard query 0xf37a  A mzonsonzpvotcdfvoeqkx.ru
131 183.145371 129.170.6.4       192.168.2.10       DNS   145 Standard query response 0xf37a No such name
132 184.643783 192.168.2.10      129.170.6.4        DNS    90 Standard query 0xa0c1  A twxvopzdbmwscytkyhfltknfce.com
133 184.685939 129.170.6.4       192.168.2.10       DNS   163 Standard query response 0xa0c1 No such name
134 186.173968 192.168.2.10      129.170.6.4        DNS    87 Standard query 0xe587  A xwswoz1zx1wdrwfih1xcjn.info
135 186.225974 129.170.6.4       192.168.2.10       DNS   147 Standard query response 0xe587 No such name
136 187.721925 192.168.2.10      129.170.6.4        DNS    91 Standard query 0x6a09  A uswtkgmozfugaaobwsktcmfutnz.biz
137 187.766482 129.170.6.4       192.168.2.10       DNS   153 Standard query response 0x6a09 No such name
138 189.268374 192.168.2.10      129.170.6.4        DNS    94 Standard query 0xe95a  A ydozonrwdnvmbincusqgdmskyhpjfm.net
139 189.386478 129.170.6.4       192.168.2.10       DNS   167 Standard query response 0xe95a No such name
```

Fig. 9 A packet capture during a DNS fluxing operation

Various metrics are derived from the collected network traces. In the DNS fluxing example, the generated queries may be differentiated from benign queries based upon inter-query timing which tends to be periodic. Additionally, the mean and variance of the query size may show a preponderance of similar queries, indicating some sort of domain generation algorithm at work.

With data traces collected, probabilities are derived as measures of how likely each observation is to be due to benign versus malicious activity. These probabilities are discussed in more depth in Sect. 4 as they feed into the Covertness Calculus.

4 Covertness Calculus

The Covertness Calculus combines a model of specific malware used by the attacker and the detection capabilities of the defender to derive a covertness measure. The covertness of a particular piece of malware is the probability that it is not detected by the defender given the malware's observables and the defense's alert logic or, equivalently, the Covertness Block Diagram. We assume that the attacker, defender, or both are able to measure and estimate the observables generated by the attacker's malware. The assumption depends on whether the situational awareness being sought is defensive or offensive.

The defender is aware of and able to modify their detector rules. However, given the complexity of these systems and the great cost of downtime caused by false positives, defenders will generally tune their systems to only trigger on high thresholds or subsets of the total possible set of appropriate rules. As has been well documented in recent reports, defenders are often overwhelmed by alerts such that even genuine attacks are missed because they're lost in a sea of false positives that are impossible to sift through [28, 29]. As a result, we believe that there is almost always a way for an attacker to accomplish their operational objectives with a low probability of detection.

More specifically, among all sets of possible observables that an attacker's malware may exhibit when accomplishing a mission, there is some combination of those observables that minimizes the probability of detection on the defender's system. The key for the attacker then is to discover a way through the defender's alert logic that minimizes the probability of detection and does so in a way that

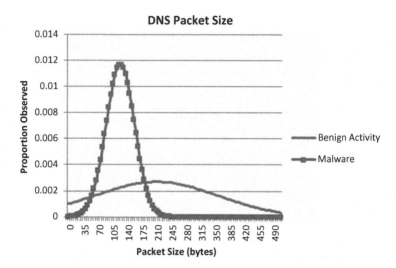

Fig. 10 A comparison of DNS packet sizes for benign activity vs. domain fluxing malware

maximizes the attacker's utility. The defender, for his part, may realize that it will take automated techniques to model how an attacker would avoid his ruleset in order to improve it.

The Covertness Calculus computes the probability of an alert triggering using the defender's Covertness Block Diagram, or equivalently their alert logic, that interacts with the observables generated by the malware. Each elementary block's probability of success is equal to the probability that the block is triggered by the malware under test. Figure 10 compares the "normal" distribution of DNS packet sizes in measured network traffic with the DNS packet size distribution obtained when malware using DNS fluxing attempts to start a command and control communications session with a remote master.

Each sensor's probability of triggering is calculated using a Mann-Whitney test [30] comparing the baseline measurement's distribution of normal behavior on the network with the measurement distributions obtained from the malware sample. In the case of DNS packet size distributions, we selected a test threshold that would correctly identify the malware 99 % of the time given the normal background DNS traffic. The observables were intentionally chosen based on the dissimilarity between baseline measurements and those taken when the malware was profiled on the previously described testbed.

Figure 11 shows the CBD with each sensor assigned a probability of triggering using a similar comparison of normal background activity with the malware's activity. The computation of the overall probability that some rule in the CBD will trigger is straightforward but possibly expensive.

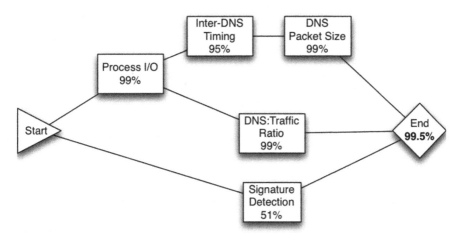

Fig. 11 The Covertness Block Diagram example with calculated probabilities for each logic node being triggered

All possible assignments of "Triggered" vs. "Not Triggered" are made for the sensors and the probability of each assignment is computed by multiplying the probabilities of the individual sensor assignments. For example, the probability of the assignment depicted in Fig. 12 is computed as:

$$0.99 \cdot (1 - 0.95) \cdot 0.99 \cdot 0.99 \cdot (1 - 0.51) = 0.0238$$

The fact that there is a path from the start state to the end state means that an alert will be triggered by the overall alert logic.

The probability, P_{detect}, that the alert logic represented by a CBD will trigger on/detect a malware instance is computed by first enumerating all 2^5 possible assignments. The probability of detection for each assignment is computed, and those probabilities are summed over all assignments. This resulting sum is P_{detect} for the malware instance. Note there is an explicit assumption that the sensors trigger independently of each other, an assumption that requires validation in each particular case of malware and sensor suite.

We have defined the probability of the alert logic triggering to be the probability that the malware is detected, P_{detect}. The resulting Covertness Score of the malware in that particular defensive infrastructure is then $1 - P_{detect}$, i.e. the probability that the malware will not be detected.

In the example first presented in Fig. 3, the malware is very likely to be detected (with probability 0.995 or 99.5 %), and thus is not covert. Covertness Score is $1 - 0.995 = 0.005$ which is very low. Given that this is the case, an intelligent attacker will try to tune the observables generated by their malware so that it will be able to operate for a longer period of time before being detected on the defender's network.

We make a few observations about this method of computing a Covertness Score. Firstly, it should be pointed out that the exhaustive enumeration of all possible

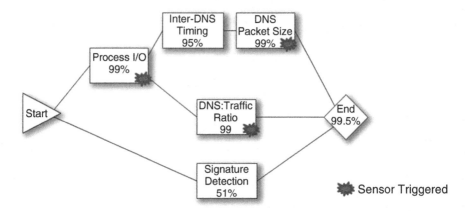

Fig. 12 The Covertness Block Diagram example showing certain sensors being triggered with the indicated probabilities for each logic node being triggered or not

sensor assignments, 2^n for the case of n distinct sensors, is exponential. Related work on evaluating reliability using general Reliability Block Diagrams using a similar formalism has been shown to be NP-Complete [31]. This indicates that there are unlikely to be more efficient ways to compute Covertness Scores as defined here without resorting to approximation or Monte Carlo methods. For $n \leq 30$ sensors, this computation should not be a problem however.

Additionally, it is worth pointing out that the Covertness Score can be computed for a range of individual sensor performance assumptions. That is, we can vary the probability that any one sensor triggers between 0 and 1 to obtain a per-sensor sensitivity assessment of covertness. In fact, inspection of the formula for the Covertness Score shows that a sensor's performance appears linearly in the score if all other sensors' performances are held fixed. This yields a simple "Differential Covertness Sensitivity" score for each sensor in the defensive sensor suite. Sensors with high scores will be candidates for tuning by both the attacker and defender.

To illustrate this construct, consider the following. Figure 13 shows the previously used CBD with an assignment of numbers to sensors, including the start and end states. Figure 14 shows the variability if the Covertness Score as Sensors 2 and 6 have trigger probabilities varied from 0 to 1. Note that each slice through the graph parallel to one the axes is a linear function but together, the function is nonlinear in the two probabilities.

We have also developed other ways to assess the sensitivity of a defensive infrastructure to malware observables, including the structural sensitivity of the overall alert logic.

Finally, we observe that while the notion of "covertness" in the present work quantifies the attacking malware's visibility with respect to the defensive infrastructure, it is also possible to develop a different notion of covertness that quantifies the information leakage of the malware as obtained by the defender. That is, what can the defender learn from the attacker's malware?

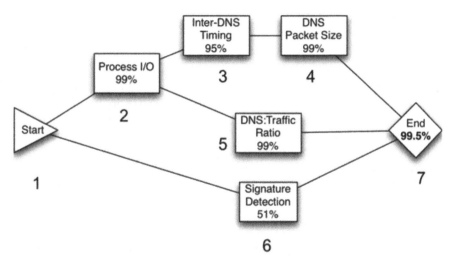

Fig. 13 The Covertness Block Diagram example with nodes labeled 1 through 7

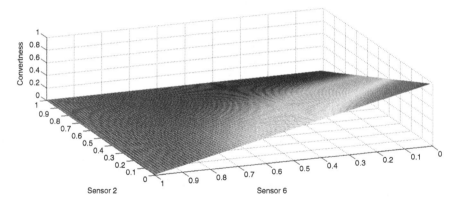

Fig. 14 The sensitivity of the example Covertness Block Diagram to variations in sensors 2 and 6 triggering probabilities. The *vertical axis* shows the Covertness Score

For example, when malware exhibits a distribution or behavior that differs from the baseline (such as discussed for DNS request frequency and depicted in Fig. 10), the size of the deviation can be used as a measure of the information leaked by the malware. Such deviations can be measured by the Kullback-Liebler divergence or other measures of the difference between probability distributions [32].

5 Conclusion

Understanding and quantifying cyber deception and covertness is becoming increasingly complicated but essential for both attackers and defenders. In many recent cyber attack cases, such as Stuxnet, Flame and APT1, the mission effectiveness of an advanced malware campaign has hinged on the covertness of activities. There is, however, little previous work in formalizing the assessment of cyber munitions' covertness as part of the overall Cyber Situational Awareness and quantification of cyber operations. This work has presented a novel technique for assessing covertness against an target defense with an advanced defensive infrastructure that is unknown but learnable by the attacker. We believe that future operations in cyberspace will rely on similar assessment of other facets of cyber munitions.

Acknowledgements This work was partially supported by Air Force Research Laboratory (AFRL) Contract FA8750-13-1-0070 . The opinions expressed in this article belong solely to the article's authors and do not reflect any opinion, policy statement, recommendation or position, expressed or implied, of the U.S. Department of Defense.

The authors thank the anonymous reviewers and the book editors for suggestions that significantly improved this chapter.

References

1. US Army. Joint technical coordinating group for munitions effectiveness program office.
2. David D Lynch and Institution of Electrical Engineers. *Introduction to RF stealth*. Scitech, 2004.
3. Dave MacEslin. Methodology for Determining EW JMEM. *TECH TALK*, page 32, 2006.
4. George Cybenko and Jason Syverson. Quantitative foundations for information operations, 2007.
5. David E. Sanger. Syria War Stirs New U.S. Debate on Cyberattacks. http://www.nytimes.com/2014/02/25/world/middleeast/obama-worried-about-effects-of-waging-cyberwar-in-syria.html, Feb 2014. Accessed: 2015-11-11.
6. US Department of Defense. The Department of Defense Cyber Strategy, 2015.
7. Mark A Gallagher and Michael Horta. Cyber Joint Munitions Effectiveness Manual (JMEM). *M& SJ*, 8:5e14, 2013.
8. US Army. Joint Publication 3–13: Information Operations, Nov 2014.
9. Molly B. Walker. New DoD program office to create cyber equivalent of the Joint Munitions Effectiveness Manual. http://www.fiercegovernmentit.com/story/new-dod-program-office-create-cyber-equivalent-joint-munitions-effectivenes/2015-10-14, Oct 2015. Accessed: 2015-11-20.
10. http://www.iseclab.org/projects/ttanalyze/. TTAnalyze: A tool for analyzing malware, 2015.
11. Clemens Kolbitsch, Paolo Milani Comparetti, Christopher Kruegel, Engin Kirda, Xiao-yong Zhou, and XiaoFeng Wang. Effective and efficient malware detection at the end host. In *USENIX security symposium*, pages 351–366, 2009.
12. Daniel Bilar et al. Statistical structures: Fingerprinting malware for classification and analysis. *Proceedings of Black Hat Federal 2006*, 2006.

13. Marko Čepin. Reliability block diagram. In *Assessment of Power System Reliability*, pages 119–123. Springer, 2011.
14. RG Bennetts. Analysis of reliability block diagrams by boolean techniques. *Reliability, IEEE Transactions on*, 31(2):159–166, 1982.
15. http://www8.hp.com/us/en/software-solutions/siem-security-information-eventmanagement/. HP ArcSight ESM, 2015.
16. http://www.splunk.com/. Splunk Operational Intelligence Platform, 2015.
17. http://www.flowtraq.com/. FlowTraq Network Security, Monitoring, Analysis, and Forensics, 2015.
18. http://www.snort.com/. Snort Intrusion Prevention System, 2015.
19. http://www.mcafee.com/us/. McAfee Intel Security Suite, 2015.
20. http://www.tripwire.com/. Tripwire Advanced Cyber Threat Detection, 2015.
21. HP Enterprise Security. HP ArcSight ESM: powered by CORR-Engine, September 2012.
22. Sandeep Yadav, Ashwath Kumar Krishna Reddy, a.L. Narasimha Reddy, and Supranamaya Ranjan. Detecting algorithmically generated malicious domain names. *Proceedings of the 10th annual conference on Internet measurement - IMC '10*, page 48, 2010.
23. Open Malware. http://openmalware.org, 2014.
24. VirusShare. http://virusshare.com, 2014.
25. The VirusWatch Archives. http://lists.clean-mx.com/pipermail/viruswatch/, 2014.
26. Cuckoo Sandbox. http://www.cuckoosandbox.org/, 2014.
27. Microsoft Developer Network. http://msdn.microsoft.com/en-us/library/, 2014.
28. Nicole Perlroth. Intelligence Start-Up Goes Behind Enemy Lines to Get Ahead of Hackers. www.nytimes.com/2015/09/14/technology/intelligence-start-up-goes-behind-enemy-lines-to-get-ahead-of-hackers.html, Sep 2015. Accessed: 2015-11-11.
29. Ben Elgin, Dune Lawrence, and Michael Riley. Neiman Marcus Hackers Set Off 60,000 Alerts While Bagging Credit Card Data. http://www.bloomberg.com/bw/articles/2014-02-21/neiman-marcus-hackers-set-off-60-000-alerts-while-bagging-credit-card-data, Feb 2014. Accessed: 2015-11-11.
30. Elizabeth R DeLong, David M DeLong, and Daniel L Clarke-Pearson. Comparing the areas under two or more correlated receiver operating characteristic curves: a nonparametric approach. *Biometrics*, pages 837–845, 1988.
31. Michael O Ball. Computational complexity of network reliability analysis: An overview. *Reliability, IEEE Transactions on*, 35(3):230–239, 1986.
32. Thomas M Cover and Joy A Thomas. *Elements of information theory*. John Wiley & Sons, 2012.

Design Considerations for Building Cyber Deception Systems

Greg Briskin, Dan Fayette, Nick Evancich, Vahid Rajabian-Schwart, Anthony Macera, and Jason Li

Abstract Cyber deception can become an essential component of organizing cyber operations in the modern cyber landscape. Cyber defenders and mission commanders can use cyber deception as an effective means for protecting mission cyber assets and ensuring mission success, through deceiving and diverting adversaries during the course of planning and execution of cyber operations and missions. To enable effective integration of cyber deception, it would be necessary to create a systematic design process for building a robust and sustainable deception system with extensible deception capabilities guided by a Command and Control interface compatible with current Department of Defense and civilian cyber operational practices and standards. In this chapter, the authors discuss various design aspects of designing cyber deception systems that meet a wide range of cyber operational requirements and are appropriately aligned with mission objectives. These design aspects include general deception goals, deception design taxonomy, tradeoff analysis, deception design process, design considerations such as modularity, interfaces and effect to cyber defenders, interoperability with current tools, deception scenarios, adversary engagement, roles of deception in cyber kill chains, and metrics such as adversary work factor. The authors expect to present the challenges and opportunities of designing cyber deception systems and to trigger further thoughts and discussions in the broader research community.

1 Introduction

Cyber deception can become an essential component of organizing cyber operations in the modern cyber terrain. Cyber defenders and mission commanders can use it as an effective instrument for protecting mission cyber assets and network infrastructure, and for deceiving and misleading adversaries. It would be very useful to create a deception design process for building a robust, sustainable, multi-layer

G. Briskin (✉) • D. Fayette • N. Evancich • J. Li
Intelligent Automation Inc., 15400 Calhoun Dr., Suite 190, Rockville, MD 20855, UK
e-mail: gbriskin@i-a-i.com; dfayette@i-a-i.com; nevancich@i-a-i.com; jli@i-a-i.com

V. Rajabian-Schwart • A. Macera
Air Force Research Laboratory, 26 Electronics Parkway, Rome, NY 13441, UK
e-mail: vahid.rajabian-schwart@us.af.mil; anthony.macera.1@us.af.mil

© Springer International Publishing Switzerland 2016   69
S. Jajodia et al. (eds.), *Cyber Deception*, DOI 10.1007/978-3-319-32699-3_4

deception system with extensive deception capabilities guided by a Command and Control (C2) interface compatible with current Department of Defense (DoD) and civilian cyber operation practices and standards. This chapter focuses on various aspects of designing a cyber deception system that meets a wide range of deception requirements and is closely aligned with mission goals and objectives.

Section 1 of this chapter outlines deception design taxonomy and general deception goals. Section 2 describes the capability requirements for deception and various C2 considerations including tradeoff analysis, deception planning and general design workflow. Section 2 also describes the general deception design process and lays down other design considerations, such as modularity of the deception framework, resiliency, agility and Moving Target Defense (MTD) applied for deception, and C2 interface and deployment. In addition, it touches on the subject of interoperability with mainstream defensive tools, which is critical to the adoption of deception techniques into current cyber security controls and practice [7]. Section 3 discusses various aspects to consider when creating a Deception Scenario, such as deception believability, certainty vs. uncertainty in deception, static vs. dynamic deception, reactive vs. pro-active deception, using deception triggers and false positives mitigation, and leveraging Software Defined Networking (SDN) in deception. This section also describes several ways to engage an attacker in a game of deception, and outlines a relationship between cyber deception and the concept of cyber-attack lifecycle or cyber kill chains. Section 4 takes a look at some of the cyber deception challenges, such as minimizing effect on mission operations and protecting deception software itself from cyber-attacks. It discusses approaches to assessing attacker's work factor and proposes a development of Deception Domain Specific Language used for deception design. Section 5 concludes this chapter.

1.1 Taxonomies

There are a number of different deception-centric taxonomies that have been proposed for the past decade. However, most of them are largely oriented towards strategic deception planning, theoretical state-of-the-art research and discussions about deception planning and positioning in overall cyber defensive strategies and mission execution. The focus of this chapter is to introduce semantics for providing a bridge between mission planning, security policy development and design of a practical cyber deception system.

1.1.1 Deception Story

The root element and the first step in the deception creation process is the *deception story*. As stated in [4] "the cornerstone of any deception operation is the deception story. The deception story is a scenario that outlines the friendly actions that will be portrayed to cause the deception target to adopt the desired perception. It is a succinct statement or narrative of exactly what the Military Deception (MILDEC) planner wants the target to believe to be the true situation, then decide and act on that basis ... The deception story identifies those friendly actions, both real and simulated, that when observed by the deception target will lead it to develop the desired perception. Deception story development is both an analytical and creative process that involves a variety of information on enemy data acquisition and processing. The deception story weaves these elements together into a coherent depiction of the situation the target will reconstruct from the information provided." Creating a narrative for the MILDEC deception story is both a prerogative and a responsibility of military mission planners [1]. In civilian environments, a deception story should be created by the combined efforts of the organization's business leadership, Information Technology (IT) and computer security management. It is also essential that the deception story is considered to be an integral part of the overall process of the organization's security policy.

Deception cyber-scenario is the second step of creating cyber deception. Deception cyber-scenario is a formal description of the desired responses of the system, based on possible attack scenarios, existing network infrastructure and specific Concept of Operations (CONOPS)/mission/Operational Security (OPSEC) requirements. Deception cyber-scenario should be generated using domain specific language specifically created for the deception design. Such scripting meta-language should be created primarily for the purpose of formally describing a deception story. The Deception Domain Specific Language (DDSL) should support ontological constructs for deception taxonomy and deception scenario mapped to available implemented state-of-the-art deception and Moving Target Defense (MTD) techniques and methods. The deception cyber-scenario is then translated into configuration data and scripts used by the software deception controls that implement deception and MTD methods. Each deception scenario should project the target(s) potentially susceptible to deception, describe the cyber-environment, system subjects of the deception (i.e., components of the system environment to become a deception of) and define various deception objects and elements.

Finally, *cyber deception plot* is a product of the deception creation process. It is comprised of a set of software cyber-deception controls, context rules, complimentary configurations and meta-protocols that prescribe how a given mission and its environment will be portrayed, in order to cause the target to adopt the desired perception. A cyber deception plot comprises of a set of deployable software modules, and configuration and deployment scripts, for each deception and monitoring unit in the deception system.

1.1.2   Cyberspace Network Environment

The cyberspace network environment is comprised of three basic notional layers, as defined in [9]:

Physical network layer uses logical constructs as the primary method of security (e.g., information assurance [IA]) and integrity (e.g., virtual private networks that tunnel through cyberspace). This is a primary target for signals intelligence

(SIGINT), including computer network exploitation (CNE), measurement and signature intelligence, open source intelligence, and human intelligence (HUMINT). The physical components of this layer include hardware, systems software, and infrastructure.

Logical network layer consists of those elements of the network that are related to one another in a way that is abstracted from the physical network, i.e., the form or relationships are not tied to an individual, specific path, or node. The logical network layer includes Uniform Resource Locator (URL), IP addresses, routable networks, etc.

Cyber-persona layer represents another higher level of abstraction of the logical network in cyberspace. It uses the rules that apply in the logical network layer to develop a digital representation of an individual or entity identity in cyberspace. Cyber-personas may relate fairly directly to an actual person or entity, incorporating some biographical or corporate network component, such as hardware, systems, software, infrastructure, data, e-mail and IP addresses, web pages, phone numbers, etc. It should be noted that one individual may have multiple cyber-personas, which may vary in the degree to which they are factually accurate. Cyber-personas can be complex, with elements in many virtual locations not normally linked to a single physical location or form.

1.1.3 Deception Profiles

When characterizing projected targets for deception, it is important to distinguish two different categories of deception profiles.

The first category is the *level of automation* an attacker could employ. In the simplest case, three different possibilities should be considered:

The first case describes *fully automated software operating autonomously*. Examples of this case are a malware uploaded to the compromised host, or a set of automated external network probes. In this case, a defender operates against a pre-defined, often known and functionally finite/limited set of attacking techniques, which is arguably easier to defeat. However, due to the pre-defined nature of the automated software, the amount of deception information that can be conveyed to an attacker is also limited. Also, in a significant number of cases, such attacking software is relatively brittle, as it is usually designed to operate in a "non-competing" environment.

A second case involves an external attacker trying (by running individual scripts or programs) to *manually* penetrate the network/system or to manipulate the compromised node. This case seems to present the most opportunities to observe an attacker, using techniques such as honeypots or honey tokens. However, it might not give a truthful representation of an attacker's intentions, since the demonstrated attacking activity might be just a "smoke screen", covering the real attack taking place someplace else.

The third case involves *semi-automated software operating under manual or automated Command and Control (C2)*. Most of the time such malware is also attacking software operating inside the target network during post-compromise phases.

This case presents a better opportunity for engaging attackers and determining their real intentions, since the C2 channel is often costly and downloaded malware can be an expensive burnable asset that is seldom used as an attacking decoy or as "chaff".

The second category for a projected attacker can be characterized as the *level of sophistication* for potential attacking techniques. We suggest the following three basic categories:

Script-kiddies level—is an attack that mainly uses well-known, open source penetration testing toolsets and well-documented techniques that are freely available on Internet. It must be noted that this does not mean that a more sophisticated adversary will not utilize these methods. On the contrary, these methods are widely used for "phishing," or initial probing for weak spots in cyber defense and for finding "low hanging fruit". This represents a big opportunity for deception designers, assuming they provide an acceptable level of realism (more details in Sect. 3.1).

Intermediate "Business" level—is an attack that is widely used by underground criminal organizations, mostly utilizing cyber tools available for purchase on "darknet" sites. Due to higher sophistication levels of these tools and use of zero-day exploits, the deception methods and techniques should provide a significantly higher degree of consistency and depth of deception (Sect. 2.1). In addition, entrapment techniques (Sect. 3.8) must be utilized as a part of the deception solution.

State Actor level—is an attack that is performed by foreign state-sponsored, high power teams with significant resources. This category accounts for a larger share of the Advanced Persistent Threat (APT) attacks. The holistic approach of combining network- and host-based deception in conjunction with the compatible MTD techniques must be considered.

1.2 Deception Goals

In this section, we distinguish two types of deception purposes:

Deception for protection: The whole purpose of this type of deception is to improve a defensive posture, and protect the cyber-assets and infrastructure of a given mission, in order to "waste the attacker's resources while permitting time to organize a better defense" [11]. In other words, the goal here is to increase the attackers' work factor, impede their efforts, deflect attention and mitigate exposure. The intention is to deter hostile actions and increase the success of friendly defensive actions [4], and also to cause "ambiguity, confusion, or misunderstanding in adversary perceptions of friendly critical information" [2]. Specifically, the objectives are to hide the existence and/or the nature of shielded systems and/or the data stored in them to:

- Prevent and significantly reduce the probability that an adversary can recover sensitive data,
- Obfuscate the value/nature of the systems and/or the data stored in them,

- Create noise around valuable information to alter adversary perception of its importance, and
- Introduce decoys and traps to the systems, in order to detect data leakage and intrusions.

The design for this type of deception should be very closely integrated with components of agile and resilient defense, such as hardening, prevention and MTD mechanisms.

Deception for Confusion and Intelligence: The purpose of this deception is to convey certain information that goes beyond the goal of mission protection. The goal is to "deliberately mislead adversary decision makers, thereby causing the adversary to take specific actions (or inactions) that will contribute to the accomplishment of the friendly mission" [4] and potentially gather intelligence on attacker methods and intent.

These types of deception are not mutually exclusive and can be designed as complimentary. Well-researched deception taxonomies, such as the ones described in [12], can be applied for both categories: dissimilation—hiding the real (such as masking, repackaging, and dazzling), and simulation—showing the false (such as mimicking, inventing, and decoying). Also there is a third hybrid approach suggested by Rowe [14]—to hide the real by showing it to be false. In addition, enticement/entrapment methods can also be applied to both categories of deception. In an alternative taxonomy [15], concealment, camouflage, false and planted information, lies, displays, ruses, demonstrations, feints, and insights can also be mapped to both deception categories [6].

2 Capability Requirements

2.1 General Considerations

Depth of deception is probably the most important property of any practical deception solution. This is especially true for network-based reconnaissance techniques where an attacker uses a set of diversified methods to obtain, infer and verify the findings. For example, in order to perform operating system (OS) fingerprinting of a particular host, an attacker can launch a set of TCP/IP probes and then confirm the findings by enumerating the host services, through banner grabbing techniques (identifying particular vendor and software version, specific for a given OS). An internal attacker can also verify her findings through various network service discovery protocols that deliver much the same information. Hence, the set of deception methods selected to convey the OS-related information for a given host, back to the attacker, must transcend different network protocols, open systems interconnection (OSI) layers, and host and networking boundaries. It must comprise of as many deception controls as needed to satisfy and cover projected attacking techniques for a given element of a constructed deception scenario.

Consistency of deception is a deception property that reflects a decision of deception planners to expose consistent or intentionally inconsistent information about a given element of deception. Inconsistency [5] can be presented either *in-depth*, e.g., different indirect response information is given back to the attacker for different probing techniques; or to have a *timing dimension* to it, with different information being exposed to an attacker who utilizes the same method at different points in time, thus creating an attacker perception of a transient environment. An inconsistent deception can also be used as a tactic when an attacker suspects that deception is being used, or it can be helpful when the deception goal is to simply frustrate, delay and confuse adversaries. Inconsistency can also raise doubts about the validity of data already obtained by the adversary.

The deception planners must also design deception scenarios to *cover multiple cyber kill chain phases* (more details in Sect. 3.9). Both an external and internal attacker must be considered and planned for. The assumption would also have to be made that an attacker may have already established a foothold inside the network or a system, and already obtained some information about the system.

The deception planners must also assume that an attacker might be observing the network from *multiple viewpoints*, from both inside and outside of the network or system. Hence, the designed deception must be consistent across the whole system or network.

Sustainability and duration of a deception plot is another important deception property. The choice of particular techniques and methods depends heavily on deception requirements: whether a given deception scenario is being designed for a short-term mission or against an APT threat with an unlimited time span. The sustainability property is also influenced by a projected attacker profile (Sect. 1.1).

Infrastructure agnosticism, or the degree of flexibility of a deception solution to be employed in a diverse and changing infrastructure environment is important. It can be defined by an ability to efficiently function by selecting and re-configuring existing software deception controls that comprise the framework, without alteration of either deception software or the underlying user infrastructure.

Hardware and Software Platforms or the range of the software and hardware platforms on which a given deception solution can be efficiently deployed is important. It is desirable for the deception scenario to be equally played out in both virtualized/cloud environment and in more traditional enterprise settings.

2.2 Command and Control (C2)

Deception Tradeoff Analysis is a persistent C2 task that spans the mission planning stages through mission conclusion/termination. In the context of a deception, the tradeoff analysis involves identifying the effect that currently used deception methods have on the attacker, versus an impact that the applied deception solution has on normal mission operations. This analysis may also include a cost component, or the amount of human and technical resources required for the deception deployment and the maintenance.

The first part of the tradeoff analysis, assessing an effect on an attacker, should start at deception planning and design stages with a clear definition of deception goals and objectives. In the cyber-context, the goals and expectations can be expressed through description of the projected attacker's follow-up actions that can be detected/tracked by available monitoring sensors. Each event may be assigned a numerical importance factor. Matching these formally expressed expectations against actual observed events should bring a solid quantified effect measurement.

The complimentary suggested approach is based on the OODA loop (Observe, Orient, Decide, and Act) concept. OODA is a cyclic process model, proposed by Boyd [13], by which an entity reacts to an event. The purpose of deception, in the context of having an effect on the attackers, is to slow down the "observe" and "orient" stages of the attacker's OODA loop. Observed changes in attack patterns can be interpreted as start/end points of the current attacker's OODA loop. By measuring the time length of each loop, we may be able to infer the degree of the effect that the applied deception methods have on the attacker.

A projected effect on the attacker can also be measured through deception testing and attack simulation. For example, in our recent deception R&D efforts sponsored by the Air Force Research Laboratories (AFRL), we use *nmap* penetration testing tools to project an effect of our anti-reconnaissance techniques on external network topology discovery. We devise an effect quantification formula that factors in the following:

- Complexity of a real network topology: a set of observable network elements and corresponding network parameters,
- Complexity of configured deception topology schema: e.g., topology that we present to an attacker,
- The difference between real and falsified topology: we define it as a distance of deception maneuver, and
- Resulting deviations: difference between network reconnaissance tools output and the configured deception schema.

We then assess deception effectiveness by calculating a ratio between distance of deception maneuver and the resulting deviation.

The second part of the tradeoff analysis, the impact on user operations, can be measured (if applicable) by assessing the loss of productivity of the mission personnel, occurring network performance degradation, service interruptions, etc. Again, it is important to set clear metrics and tolerable mission thresholds for impact measurement during the deception planning and design stage.

2.2.1 Deception Design and Planning

Joint Publication 1–13.4 [4] defines *deception planning* as "an iterative process that requires continual reexamination of its objectives, target, stories, and means throughout the planning and execution phases … The MILDEC planning process consists of six steps: deception mission analysis; deception planning guidance; staff deception estimate; commander's deception estimate; Chairman of the Joint Chiefs of Staff estimate review; deception plan development; and deception plan review and approval."

This chapter focuses on *deception design*. The workflow of the deception design process is described in Sect. 2.3. This section outlines the expectations and requirements for deception planning and design. One of the most important inputs to the deception design is a *mission context* that is a result of the mission analysis performed as a part of deception planning. Joint Publication 3-12(R) [9] defines a mission context as a combination of "current and predictive knowledge of cyberspace and the Operation Environment (OE) upon which Cyber Operations (CO) depend, including all factors affecting friendly and adversary cyberspace forces." Important elements of the mission context are the mission's cyber assets, network assets for defense and recovery, communication channels, data feeds, user interfaces, mission capabilities (e.g., critical systems, alternative paths, backups, etc.), constraints, and boundaries and limitations that affect potential ranges of deception, stemming from the needs of providing transparent mission operations. In our deception design paradigm, we also consider factors that include:

- a specific mission network and application context,
- enterprise network setup (scale, device types, topology, typical configurations),
- dependencies and operations use cases (command and control context, mission task list),
- computing environment (operating systems, anticipated services, public and private cloud, etc.),
- cyber asset and importance factors,
- installed security modalities (firewall and configurations, intrusion, etc.),
- metrics (security metrics for analysis and test and evaluation, performance metrics, etc.), and
- workflow (where deceptions reside in the Air Task Order (ATO) production, for example).

2.2.2 Situation Awareness and Run-Time Centralized Configuration Control

Once the planning and design process is complete, the Deception Plot is developed and deployed, defining where and when the deception execution cycle must take place. The process of execution involves two basic functions, assessing and control. "Assessing involves the receipt and processing of information concerning the MILDEC operation, and control entails making iterative decisions and issuing instructions until termination. The deception plan is the basis for execution, but execution may take place in conditions that are more dynamic than the plan anticipated" [4]. One of the key requirements of situational awareness is a clear understanding of the existing Operational Environment during mission operations. Joint Publication 3-12(R) [9] defines the Operational Environment (OE) as a "composite of the conditions, circumstances, and influences that affect the employment of capabilities and bear on the decisions of the commander". Some of the cyber OE elements include but are not limited to: network outages or degradation, detected intrusions/attacks/indicators of compromise (IoCs), unauthorized activity, alerts/threat information, current network traffic analysis, etc.

Joint Publication 1–13.4 [4] defines the six principles of MILDEC that provide guidance for planning and executing MILDEC operations. These six principles are:

- Focus: The deception must target the adversary decision maker capable of causing the desired action(s) or inaction(s).
- Objective: Cause an adversary to take (or not to take) specific actions, not just to believe certain things.
- Centralized Planning and Control: MILDEC operations should be centrally planned and directed.
- Security: Deny knowledge of a force's intent to deceive and the execution of that intent to adversaries.
- Timeliness: A deception operation requires careful timing.
- Integration: Fully integrate each MILDEC with the operation that it is supporting.

Mission commanders continuously monitor changes in the OE and make adjustments. Hence, deception design should implement the capabilities to perform dynamic changes in the deception story, according to changing conditions. Such capabilities can only be efficiently provided through a centralized deception command and control infrastructure that will enable:

- Deployment and dynamic configurations for deception scenarios, and
- Dissemination, synchronization and coordination of network deception events and attack responses across the whole mission environment.

Joint Publication 3-12(R) [9] states that the "C2 of . . . defense cyber operations (DCO) may require pre-determined and preauthorized actions based on meeting particular conditions and triggers, executed either manually or automatically if the nature of the threat requires instantaneous response." The design of an effective cyber deception system should include capabilities to perform all three types of C2 operations:

- Man-in-the-loop—allowing a manual run-time deception scenario update.
- Autonomous mode—using pre-configured rule-based deception policy for triggering deception responses.
- Man-on-the loop—supporting context-aware capabilities whereas different rules might be applied to the same input/stimuli depending on current operational context and situational awareness considerations. Man-on-the-loop allows for run-time context rule updates.

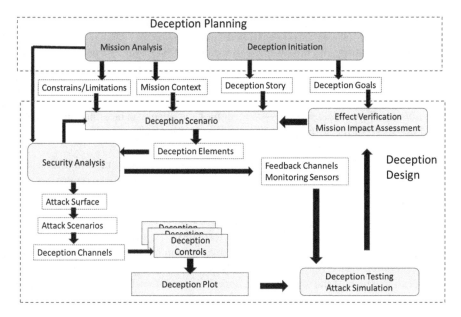

Fig. 1 Deception design work flow loop

2.3 Deception Design Process

The Deception Design process workflow is shown in Fig. 1.

As shown in Fig. 1, there are two important outputs from the *Mission Analysis* [3] described in Sect. 2.2:

- *Constraints/Limitations*—inferred descriptions of possible deception limitations based on the requirement of user transparency in mission operations
- *Mission Context*—also described in Sect. 2.2, is comprised of descriptions of cyber-assets and cyber-personas, mission operations use cases, topology, network and application contexts, dependency chains, etc.

Deception Story described in Sect. 1.1 and *Deception Goals* described in Sect. 1.2 are the other two important outputs of each Deception Planning process that are used to generate the Deception Scenario.

Deception Scenario is defined in Sect. 1.1. It is comprised of a number of *Deception Elements*. Each Deception Element formally describes a particular deception goal. For example, there might be a Deception Element that states that a falsified "topological location of host A should be shown as located behind gateway X".

Security Analysis is performed for each Deception Element of the Deception Scenario. Security Analysis determines all possible methods with which an attacker may obtain a topology of the host A in the example above. This includes analysis and determination of the *network and software attack surface* of the host itself, plus all other nodes and devices that may directly or indirectly relay the relevant information about topological location of host A (routers, Active Directory DNS, services, other hosts communicating with host A through variety of OSI Layer 2–7 protocols, etc.). The network attack surface can be defined as the sum of all potential vulnerabilities and all points of network interaction in connected hardware and software that are accessible to external (to a given node) users. The software attack surface is, in fact, a *useable software attack surface*—a portion of the attack surface that an attacker could use to crash the software, access sensitive information, or gain access to the host machine remotely.

Security Analysis is also performed for a given Mission from which relevant *Attack Scenarios* are developed. The Attack Scenarios are used for two purposes:

- To determine and define communication *Deception Channels* used to directly or indirectly relay deception information back to the attacker. As stated in [4], "within MILDEC, conduits are information or intelligence gateways to the deception target. Conduits may be used to control flows of information to a deception target." The Deception Channel will contain the context required to determine which of the available *Deception Controls* to select, in order to satisfy the requirements of a given Deception Element. The Deception Controls are software modules which, when used with accompanied configuration data and scripts, implement deception techniques and methods. The context might contain the list of required network protocol(s) and specific details of protocol message exchanges, and required deception actions, such as modification of particular network message fields, manufacturing protocol replies, blocking certain types of messages, tracking specified behavior patterns, listing the hosts/devices involved, etc.

- Perform *Attack Simulation and Deception Testing*, in order to determine the effectiveness of the selected Deception Controls and to assess a deception impact on normal user mission operations.

The Security Analysis will also determine requirements for the *Feedback Channels* that describe the type and topology of *monitoring sensors* used to detect network and system events needed for tracking an adversary's behavior and for assessing the state of the systems involved in mission operations. These channels will be used during deception design testing, as well as during mission operations.

The *Deception Plot* described earlier in Sect. 1.1, is synthesized by combining various Deception Controls. The generated Deception Plot components are then distributed among participating deception appliances and software modules.

As shown in Fig. 1, deception design is an iterative process. When the testing results are analyzed and the *Effect Verification and Mission Impact Assessment* is calculated against the stated deception goals generated during the *Deception Planning Process*, the deception scenario is refined and the whole deception design loop is repeated until an acceptable compromise between the desired effect and the mission impact is found.

2.4   Other Design Considerations

2.4.1   Modularity

As mentioned earlier, deception planning and design is an iterative process that involves continuously "checking the logic and consistency of the internal elements of the deception. This allows the deception planner to identify desired perceptions, observables, and executions that may need refinement, and to add supporting observables as needed to strengthen certain elements of the deception story or diminish the impact of troublesome competing observables. Each element of the deception story should have associated deception means that can credibly portray the data, plus identified conduits that transfer this information into the enemy's information processing system" [4]. In order to comply with those requirements, especially in the case of complex expanded missions' deception scenarios, it is necessary to provide a *modular deception framework* to enable deception designers to perform deception plot synthesis with facilities to operate and manipulate various deception elements to resolve conflicts, provide agility and build a holistic deception solution for a given mission. One approach that Intelligent Automation Inc. (IAI) is developing is to use deception controls as building blocks for a deception solution. However, there exists the challenge of combining various deception controls to work synergistically for depth and consistency in deception, and to detect and resolve possible conflicts between different deception controls that may arise during deception story presentation. A key part of this approach is to perform verification and validation during the deception design when a deception plot is being synthesized. We are currently experimenting with different ways of combining deception controls using a combination of merging and pipelining.

2.4.2   Resiliency, Agility and MTD in Deception

Deception and MTD capabilities are two integral parts of holistic agile cyber-defense posture. MTD is a cyber-defense strategy in which a set of system configurations is dynamically changing at network and/or host level to increase uncertainty and complexity for adversaries seeking to discover and exploit vulnerabilities in a target network or host environment. In IAI deception implementation, by combining MTD and deception concepts, we create a perception of a transient network environment, perform randomization and pollution of the network attack surface, and increase the work factor for an attacker to determine the direction, volume and importance of network traffic. We also introduce non-concurrency in network parameter modification schemes for unpredictable network parameter transitions and for minimization in interruption of network connectivity.

The deception design process should be integrated with planning and deployment of MTD technologies, in order to create a coherent and consistent presentation of the network and host environment to potential attackers, to reduce the cost of deployment and maintenance, and to minimize impact on user/mission operation. In our approach, we are aiming at an integrated planning, design and common configuration environment, in order to achieve mutual compatibility and effectiveness. Moreover, we believe that an integrated MTD and deception software base for network- and host-based components respectively, is the most practical and effective way to a fast and successful transition of a R&D solution into the user environment.

Deception should be an integral part of any agile software system [10] since it aligns with the main goals for achieving cyber resiliency: increasing cost to the attacker, increasing chance of detection, minimizing effect of the attacker and increasing the uncertainty that the attack was successful. While the first three stated objectives are factored largely through the trade-off analysis, the last objective should be implemented as a part of the deception story, especially in case of deception for confusion and intelligence.

2.4.3   C2 Interface

It is stated in [2] that "C2 functions are performed ... by a commander in planning, directing, coordinating, and controlling forces and operations in the accomplishment of the mission". Our deception design paradigm supports all three main C2 Interface modes. Figure 2 shows notional interfaces for these modes of operations.

For the *Man-in-the-Loop mode*, a mission commander creates deception scenarios, followed by a deception plot generation and deployment. Upon receiving orders, the mission commander updates the original deception scenarios, re-generates and deploys a new deception plot based on the Situational Awareness (SA) data and feedback from the deception system in the Mission Domain.

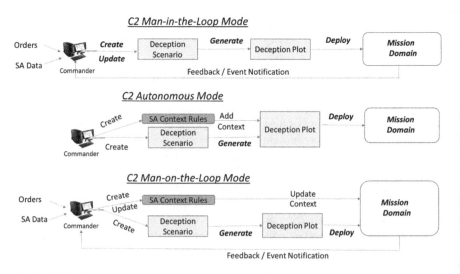

Fig. 2 C2 interface modes

For the *Autonomous mode*, the mission commander creates deception scenarios and context rules that are included in the generated deception plot being deployed. The built-in context rules will govern the triggering of different deception scenarios based on the occurrence of network events within the mission domain. In this case, the mission commander does not act based on the feedback channel information. In practice, these two modes should be combined to provide more effective deception management.

For the *Man-on-the-Loop mode*, the mission commander has an option of creating and deploying deception scenarios and context rules independent from each other. This allows for changing the deception story without re-generation and re-deployment of the deception plot. Instead, upon receiving the new SA information and the feedback from the deception system in the mission domain, the mission commander modifies a deception story by simply updating and deploying the SA context rules that govern execution of the deception plot.

2.4.4   C2 Coordinated Response and Deployment

Attack scenarios might be fairly complex and include multiple steps. Accordingly, deception scenarios should reflect this fact. For example, consider an attack scenario where an attacker establishes a foothold on one of the compromised hosts and attempts to obtain a password, and discovers and gets access to an important C2 server. The deception scenario and generated deception plot might include host-based deception controls for the entrapment (marked/trackable security credentials left for the attacker to pick-up), network authentication protocol monitoring proxy to control and intercept an attempt to use the marked password hash, and to redirect to an installed decoy and the decoy control itself. In addition, an impediment deception control may also be used to simulate a bad network connection, in order to further increase the attackers' work factor. In order to deploy a scenario of such complexity, a multi-layered and multi-staged deception plot must be created with various deception controls deployed on several hosts and deception units, which must be activated synchronously. In order to coordinate the deployment, activation and monitoring of various deception control and feedback sensors, the *Deception Management Network Protocol* must be created to facilitate dynamic configuration dissemination, deception coordination, and deployment of deception controls.

2.4.5   Interoperability with Mainstream Defensive Controls

Deception can be an effective tool for defending computer systems as a third line of defense when access controls and intrusion-prevention systems fail. The deception solution can benefit from already available network and host-based monitoring and alert generating capabilities of deployed intrusion detection system/intrusion prevention system (IDS/IPS) and security information and event management (SIEM) systems. Since it is not feasible to develop an interface to each mainstream

IDS/IPS/SIEM system, it seems more practical to develop open interfaces (southbound or northbound) to allow receiving of alerts, to process network events, and to analyze network flow generated by third party products.

Another significant topic is interoperability with access control systems. Efforts should be made to ensure that internal firewalls do not block communication between various deception units deployed throughout the network, or avoid discarding configuration dissemination updates between the centrally located deception management entity and peripheral deception units. Moreover, deception channels, used to relay deception information back to an attacker, may be impeded by the firewall that is configured to block egress Internet control message protocol (ICMP) traffic or to block all network packets with enabled IP option. Since some third party access control systems might also be subjects of an attack, they should also be part of the deception story. For example, an attacker may conduct a firewalking attack—an active reconnaissance network security analysis to determine access control list (ACL) filters and internal mappings behind a firewall. If firewalking is a part of the deception attack scenario, a deception unit should be installed in front of the firewall to provide deception coverage. Alternatively, wherever possible, the deception module should be integrated with a firewall that would defer to the deception module for all ingress and egress packets the firewall is configured to block.

3   Deception Factors to Consider for the Deception Scenario

3.1   What Is Believability?

According to [4], "The story must correspond to the deception target's perceptions of the friendly force's mission, intentions, and capabilities." In our opinion, there are several believability factors that must be considered when constructing a deception story:

- Firstly, the deception scenario under construction must include deception controls that together make the deception story *verifiable*. As stated in [4], "the adversary should be able to verify the veracity of the deception story through multiple channels and conduits." In our deception design paradigm, the synthesis of deception controls is aligned with the security analysis that aims to discover all possible channels and conduits through which a deception story is revealed and verified by an attacker.
- Secondly, as stated in [4], "the deception target must believe that the friendly force has the capability to execute the operations that are being portrayed by the deception story." In our deception design paradigm, it means that deception elements that comprise a deception scenario should be designed in a way that provides an *internal integrity* for a deception story. The deception elements exposed to an attacker should be perceived as pieces of a puzzle that, when being

put together, relay fictitious mission capabilities that will hopefully match an attacker's expectations.

- Lastly, when designing for believability, the *adversaries' cognitive biases* must be considered. There are two types of biases that should be addressed:

  - *Confirmation bias* which is defined as "the seeking or interpreting of evidence in ways that are partial to existing beliefs, expectations, or a hypothesis in hand" [16]. This is also tied with the verifiability property discussed above. In our deception design paradigm, multi-layer deception sets provide multiple confirmations for a given deception element.
  - *Conjunction fallacy bias* described in [17], which postulates that people do not subjectively attach probability judgments to events. Instead, they attach probabilities to the description of these events. Moreover, the more explicit and detailed the description of the event is, the higher the probability people assign to it. In addition, unpacking the description of the event into several disjoint components increases the probability people attach to it. Our deception design paradigm supports this concept, by combining multiple smaller deception elements in way to make the story more believable.

3.2 Projecting Certainty vs. Uncertainty in Deception (Equivocation)

Most deception designers would prefer engineering deception scenarios that project certainty and consistency. There are three main reasons for this. Firstly, it is easier and more straightforward to map an incoming mission and its deception goals and objectives. Secondly, if the uncertainty principle is used, the resulting deception plot would be harder to test and verify initially, and to maintain after deployment. Finally, it is arguably harder to guarantee the believability of a deception story that projects any uncertainty. We do believe, however, that projecting uncertainty is a valuable deception strategy for protection using deception (Sect. 1.2), as it slows down the attacker's OODA cycle and increases the attacker's work factor. In our deception work sponsored by AFRL we have expanded the concept of uncertainty in the deception story to implement a set of impediment techniques used while under attack. This set of techniques, called *protocol equivocation*, creates uncertainty, ambiguity, and inconsistency in network responses. It introduces delays, intermittent errors, and unexpected breaking of protocol sequences in network message exchanges. These techniques allow for monitoring and manipulating success and failure messages, allowing some to pass, faking some, and blocking or altering the response to confuse the attacker. Attackers expect protocol failures, because failures are abundant in attack situations, due to the attacker often not having perfect intelligence about the network. Inserting and removing failures herds the attacker to network resources at the defenders' whims. In the case of reconnaissance, in addition to getting inconsistent results from different network probes, an attacker will have intermittent network probe failures, which in theory should prevent the attacker from arriving at a definitive conclusion. This technique is especially effective in the case of internal reconnaissance that is usually performed by automated malware installed on an already compromised host. In reality, most of the known malware is not "bullet-proof" since it is written for minimum size, is fragile and is intended to operate in a non-contested environment. This implies that malware is also vulnerable to attacks and exploitation. In cases when a compromised host is identified, protocol equivocation techniques may obstruct the attacker's C2 and data exfiltration channels by utilizing protocol equivocation methods in conjunction with Man-in-the-Middle (MITM) techniques.

3.3 Is Explicit Deception Beneficial?

We define an explicit deception as a deception story with an intentionally low believability factor. Although it might appear that an explicit deception has very low usability value, we do see a couple of cases where explicit deception is important and must be applied.

The first case is the situation when an attack is detected and deception planners suspect that an attacker already gathered a certain amount of mission data. If an explicit deception ploy is then deployed and an attacker has detected it, the defenders might be able to raise doubts about the validity of data already obtained by the adversary. Note that an attacker should *infer* the deception, in other words, the deception plot should be designed to show a certain amount of effort to obfuscate the fact of the deception.

The second use case for an explicit deception is the third hybrid approach described in Sect. 1.2: to hide the real by showing it to be false. For example, the defender can hide an important C2 server by masquerading it to be a honeypot. Here, a deception designer explicitly shows a deception attempt in order to mislead an attacker about the real target.

3.4 Static vs. Dynamic Deception

In our deception design paradigm, the deception elements that a deception scenario is comprised of contain rules that govern presentation of deception information (deception action) to an attacker. These rules are based on initial deception objectives and also depend on network events received by deception systems during deployments. Examples of such events include detecting a TCP SYN packet arriving from a particular host, or detecting a sequences of packets that can be identified as reconnaissance probes. In this case, we define static deception as generating the identical responses to the same stimuli over a period of time within which the current deception configuration is in effect. However, for dynamic deception, we use a configured algorithm to change responses to the same stimuli over a period of time, thus creating uncertainty and a perception of a transient network and system environment. With our deception framework, deception designers can combine static and dynamic deception based on mission objectives, achieving a high level of believability.

3.5 Pro-active vs. Reactive Deception

One of the biggest questions facing a deception designer is how and when to deploy a particular scenario. In our deception design paradigm, a pro-active deception scenario can be deployed *before* malicious actions are detected and the identity of an attacker is determined. A reactive scenario, on the other hand, is only deployed when an attack is detected and attacker(s) are identified. With our deception framework, the deception designer should combine pro-active and reactive deception approaches, as the two approaches are complementary. A pro-active deception is usually lightweight and does not rely on attack detection. Pro-active deception should be configured to avoid impact on normal user operations. Reactive deception usually does not have an effect on user operations, and it is disabled when an attack is not detected.

3.6 Deception Triggers and False Positive Mitigation

As mentioned previously, a particular deception scenario can be triggered or activated even for a pro-active deception setup, for example, when a particular network or system resource has been "touched"/tampered. For reactive systems, a deception scenario is activated due to either attack detection by the deception system itself, or due to a received alert from third-party deployed IDS/IPS system. With all these alerts, the challenge becomes the mitigation of false positives. Our approach is to determine the threshold for the triggering rules based on a tradeoff analysis performed during the deception design phase, between the severity of alerts and the impact on user operations due to triggering of a given deception scenario. Other ways to improve the false positive ratio is to implement built-in direct threat intelligence mechanisms such as IoCs, signatures, and white and black listing. Another reliable deception trigger approach is using traps as described in Sect. 3.8.

3.7   Software Defined Networking (SDN) vs. Stand-Alone Appliances

Various deception solutions rely on the SDN concept that allows a software program to control the behavior of an entire network segment. Separating a network's control logic from the underlying data-forwarding plane allows for deployment of high-level control applications that specify the behavior of an entire network. High level network control makes it possible to specify compound tasks that involve integrating many disjoint network functions (e.g., security, resource management, prioritization, etc.) into a single control framework, which enables (1) mapping deception mission/application level requirements to tangible network constructs, (2) robust and agile network reconfiguration and recovery, (3) extremely flexible deception network planning, and in turn, (4) dramatic improvements in network efficiency, controllability, and survivability. SDN switches in the network configuration that features deception capabilities will replace the network switches inside of selected network segments. Such deception–enabled switches combine common bridge-routing capabilities with deception specific functionality through its control plane that includes monitoring, tracking and modifying network traffic, according to pre-configured deception scenarios; intercepting and controlling all local network traffic including multicast and broadcast queries and responses. It can manufacture fictitious replies on behalf of real or fake local hosts and generate synthetic messages, as well as superfluous queries, in order to entice the malicious software to initiate a spoofing attack. Though the SDN solution does scale up well, it has two significant drawbacks: (1) it has a rather large network and software attack surface, hence it represents an attack target; and (2) it requires replacement of all internal network switches with SDN-enabled switches.

The alternative to SDN-based solutions is stand-alone appliances. In a project sponsored by AFRL, we have implemented a deception solution based on stand-alone software deception units deployed throughout a network, at the edge of each network segment. These units collectively thwart reconnaissance propagation, C2 and exfiltration attempts and provide feedback through monitoring. This solution achieves deception capabilities by manipulating intercepted network packets, and selectively crafting responses on behalf of real and fake nodes inside the shielded segment. Each unit has a very small attack surface, and presents itself as a transparent network bridge, thus enabling stealth and hardening capabilities absent from SDN solutions. It is a "bump-in-the-wire" solution that runs on commercially off the shelf (COTS) software and hardware platform. It is infrastructure-agnostic and does not require changes in existing network infrastructure.

Note that the two approaches are not mutually exclusive as they can complement each other in a single deployment. For example, SDN-based switch can perform dissimilation functions of hiding or masking the real nodes on the network, while stand-alone appliances positioned "in front of" the switch would perform simulation functions such as decoying and protocol equivocation.

3.8 Engaging the Attacker

Attacker entrapment is one of the most popular cyber defense techniques that should be integrated in the deception solution design. Traps are usually used to detect an attacker. However, they are very effective at getting an attacker engaged, which enables the delivery of a variety of realistic deception information. Our approach utilizes deception objects like fake network or system entities that have no real function at the node, device or server. When such an object is accessed and/or altered, either manually by an adversary or automatically by automated tools, a corresponding deception scenario is triggered. Such objects include but are not limited to:

- Using *Honey Tokens*, or "marked" security authentications credentials like passwords, password files, security certificates, etc. to deceive attackers. An example of implementing this approach is using fake passwords ("ersatzpasswords.") or password hash files, such that when attackers use traditional cracking tools to recover users' passwords, they will discover fake passwords. When such passwords are used to login to the targeted systems, they will trigger an alarm [8]. Another approach is to use many passwords associated with a single username, where all of them are fake except one.
- Using *Indirect Traps* to generate, for example, a fictitious session initiation protocol (SIP) message that references a decoy as a voice conferencing server. Access to such a decoy would trigger a deception scenario.
- Exposing a *pre-existing vulnerability* by pretending to be unfixed to entice an attacker to use the attack surface that leads to an access to such vulnerability.
- Using *decoy documents or a data distributor* as a mechanism to monitor whether the insider accesses the decoy.

3.9 APT Cyber Kill Chains and Mission Deception Focus

In our approach, we use the Cyber Kill Chain model developed by Lockheed Martin [18] as a referencing baseline to describe different phases of a cyber-attack. A deception is especially effective in early stages of cyber kill chains, such as external reconnaissance, since it reduces risk to defenders and better exploits the attacker's unfamiliarity with the target system. During the early stages of a cyber-attack, an attacker is often "phishing", using well-known reconnaissance techniques to look for "low hanging fruits". Our deception framework provides multiple pro-active deception scenarios to mislead attackers, deflect their attention from important subjects, make the system look very unreliable by introducing deliberate delays and error messages or make the system look like a honeypot.

For the purpose of deception design, we distinguish three major phases of a cyber-attack (Fig. 3):

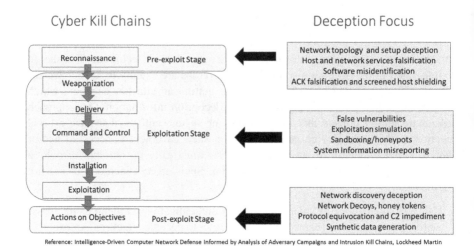

Fig. 3 Cyber kill chains and deception focus

- Pre-exploit stage: Here, the attacker conducts reconnaissance to perform host detection, service enumeration, network topology mapping, ACL and packet filter detection, and OS fingerprinting. Deception scenarios for this stage are centered mostly on network topology and network setup deception (false topology, hidden and fictitious nodes/subnets, false network parameters), providing false data on host discovery and network services, OS misidentification gateway/firewall shielding and ACL falsification.
- Exploitation stage: This covers weaponization, delivery, exploitation and installation phases of a cyber kill chain. Deception scenarios for this stage are focused on exposing false vulnerabilities and simulation exploitation processes within a decoyed sandboxing environment, misreporting system information.
- Post-exploitation stage: Command and Control and Action on Objective (internal reconnaissance, propagation, data gathering and exfiltration, denial of service, data and system destruction and alteration). Deception scenarios for this stage are mostly devoted to post-exploit deception protection against internal threats emanating from compromised nodes located inside network segments. Malicious software planted by an attacker can attempt an automated network discovery of nodes, devices, hosts, services, users, groups and network shares, as well as data harvesting from various network hosts. Other potential threats and hence opportunities for deception include horizontal malware propagation, attempts to access, modify or delete data and software, vital data collection, C2 and data exfiltration. Deception design for this stage should include network protocol manipulation and protocol exploitation, deflection of attention from vital mission cyber assets and redirection to fake services and network shares, generation of synthetic data and network information, implementation of attacker behavior monitoring, tracking and entrapment.

4 Deception Challenges

4.1 Minimizing Effect on Mission Operations

In our deception solution, we utilize a number of different approaches to minimize deception effect on mission operations. The scope of this discussion is limited to network-based deception.

We classify three behavior types of network activity.

- *Normal user event*: This type of network activity can be attributed to a normal user operation. An example is detection of a three-step TCP connection imitation sequence. Despite the fact that a potential attacker can also initiate a connection that looks legitimate, in the absence of any additional context, we treat this event as a normal user activity.
- *Dual use network event*: This type of activity can be attributed to either a potential attacker or to a power user, such as a system administrator. This type of event is considered uncharacteristic for a normal user, and not required for normal user operations. An example of this type of event is detected traceroute (ICMP) packets.
- *Plausible attacking activity*: This type of event is uncharacteristic for either normal user or for a power user. Examples include detected Xmas tree scans, reverse mapping, and web crawling attempts.

The first challenge is classification semantics to describe configuration rules for applying deception techniques as they pertain to specifics of the network environment. For example, if plausible attacking activity is detected, the deception can be triggered unconditionally. However, if a dual use network event is detected, additional conditions must be present for deception triggering. Example of such required additional information is a source IP address of the host that initiated a network connection requesting DNS zone transfer. The deception will be triggered if the request came from a host that is not expected to perform network administrative activities.

Second, different deception techniques have different impacts on different types of users. For example, presenting fictitious services on the network has no impact on normal user operations, although it might create some problems from the network management point of view. At the same time, employing entrapment techniques such as fake passwords or honey objects, such as unused data files or unmapped web pages, should not have any impact on normal or power user activities. However, using impediment techniques such as protocol equivocation may seriously impact normal user operations, and should only be triggered when an attack is detected and an attacker is identified and can be directly targeted.

Third, a deception can be triggered if the detected activity commonly attributed to normal user operations, is trying to perform the action beyond the scope of the assigned authority. For example, if a certain host is trying to initiate a TCP connection using a port that is blocked by an internal firewall, or if what appears to be a legitimate internal user is trying to access a forbidden network share.

To address these challenges, a possible alternative approach (which might, however, require alteration of existing practices) is to utilize a multi-factor authentication for all power user activity associated with network requests/queries. The impact of deception activities on normal operations is best addressed by careful selection of appropriate deception techniques aligned with security policies. It is also important to note that the provenance of each analyzed packet must be verified by either using capabilities of the deception system itself or by relying on other deployed third-party defense controls such as firewalls.

4.2 Deception Controls as Subjects of Potential Attacks

As with any other network or host-based components, deception controls may become the subject of an attacker's attention, and therefore must be hardened to prevent a compromise or denial of service (DoS). We see this as a dual objective of reducing network attack surface of the software module that contains deception controls, and of reducing software attack surface for each deception control. Both network and software attack surfaces increase when adding more deception controls to each deception unit. This is especially true for network-based deception since a network-based deception unit is supposed to process all egress/ingress and internal network traffic for a given network segment. Security analysis that determines network and software attack surfaces must be performed for each deception control added to a deception plot. To overcome this challenge, we have implemented each deception unit as a standalone appliance that presents itself externally as a network bridge, thus significantly reducing the attack surface, in comparison with router-switches, while also providing a certain level of stealth in all deception activities.

4.3 Attacker's Work Factor Assessment

As stated in [4] it is important to cause "the adversary to misallocate personnel, fiscal, and material resources in ways that are advantageous to the friendly force." The challenge is in how a credible work factor assessment can be performed. We believe that feedback channels must be identified and enabled, and used to monitor attacker activities and to measure an adversary's perceptions and actions. Feedback channels are critical in assessing the success of any deception operation or component. Hesketh [19] defines three general categories of signals that can be used to know whether a deception had an effect on an attacker:

- The target acts in the wrong time and/or place.
- The target acts in a way that is wasteful of its resources.
- The target delays acting or stops acting at all.

Some examples of reliable pieces of evidence to confirm that an attacker did invest significant resources include:

- Detected use of zero-day exploits and other stages 1 and 2 "burnable" high-value malware for installation, persistence downloading and launching backdoor programs (scripts or executables) or privilege escalation exploits
- Utilization of distributed denial of service (DDoS) capabilities, such as botnets attacking fictitious targets

Other work factor indicators, although less reliable, include the time an attacker spent conducting active reconnaissance, or an attacker switching back to reconnaissance phase.

4.4 Deception Domain Specific Language

Domain Specific Languages (DSLs) are high-level languages for design capture in particular problem areas. Our goal for Deception Domain Specific Language is creating a framework that would serve as an unambiguous specification that:

- guides and documents design process and deception scenario implementation,
- provides semantics that capture and map mission context,
- provides capabilities for a formal description of a given deception story that would allow for reasoning at a high level about a deception scenario, and for defining properties for deception elements applicable to the low-level implementation,
- enables deception scenarios to be used as a base to generate test vectors, and
- translates to the deception plot implementation.

Deception Domain Specific Language (DDSL) is a scripting specification meta-language used to create a deception scenario from a given deception story and a mission context. DDSL supports ontological constructs for deception taxonomy and deception scenarios mapped to deception and MTD techniques and methods of deception solutions. It formally describes the desired responses of the system based on possible attacker scenarios, existing network infrastructure and the specific CONOPS, mission, and OPSEC requirements. DDSL contains a *Parser* that translates formal definitions and properties of each deception element of the deception scenario into intermediate meta-data used by the *Deception Plot Generator* to select appropriate deception controls and to generate configuration and deployment scripts that comprise Deception Plot.

5 Conclusions

In this chapter we have described various design considerations for building cyber deception systems. The objective of this chapter is to provide a comprehensive introduction of various cyber deception topics, including deception taxonomy, goals, and general requirements, and to discuss important design aspects, such as the deception design process and various deign factors (e.g., believability, engagement, static vs. dynamic deception, command and control, cyber kill chain and mission context, etc.). Due to the page limit, we have only briefly described our current implementation of a holistic cyber deception system, which adopts the design principles and process depicted in this chapter. Initial experimentation has demonstrated promising future for incorporating cyber deception techniques in military defense scenarios. To see the full potential and its operational effectiveness and relevance, much work needs to be accomplished, which includes development of techniques, interoperability validation, integration with C2 framework, minimum impact on missions, rigorous test and evaluation, etc.

References

1. FM101-5_mdmp. "The Military Decision-Making Process"
2. "THE JOINT OPERATION PLANNING PROCESS FOR AIR," Last Updated: 09 November 2012
3. "Command and Control of Joint Air Operations", Joint Publication 3–30, 10 February 2014
4. Joint Publication 1–13.4 "Military Deception", 26 January 2012
5. "Thwarting Cyber-Attack Reconnaissance with Inconsistency and Deception", by Neil C. Rowe and Han C. Goh
6. http://faculty.nps.edu/ncrowe/mildec.htm
7. NIST Pub 800.53, Rev.4 SC26-SC30
8. CERIAS Tech Report 2015–11 "Using Deception to Enhance Security: A Taxonomy, Model, and Novel Uses", by Mohammed H. Almeshekah, Center for Education and Research Information Assurance and Security, Perdue University
9. Joint Publication 3-12(R) "Cyberspace Operations", 5 Feb 2013
10. "Cyber Resiliency & Agility – Call to Action", by Suzanne Hassell, MITRE Resiliency Workshop May 31, 2012
11. "Planning Cost-Effective Deceptive Resource Denial in Defense to Cyber-Attacks", by Neil Rowe. In Proceedings of the 2nd International Conference on Information Warfare & Security, page 177. Academic Conferences Limited, 2007
12. "Cheating and Deception", by J. Bowyer Bell and Barton Whaley. Transaction Publishers New Brunswick, 1991.
13. "The Essence of Winning and Losing", by Boyd, John, R., 28 June 1995.
14. "Defending Cyberspace with Fake Honeypots", by Neil Rowe, E. John Custy, and Binh T. Duong. Journal of Computers, 2(2):25–36, 2007.
15. "Victory and Deceit: Deception and Trickery at War", by James F. Dunnigan and Albert A. Nofi. Writers Club Press, 2001.
16. "Confirmation Bias: A Ubiquitous Phenomenon in Many Guises", by Raymond S. Nickerson. Review of General Psychology, 2(2):175–220, June 1998

17. "Extensional Versus Intuitive Reasoning: The Conjunction Fallacy in Probability Judgment", by Amos Tversky and Daniel Kahneman. Psychological review, 90(4):293–315, 1983.
18. "Intelligence-Driven Computer Network Defense Informed by Analysis of Adversary Campaigns and Intrusion Kill Chains", by Eric M. Hutchins, Michael J. Cloppert, and Rohan M. Amin, Leading Issues in Information Warfare & Security Research, 1:80, 2011.
19. "Fortitude: The D-Day Deception Campaign", Roger Hesketh. Overlook Hardcover, Woodstock, NY, 2000.

A Proactive and Deceptive Perspective for Role Detection and Concealment in Wireless Networks

Zhuo Lu, Cliff Wang, and Mingkui Wei

Abstract In many wireless networks (e.g., tactical military networks), the one-to-multiple communication model is pervasive due to commanding and control requirements in mission operations. In these networks, the roles of nodes are non-homogeneous; i.e., they are not equally important. This, however, opens a door for an adversary to target important nodes in the network by identifying their roles. In this chapter, we focus on investigating an important open question: *how to detect and conceal the roles of nodes in wireless networks?* Answers to this question are of essential importance to understand how to identify critical roles and prevent them from being the primary targets. We demonstrate via analysis and simulations that it is feasible and even accurate to identify critical roles of nodes by looking at network traffic patterns. To provide countermeasures against role detection, we propose role concealment methods based on proactive and deceptive network strategies. We use simulations to evaluate the effectiveness and costs of the role concealment methods.

1 Introduction

Many mission-critical wireless networks, such as tactical military networks [7, 8, 14], are formed based on mobile ad-hoc networks that involve unique challenges due to mission or tactical requirements, such as reliability and security in hostile environments. In those wireless networks, the roles of nodes are non-homogenous; i.e., they are not equally important. For example [1], Intelligence, surveillance, and reconnaissance (ISR) operations are used to collect operational information, such as status of the enemy, terrain, and weather. Such information will be delivered to the commander, who will make the best judgement to task ISR assets and soldiers. Therefore, commanding and control roles are pervasive

Z. Lu (✉)
The University of Memphis, Memphis, TN, USA
e-mail: z.l.lu@ieee.org

C. Wang
Army Research Office, Durham, NC, USA

M. Wei
North Carolina State University, Raleigh, NC, USA

© Springer International Publishing Switzerland 2016
S. Jajodia et al. (eds.), *Cyber Deception*, DOI 10.1007/978-3-319-32699-3_5

in mission-critical or tactical wireless networks. At the network level, such roles lead to the one-to-multiple communication model in the network, which is usually facilitated by either unicast or multicast protocols [11–13].

The non-homogeneity of roles in the network can be taken advantage of by an adversary to attempt to identify the role of a node before performs attacks. This enables the adversary to find and target the most valuable node inside the network, such as the commanding node that sends control commands to other nodes for mission operations. Hence, security approaches must be deployed to prevent the adversary's efforts to identify all roles in the network.

In this chapter, we focus on investigating an important question: *how to detect and conceal the roles of nodes in wireless networks?*. This question is a very important security question, and however has not been well explored in the literature. To be more specific, we say that a node has a commanding role in a wireless network if the number of its active network flows with other nodes exceeds a given threshold and say that it has an acting role otherwise. We note that the concept of a node being commanding does not necessarily mean that the node is sending real operational commands to others, but indicates that it is actively interacting with others, thereby playing an important role in the network by definition. Accordingly, the research problem becomes two-fold.

1. The role detection problem, i.e., whether we can accurately identify such critical commanding nodes in a network from an adversary's point of view.
2. The role concealment problem, i.e., whether we can protect such nodes from being identified from a defender's points of view; in addition, we also need to answer what kind of defense is good with low cost the conceal the role of a node in the network.

At the first glance, the role detection and concealment problems seem to be an endless arms race. If the adversary figures out a good detection strategy, based on which the defender can provide corresponding defense; then, the adversary can in turn change the original strategy to beat the new defense strategy, and so on. We also notice that both role detection and concealment are not well investigated in the literature. In the chapter, we focus on both problems, and present and extend the research results in our previous study in [16]. In particular, we aim to address the problem from a proactive and deceptive perspective. That is, we allow the defender to proactively and deceptively change network dynamics to make the network state hard to be accurately observed to minimize the detection performance under the adversary's current best possible detection strategy.

We first show that network flow analysis [2, 5, 6, 10, 15, 18–20] servers as a foundation for role detection, based on which we develop our role detection method. Simulation results demonstrate that it is quite feasible and accurate to identify the roles of commanding nodes. These nodes can become an adversary's primary targets, thus must be protected by role concealment methods. Accordingly, we propose a line of network strategies, which proactively and deceptively cause network traffic dynamics to counter network flow analysis based role detection. We also use simulations to evaluate the performance and cost of such proactive strategies.

The rest of this chapter is organized as follows. In Sect. 2, we introduce system and network models, as well as state the research problems. In Sects. 3 and 4, we use analysis and simulations to present our findings in role detection and concealment, respectively. Finally, we conclude in Sect. 5.

2 Models and Problem Statement

In this section, we first introduce system and network models, then formally state the research problems of role detection and concealment.

Notations. We denote by $\mathbf{A}^T$ the transpose of matrix $\mathbf{A}$. We use $\mathbb{R}^{n \times m}$ to represent the set of all n-by-m real-valued matrices.

2.1 Network Model

We consider a wireless network with n nodes (indexed by $\mathcal{N} = \{1, 2, \cdots, n\}$) distributed independently and uniformly on region $\Omega = [0, \sqrt{n/\lambda}]^2$ for a large node density λ such that the network is connected (asymptotically almost surely) [17]. We say two nodes have a network link if they are in each other's wireless transmission range r.

We also assume that nodes are move around the network and the distribution of their appearance follows the uniform distribution on the network region Ω.

2.2 Node and Role Model

Based on network activities, there two types of nodes in the network:

- Nodes that maintain many active end-to-end connectivities to other nodes throughout the network. We define such nodes to be serving as the *commanding* role in the network.
- Nodes that help forward data and only maintain a limited number of active end-to-end connectivities to other nodes. We say such nodes serving *acting* role in the network.

We can see that the role of a node (either commanding or acting) is based on the number of end-to-end connectivities. In addition, since a connectivity between two nodes means that there is a network flow with non-negligible data rate from a node to the other.

Definition 1 (Commanding and Acting Roles). We say a node is *commanding* if it has network flows with rates in rate region Σ to/from at least n_c nodes (where $n_c > 1$ is said to be the threshold for commanding); and say it is *acting* otherwise. Mathematically, we define that the role of node i ($i \in \mathscr{N}$), denoted by R_i, has value 1 if it is commanding, and value 0 otherwise; i.e.,

$$R_i = \begin{cases} 1 \text{ if node } i \text{ is commanding,} \\ 0 \text{ if node } i \text{ is acting,} \end{cases}$$

for $i \in \mathscr{N}$.

Then, we define a role vector $\mathbf{R} = [R_1, R_2, \cdots, R_n]^T$. Accordingly, the roles of all nodes can be characterized by the role vector $\mathbf{R}$ in the network. We note that $\mathbf{R}$ contains important, sensitive information in the network and should never be revealed. If $\mathbf{R}$ is disclosed, an adversary can immediately know which node plays an important (commanding) role in the wireless network.

It is worth noting that the rate region Σ is a set of allowed rates. It can be a generic region, such as $[\sigma, +\infty)$ to take into account any network flow as long as the flow rate is larger than a threshold σ. It can also be a specific region, such as $[\sigma - \epsilon, \sigma + \epsilon]$ to only consider network flows generated by a military standard with a fix communication rate σ, where ϵ is the allowed error margin.

2.3 Adversary Model

The goal of the attacker is to successfully detect the role of each node in the network; i.e., decide whether a node is commanding or acting. We assume a relatively strong attacker existing in the network. Specifically, the attacker can overhear the data transmissions on each link and estimate the transmission rate at each link (e.g., by placing eavesdroppers all over the network). The attacker is aware of the network topology; hence, given a routing protocol used in the network (e.g., shortest path routing), the attacker knows the routing path between any source-destination pair. Although this attacker model appears to be strong in practice, but it is always desirable to develop security strategies based on a strong attacker model.

The attacker will observe the network for a sufficiently-long observation period; then attempt to detect the role of each node. In this chapter, the role detection and concealment methods, and their associated operations will all happen within this observation period, unless otherwise specified.

2.4 Problem Statement

Given the network, role and adversary models, we state our research problems of both role detection and concealment. For role detection, the adversary wants to know who is serving a commanding role and communicating with a non-negligible number of nodes in the network. In order to do so, the adversary has to collect all link traffic data and attempt to use a algorithm to correlate all data with the network-wide connectivity information. We formally define role detection as follows.

Definition 2 (Role Detection). The goal of the adversary is (by observing network traffic transmissions) to find a role vector estimate $\hat{\mathbf{R}}$ such that $\hat{\mathbf{R}}$ is in close value to the real role vector $\mathbf{R}$. In the best case, $\|\hat{\mathbf{R}} - \mathbf{R}\|$ should be minimized.

For role concealment, the network designer or defender wants to design and deploy a network strategy such that the adversary is almost impossible to use the data it observes in the network to perform accurate role detection. We define role concealment as follows.

Definition 3 (Role Concealment). The goal of the network defender is to make the real role vector $\mathbf{R}$ difficult to detect. In the best case, for any node i, the adversary's estimate $\hat{R}_i$ in $\hat{\mathbf{R}}$ should be equal to real value R_i in $\mathbf{R}$ with probability 0.5 (i.e., equivalent to a random 0/1 guess).

As aforemention, it appears that the duel between role detection and concealment is an endless arms race: a concealment method can be developed based on attacking a role detection method, and vice versa. As both role detection and concealment are not well studied in the literature, we present the initial study on the two problems, which is based on our previous results in [16]. In particular, we first show that the state-of-the-art on network flow analysis makes it feasible to detect node roles in a network, then exploit proactive design of countermeasures to conceal node roles, which leads to substantial difficulty for any role detection based on network flow analysis.

3 Role Detection

In this section, we describe our design of role detection methods based on network flow analysis. We first introduce the backgrounds on network flow analysis, then design our methods, and finally use simulations to show the effectiveness of our method.

3.1   Backgrounds on Network Flow Analysis

According to our definitions, the role R_i of node i is based on the number of its network flows to other nodes. Therefore, our design of role detection must be based on network flow analysis, i.e., estimating the rates of all possible flows in the network. Recent advances in network flow analysis have already established a research line called network tomography, which is an effective way to infer end-to-end flow or link rates from network measurements [2, 5, 6, 10, 18–20]. Thus, it is necessary to briefly introduce network tomography before moving to the design of network flow analysis based on role detection.

In the network with n nodes, there are at most $\frac{n(n-1)}{2}$ undirected flows.[1] All of them are associated with a flow rate vector $\mathbf{x} \in \mathbb{R}^{\frac{n(n-1)}{2} \times 1}$, whose entry represents the rate of each flow. The attacker aims to get an estimate $\hat{\mathbf{x}}$ in close value to $\mathbf{x}$. However, the attacker cannot directly see $\mathbf{x}$, but can only observe the data transmission on each link. Therefore, the attacker has to estimate the flow rate vector from a link rate vector, which belongs to network tomography. In particular, the objective of the attacker is to compute an estimate of $\mathbf{x} \in \mathbb{R}^{\frac{n(n-1)}{2} \times 1}$ from the observed link rate vector $\mathbf{y} \in \mathbb{R}^{L \times 1}$, where L is the number of point-to-point links in the network. Each entry of $\mathbf{y}$ is the data transmission rate at each link.

It has been shown in the literature (e.g., [5, 6, 10, 19]) that $\mathbf{x}$ and $\mathbf{y}$ exhibit a linear relationship, i.e.,

$$\mathbf{y} = \mathbf{Ax}, \tag{1}$$

where $\mathbf{A} = \{a_{i,j}\}$ is called the routing matrix in the network, whose element $a_{i,j}$ has value 1 if the i-th link is on the routing path of flow j, and value 0 otherwise.

Figure 1 shows a simple example to determine the routing matrix $\mathbf{A}$ is a network. In Fig. 1, there are only 4 nodes A, B, C, D and 4 undirected links 1 (A-B), 2 (A-C), 3 (B-D), 4 (B-C) in the network. There can be 6 potential end-to-end flows in the

link indexing: 1-4	A-B	A-C	A-D	B-C	B-D	C-D	flow indexing: 1-6
	1	2	3	4	5	6	
1	1	0	1	0	0	0	
2	0	1	0	0	0	0	
3	0	0	1	0	1	1	
4	0	0	0	1	0	1	

routing matrix A

Fig. 1 A simple example of formulating network flow analysis

---

[1]For the sake of notation simplicity, we consider undirected links in this chapter. We note that the directed link case does not affect any formulation in this chapter and thus is a straightforward extension.

network: 1 (A-B), 2 (A-C), 3 (A-D), 4 (B-C), 5 (B-D), and 6 (C-D). The routing matrix **A** is a 4-by-6 matrix representing how point-to-point links form end-to-end flows. In particular, $a_{i,j}$ is 1 if the j-th flow is routed over the i-th link, and is 0 otherwise. For example in Fig. 1, flow 3 (A-D) will be routed over link 1 (A-B) and link 3 (B-D). Therefore, we can see that $a_{1,3} = 1$, $a_{2,3} = 0$, $a_{3,3} = 1$, and $a_{4,3} = 0$ in **A**.

Various methods have been developed to solve (1) in an effective way (e.g., [2–6, 10, 18]). In this section, we do not intend to develop any method to solve (1), but aim to leverage existing solutions to (1) for building a role detection method.

3.2 Detection Method Design

We design a detection method to detect the roles defined in Definition 1. As the role of a node is defined based on how many network flows it has, the method consists of two steps in the following.

1. Flow rate estimation. Use a network tomography method (or choose the best tomography method) to estimate all rates of possible flows in the network, denoted by a vector

$$\hat{\mathbf{x}} = [\hat{x}_1, \hat{x}_2, \cdots, \hat{x}_{\frac{n(n-1)}{2}}]^T. \tag{2}$$

The estimate $\hat{\mathbf{x}}$ should be in close value to **x**.

2. Role detection. For each node i, estimate its role $\hat{R}_i$ as

$$\hat{R}_i = \mathbf{1}_{\left\{\left(\sum_{f \in \mathscr{F}_i} \mathbf{1}_{\{\hat{x}_f \in \Sigma_1\}}\right) \geq \sigma_2\right\}}. \tag{3}$$

where $\mathscr{F}_i$ denote the set of indexes of all network flows from/to node i, Σ_1 is the rate threshold range for flow detection, σ_2 is the threshold for role detection, and $\mathbf{1}_E(x)$ is the indicator function that has value 1 if event E happens and value 0 otherwise.

To be more specific, the second step in the role detection method is to first compute the number of network flows from/to node i with rate within the threshold range Σ_1, then compare the number with the threshold σ_2 to decide whether the role is commanding or acting. It is obvious that if the first step can estimate the rates of all possible network flows with small error, the second step will then accurately detect all roles. In the literature, there are a wide range of tomography methods available for the first step. In this chapter, we use an efficient basis pursuit denoising method in [9].

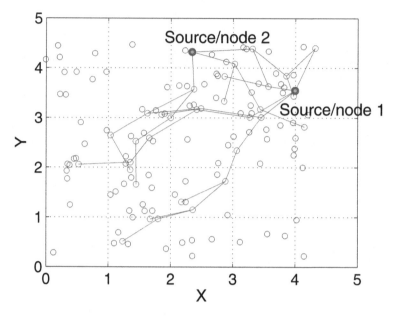

Fig. 2 Network topology and flows initiated by two commanding nodes (nodes 1 and 2) in the 2-D network region (wireless transmission range is normalized to 1)

3.3   Performance Evaluation

We use simulations to evaluate the effectiveness of the proposed role detection method. In our simulations, the transmission range of each node is normalized to 1. We generate a 100-node network with density 5 (i.e., there are on average 5 nodes in a unit area). All nodes are uniformly distributed in the network. There are 2 commanding nodes and 98 acting nodes in the network. Each commanding node is communicating with 10 other random nodes. Among all acting nodes, there are 10 random source-destination node pairs. The rate of each network flow is randomly distributed from 1 to 2 Mbps.

Figure 2 shows a network topology in one simulation run. In Fig. 2, two commanding nodes are sources/nodes 1 and 2. Each commanding node has 10 network flows to other random destinations in the network. The links with active network traffic induced by these flows are shown in solid lines. Note that for better illustration, random network flows between acting noes are not shown in Fig. 2.

Given the network topology and flow setups in Fig. 2, we evaluate the performance of the role detection algorithm. In particular, we first show the performance of flow rate estimation, as it is the basis for role detection. We use a basis pursuit denoising algorithm proposed in [9] for flow rate estimation. Figure 3 depicts the comparison between estimated and real flow rates for commanding node 1 to all possible destinations (i.e., nodes 2–100). It is observed from Fig. 3 that among all

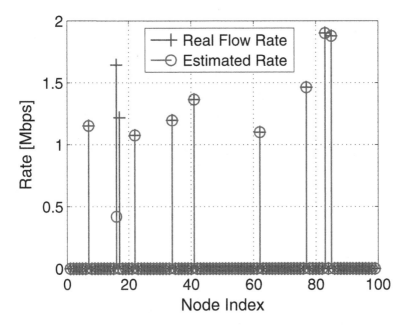

Fig. 3 Estimated flow rates from node 1 to nodes 2–100 in comparison to real flow rates

Fig. 4 Role detection performance

ten flows of node 1, eight are accurately estimated, one has a substantial error, and the one is completely miss-detected. This indicates the following.

- Most of the flow rates can be accurately estimated, providing a good foundation for the next role detection step;
- The thresholds Σ_1 and σ_2 in role detection algorithm (3) should be properly set to achieve a good balance between detection ratio and false alarm.

Next, we run role detection based on the flow rate estimation with network setups in Fig. 2. We set $\Sigma_1 = [700, +\infty]$ (i.e., we want to detect all flows with rate no less than 700 Kbps) and $\sigma_2 = 7$. Figure 4 shows the following four performance metrics.

- Flow detection error rate, which is the probability that the existence of a flow is not detected.

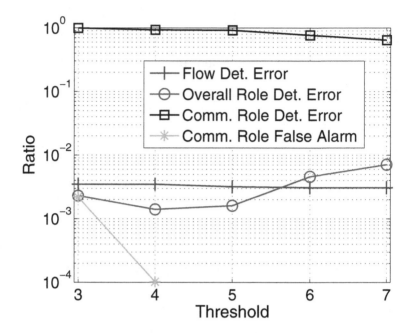

Fig. 5 The performance metrics of role detection for different threshold σ_2 (from 3 to 7)

- Commanding role detection rate, which is the probability that a commanding node is indeed detected as a commanding node.
- Commanding role false alarm, which is the probability that an acting node is mistakenly detected as a commanding node.
- Overall role detection error rate, which is the probability that the role of a node (either commanding or acting) is correctly detected.

We see from Fig. 4 that the flow detection error rate is 1.4 %, indicating that most network flows can be identified in the network with rate threshold region $\Sigma_1 = [700, +\infty]$. Commanding role detection rate and false alarm are very important metrics for detecting critical roles in wireless networks. As most of the time, the adversary may be only interested in these important nodes and consider them as the primary targets. We see that the role detection method can detect these roles with 100 % accuracy and 0 % false alarm for the network setups in Fig. 2. And the overall role detection error rate is also 0 % shown in Fig. 4.

Results from Fig. 4 are obtained from one simulation run with a particular network topology. Therefore, we also comprehensively evaluate the role detection performance by averaging 100 random network topologies, each of which also includes randomly generated network flows, commanding and acting nodes.

In all simulation runs, we set $\Sigma_1 = [700, \infty]$ and vary the value of σ_2 from 3 to 7 in the network. Figure 5 shows the performance of the role detection method. We can observe from Fig. 5 that as the threshold increases, the commanding role detection rate decreases; at the same time, the commanding role false alarm also decreases. This is intuitively true because a higher threshold means a tougher detection standard, which decreases the detection ratio and false alarm at the same time. From Fig. 5, it is observed that around 50 % of n_c is a good threshold to achieve good performance in commanding role detection. Note that this can also depend on conditions in applications, such as how many network flows usually a command node has.

In summary, our simulations show that role detection is not only feasible but also accurate, which poses a challenging issue against protecting critical nodes from being exposed in wireless networks.

4 Role Concealment

4.1 Design Methodology

We have shown the feasibility and effectiveness of the proposed role detection method in the previous section. From a network defender's perspective, it is critical to design strategies to make sure that an adversary can conclude nothing or wrong information from role detection, which we call *role concealment*. It has a great potential to be deployed in wireless networks where role detection must be prevented.

Nonetheless, an adversary can perform role detection via only passive observation or overhearing, which means that the presence of the adversary may be never correctly known. Therefore, role concealment should be proactive (i.e., always actively online) rather than following a wait-and-detect-then-act paradigm.

To systemically develop proactive strategies for role concealment, a natural starting point is to take a close look at the role detection process, then attempt to break its underlying conditions to make it not work. Apparently, role detection is based on network flow detection. Therefore, we first need to understand how network flow detection works. Figure 6a shows an illustrative example of how flow detection works: node A has an end-to-end flow with rate 100 Kbps to node H and node B also has a flow with rate 50 Kbps to node H. Suppose that there is an adversary that attempts to use flow analysis to deduce from all link observations the facts that nodes A and B have 100 and 50 Kbps flows to node H, respectively. If the adversary is aware of routing paths and overhears all link transmissions, i.e., A→C: 100 Kbps, C→F: 100 Kbps, B→D: 50 Kbps, D→F: 50 Kbps, and F→H: 150 Kbps (because two flows use the same F→H link). It is easy to use a method to get the facts: A→H is 100 Kbps, B→H is 50bps, and there is no other flow in the network.

There are two conditions that a flow analysis method relies on to successfully deduce the facts of A→H and B→H. (1) The reason that the adversary observes link transmissions is only because there exist some end-to-end flows in the network. In other words, if there is no network flow, no data transmission should be observed.

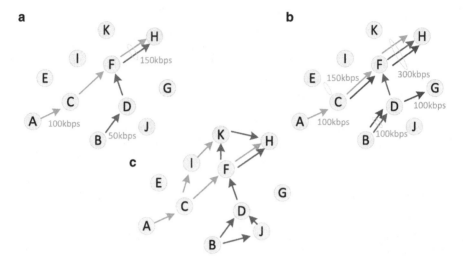

Fig. 6 Simple examples of proactive strategies: (**a**) normal network operation, (**b**) transmitting deception traffic, and (**c**) changing routing strategies

(2) The adversary is aware of how data is routed from a source to a destination, which is determined by the routing protocol used in the network.

Both conditions usually hold in a network because optimized data transmission and routing mechanisms are widely used in a network. As one major objective in network design is to optimize the performance (e.g., maximizing the throughput or minimizing the delay), it is apparent that nodes should not transmit anything if they have no data to transmit or forward. In addition, a routing path should always be optimized from a source to a destination. Such design can lead to static or predictable routing paths, which can be in turn taken advantage of by the adversary to infer information.

Accordingly, in order to make flow analysis inaccurate, strategy design should be focused on breaking the two conditions that it depends on.

- Transmitting redundant traffic into the network to break condition (i). We call such traffic *deception traffic*. In this case, the link observations are due to either deception traffic or real network flows. For example, as shown in Fig. 6b, nodes B, C, D, and F all transmit an amount of deception traffic into the network, which acts like camouflage over the real network flows. Thus, it becomes difficult for the adversary to infer the real network flow information.
- Keeping changing routing to break condition (ii). We call such a strategy *routing changing*. For example, as shown in Fig. 6c, if nodes A and B no longer use the static routing paths, but vary their paths to node H, it is difficult for the adversary to correctly acquire the exact routing path information that varies over time.

Both deception traffic and routing changing strategies are proactive and can cause more dynamics in the network to make flow analysis inaccurate, and accordingly failing role detection. On the other hand, however, they also break the requirement of the optimized network design (e.g., not always choosing the optimal routing path), thereby resulting in potential performance loss. A key question is under limited costs, what we can do for role concealment.

On the other hand, we can also adopt the following strategy to make the adversary difficult to analyze who is of the commanding role.

- Creating more commanding roles in the network. In other words, some acting nodes will pretend that they are commanding nodes and create redundant end-to-end flows to other nodes. We call such a strategy *deceptive role-making*. Apparently, deceptive role-making allows the adversary to perform accurate network flow analysis, but intentionally make more command roles hide the real ones in the network.

In this chapter, our objective is to design these three strategies with the simplest form to avoid incurring substantial operational complexity in an already complicated network environment. Thus, we consider and evaluate the following strategies.

- A deception traffic strategy, in which each node transmits deception traffic independently to its one-hop neighbor. The amount of deception traffic on each link is always bounded above such that the performance degradation is also limited.
- A routing changing strategy, in which each node will randomly select a different (longer) path for data delivery to a destination.
- A deceptive role-making strategy, in which several randomly chosen nodes create many redundant network flows to other nodes.

4.2 Simulations

We use simulations to evaluate the effectiveness and cost of role concealment methods. We generate a similar 100-node network with density 5. All nodes are uniformly distributed in the network. There are 2 commanding nodes and 98 acting nodes in the network. The rate of each network flow is randomly distributed from 1 to 2 Mbps. There exists an adversary that attempts to use the role detection method discussed in Sect. 3 to detect the role of each node in the network.

We first evaluate the performance of the deception traffic strategy. The deception traffic rate on each link is uniformly distributed from 0 to a given rate. Figure 7 shows the performance metrics of role detection affected by deception traffic with limited average traffic rate from 100–900 Kbps. We can see from Fig. 7 that as the deception traffic rate increases, the commanding role detection ratio decreases and approximately remains at 75.0%; and the commanding role false alarm increases sharply to 43.5%. This means that deception traffic significantly reduces the performance of role detection, particularly increasing false alarm as all nodes are transmitting in the network (so they are very likely to be considered as

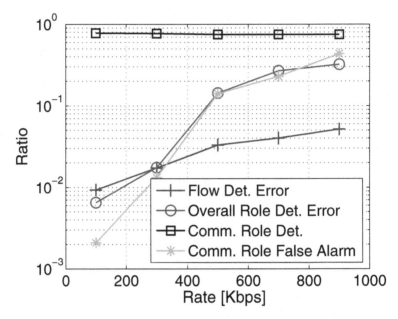

Fig. 7 The performance metrics of role detection under the deception traffic based role concealment method

commanding). It is obvious that the overhead cause by deception traffic is large as
it requires every node to transmit in the network.

We then evaluate the performance of the routing changing strategy. Each node
will use K-th shortest path routing for data delivery ($K \geq 2$); however, we assume
that the adversary has no such knowledge and thinks everyone still uses the (1st)
shortest path. This, obviously, will lead to routing information mismatch in role
detection. Figure 8 shows the performance metrics of role detection affected by
routing changing with $K = 2, 3, 4, 5, 6$. it is noted from Fig. 7 that as long as $K \geq 2$,
the commanding role detection ratio is approximately 50.0 %, which indicates that
commanding role detection now becomes a random 0/1 guess; and the commanding
role false alarm increases slightly to 1.41 %, which is much smaller compared with
the false alarm induced by deception traffic in Fig. 7. This is because when using
routing changing, no node will transmit redundant traffic (so they are less likely
to be considered having more network flows). The role detection error is due to
information mismatch.

In addition, we also measure the delay performance degradation due to routing
changing as each node will use a longer routing path, causing more delivery delay.
Figure 9 shows the delay cost for $K = 2, 3, 4, 5, 6$. For example, $K = 6$ will cause
12.2 % more delay in average message delivery in the network.

Lastly, we measure the performance of the deceptive role-making strategy.
In our simulations, there are still 100 nodes in the network, and 2 of them
are commanding nodes and the others are acting nodes. To deploy deceptive

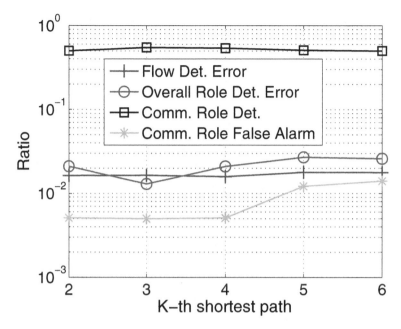

Fig. 8 The performance metrics of role detection under the routing changing based role concealment method

role-making, we randomly select s acting nodes to play the commanding role. Each of they randomly selects a number of destinations to maintain active end-to-end flows. Figure 10 illustrated a snap-shot of all network connectivities when $s = 8$, in which solid nodes are either commanding nodes or deceptive commanding nodes (acted by acting nodes). We can observe from Fig. 10 that even if the adversary can accurately identify all connectivities in the network, it still has to guess who are the real commanding nodes. If the adversary decides to choose one to attack, the best case is to hit a real commanding node with probability $2/10 = 1/5$. Apparently, the probability will decrease when we deploy more deceptive commanding nodes in the network.

4.3 Discussions

From our simulations, we can see that deception traffic is effective in causing false alarm in role detection, but requires all network nodes transmitting deception traffic, causing throughput degradation in the network. Routing changing is a good strategy to counter the commanding role detection at the cost of the delay performance. Deceptive role-making can make the adversary hard guess which node is really commanding even if the adversary can 100% identify all network

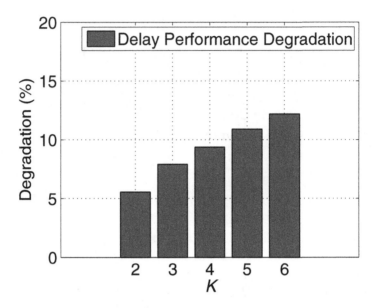

Fig. 9 Delay performance degradation

connectivities in the network. The strategy can be quite effective if the number of deceptive commands nodes is large. However, on the other hand, it brings more cost (i.e., redundant data transmissions) in the network. All of these strategies can be chosen based on different application requirements and conditions in wireless networks. We note that more sophisticated strategies (e.g., combination of two or more) can be developed upon the deception traffic, routing changing, and deceptive role-making strategies discussed in this chapter.

5   Summary

In this chapter, we studied the role detection and concealment problems from a proactive and deceptive perspective. We showed that our role detection methods can identify critical roles of nodes. Then, we proposed three proactive strategies (deception traffic, routing changing, and deceptive role-making) for role concealment, and used simulations to show the effectiveness and cost of the proposed strategies. Our results did demonstrate that it is vital to be proactive and deceptive to protect critical nodes from being identified in wireless networks.

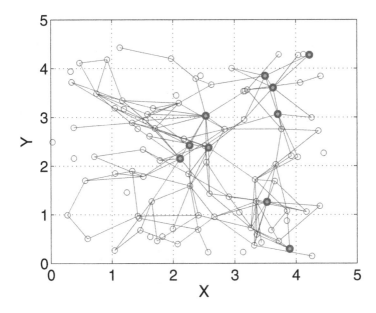

Fig. 10 The connectivity graph of the network

References

1. Bar-Noy A, Cirincione G, Govindan R, Krishnamurthy S, LaPorta T, Mohapatra P, Neely M, Yener A (2011) Quality-of-information aware networking for tactical military networks. In: Proc. of IEEE IEEE International Conference on Pervasive Computing and Communications (PERCOM) Workshops, pp 2–7
2. Bu T, Duffield N, Presti FL, Towsley D (2002) Network tomography on general topologies. In: Proc. of ACM SIGMETRICS
3. Candes E, Romberg J, Tao T (2005) Stable signal recovery from incomplete and inaccurate information. Communications on Pure and Applied Mathematics pp 1207–1233
4. Candes E, Romberg J, Tao T (2006) Robust uncertainty principles: exact signal reconstruction from highly incomplete frequency information. IEEE Trans Information Theory 52:489–509
5. Castro R, Coates M, Liang G, Nowak R, Yu B (2004) Network tomography: Recent developments. Statistical Science 19:499–517
6. Chen A, Cao J, Bu T (2010) Network tomography: Identifiability and fourier domain estimation. IEEE Trans Signal Processing 58:6029–6039
7. Elmasry GF (2010) A comparative review of commercial vs. tactical wireless networks. IEEE Communications Magazine 48(10):54–59
8. Elmasry GF, McCann CJ, Welsh R (2005) Partitioning QoS management for secure tactical wireless ad hoc networks. IEEE Communications Magazine 43(11):116–123
9. Friedlander MP, Saunders MA (2008) Active-set methods for basis pursuit. In: Proc. of West Coast Optimization Meeting (WCOM)
10. Horton JD, Lopez-Ortiz A (2003) On the number of distributed measurement points for network tomography. In: Proc. of ACM SIGCOMM Internet Measurement Conference (IMC), pp 204–209
11. Kidston D, Shi M (2012) A multicast routing technique for tactical networks. In: Proc. of IEEE Conference on Military Communications (MILCOM), pp 1–6

12. Kunz T, Li L (2010) Broadcasting in multihop mobile tactical networks: To network code or not. In: Proc. of The International Wireless Communications and Mobile Computing Conference (IWCMC), pp 676–680
13. Kunz T, Li L (2014) Robust broadcasting in tactical networks using network coding. In: Proc. of IEEE Conference on Military Communications (MILCOM), pp 1213–1222
14. Lee SH, Lee S, Song H, Lee HS (2009) Wireless sensor network design for tactical military applications: remote large-scale environments. In: Proc. of IEEE Conference on Military Communications (MILCOM), pp 1–7
15. Lu Z, Wang C (Apr. 2015) Network anti-inference: A fundamental perspective on proactive strategies to counter flow inference. In: Proc. of IEEE International Conference on Computer Communications (INFOCOM)
16. Lu Z, Wang C, Wei M (Oct. 2015) On detection and concealment of critical roles in tactical wireless networks. In: Proc. of IEEE Conference on Military Communications (MILCOM)
17. Penrose M (2003) Random Geometric Graphs. Oxford Univ. Press
18. Soule A, Lakhina A, Taft N, Papagiannaki K, Salamatian K, Nucci A, Crovella M, Diot C (2005) Traffic matrices: Balancing measurements, inference and modeling. In: Proc. of ACM SIGMETRICS
19. Yao H, Jaggi S, Chen M (2010) Network coding tomography for network failures. In: Proc. of IEEE International Conference on Computer Communications (INFOCOM)
20. Zhao Q, Ge Z, Wang J, Xu J (2006) Robust traffic matrix estimation with imperfect information: Making use of multiple data sources. In: Proc. of ACM SIGMETRICS, pp 133–144

Effective Cyber Deception

A.J. Underbrink

Abstract Cyber deception may be an effective solution to exposing and defeating malicious users of information systems. Malicious users of an information system include cyber intruders, advanced persistent threats, and malicious insiders. Once such users gain unobstructed access to, and use of, the protected information system, it is difficult to distinguish between legitimate and illegitimate users.

We view cyber deception as comprised of two broad categories: active deception and passive deception. Active deception proactively applies strategies and actions to respond to the presence of malicious users of an information system. Actions of a malicious user are anticipated prior to their execution and counter actions are predicted and taken to prevent their successful completion or to misinform the user. Active deception may employ decoy systems and infrastructure to conduct deception of malicious users and sometimes assumes that a malicious user has already been detected and possibly confirmed by sensing systems.

Passive deception employs decoy systems and infrastructure to detect reconnaissance and to expose malicious users of an information system. Decoy systems and services are established within the protected boundary of the information system. Interactions with decoy systems and services may be considered suspicious, if not conclusively malicious. Since reconnaissance and exploration of the information system are the first steps in the process of attacking an information system, detecting reconnaissance enables an active defense system to quickly identify a malicious user and take action. Like active deception, passive deception can provide misinformation to the malicious reconnaissance. We argue that effective cyber deception includes both active and passive techniques.

1 Introduction

Malicious users include cyber intruders, advanced persistent threats (APT), and malicious insiders. All types of malicious users are a major threat to critical information systems. Although cyber deception has been explored as an effective

A.J. Underbrink (✉)
Sentar, Inc., 315 Wynn Drive, Suite 1, Huntsville, AL 35805, USA
e-mail: al.underbrink@sentar.com

© Springer International Publishing Switzerland 2016   115
S. Jajodia et al. (eds.), *Cyber Deception*, DOI 10.1007/978-3-319-32699-3_6

solution to detecting and defeating malicious users, we view cyber deception as comprised of two broad categories which may be used in cyber operations to counter malicious users: active deception and passive deception. We define active deception as proactively applying strategies and actions to interactively engage malicious users of the information system. Actions of a malicious user are anticipated prior to their execution and predictive counter actions are executed to prevent their successful completion. Active deception may employ decoy objects and infrastructure to conduct deception of malicious users. Active deception typically assumes that the presence of a malicious user has been detected and possibly confirmed by cyber sensing systems.

We define passive deception as employing decoy products and infrastructure to sense and detect suspicious use and to expose malicious users of an information system. Decoy systems and services may be proactively instantiated within the protected boundary of the information system. Since reconnaissance and exploration of the information system are the first steps in the process of attacking a protected network system, detecting reconnaissance enables a computer network defense (CND) system to be proactive rather than reactive. Passive deception can play an important role in the early detection of reconnaissance and the disruption of the reconnaissance process. In this article, passive deception is advocated as a large-scale sensing system for active cyber deception.

Effective cyber deception therefore may include elements of both active and passive techniques. From the perspective of active cyber deception, an architecture for command and control (C2) is needed. C2 may be centralized or distributed to proactively deceive malicious users. Passive cyber deception is arranged as a sensor system for detecting the presence and activity of malicious users.

2 Survey of Related Work

Deception has been used extensively in cyber research, particularly in the use of honeypots and honeynets constructed for the purpose of gathering information about attackers and their techniques (e.g. [1, 7, 10, 16]). Although as a strategic tool for defending information assets, deception has received less attention. A number of researchers have investigated the possibilities, with the result that a number of interesting techniques have been proposed.

Early work in the area of deception includes exploring the use of deception in cyber warfare; see Tirenin and Faatz [17]. Although deception has sometimes been viewed as an offensive activity, deceptive information operations may be used to support defensive functions. In Tirenin and Faatz's view, the deception objective of cyber defense is to create confusion and uncertainty for potential attackers regarding the value and location of critical information systems and resources. They recommended that for the deception to be effective, it must be dynamic, presenting a continually-changing situational picture to the attacker. These changes must occur rapidly, preventing the attacker from constructing a valid understanding of the systems and networks. However, these changes must be transparent to legitimate users. This requires considerable knowledge of the friendly systems and networks, as well as a high degree of cooperation among the elements participating in the deception; moreover these elements are likely to be physically and logically dispersed and decentralized, and possibly belonging to coalition members of the total friendly force, as well as the commercial infrastructures.

From a strategic perspective, Gerwehr and Anderson [6] proposed that deception could be used in deception in two general ways: (1) as protective measures and (2) as intelligence-gathering measures against a range of attacks on information infrastructure. In the view of these researchers, effective deception requires a well-defined CONOPS, and they propose a prototypical concept of operations which the authors they call *deception in depth*. In this approach, defensive deceptions are implemented in concentric layers, with the weakest at the periphery and the strongest at the core. This is similar to operating system security models. A consequence of this is that attackers are deceived differentially, depending on their capabilities. This provides the defenders with differential outcomes, enabling them to assess the intentions and capabilities of the attackers. Gerwehr and Anderson argue that through a well-developed theoretic framework, tool development, wide-ranging experimentation, and thorough analysis it is possible to hope to provide operational guidance and instruments for both employing and combating deception in information security.

Michael and Wingfield [8] investigated policy implications for the potential for abuse in the use of cyber decoys as a means for automating counterintelligence activities and responses to cyber attacks. Noting that not just states, but nongovernmental entities and individuals may employ cyber decoys, they presented a principled analysis of the use of cyber decoys, and they explored the absolute minima in terms of customary principles for what might be considered to be acceptable use of deception. In their view, a decoy anticipates some type of inappropriate interaction between a calling process and a protected unit of software, providing in advance rules for learning about and evaluating the nature of the interaction, in addition to rules for response. Policies are needed that would place boundaries on the extent and type of deception to be employed, but providing some degree of latitude to the user of decoys to inject creativity into deceptions so as to increase the likelihood that the deceptions will be effective. The boundaries could be used to delineate the thresholds that if breached could result in the misuse or unlawful use of decoys.

Yuill [18] investigated the processes, principles, and techniques that are involved in deception operations for computer security defense. Computer security deception operations are defined as the planned actions taken to mislead attackers and thereby cause them to take (or not take) specific actions that aid computer-security defenses. An objective of this research was to systematically model and examine computer security deception operations. The research addressed these issues by focusing on deception for computer-security defense. The four main contributions of this research were: (1) a process model for deception operations which would provide deception planners with a framework for conducting deception operations; (2) a process model of deceptive hiding which would aid the defender in developing new hiding techniques and in evaluating existing techniques; (3) deception-based intrusion detection systems; and (4) experiments and evaluation. This research provides exploratory and confirmatory assessment of the deception models.

Yuill et al. [19] defined a model for understanding, comparing, and developing methods of deceptive hiding. The model characterized deceptive hiding in terms of how it defeats the underlying processes that an adversary uses to discover the hidden thing. Yuill et al. posited that an adversary's process of discovery can take three forms: direct observation, investigation, and learning from other people or agents. Deceptive hiding works by defeating one or more elements of these processes. They applied their model to computer security. They believe their process model offers a conceptual framework for developing new deceptive hiding techniques and for evaluating existing techniques, and that it provides a common frame of reference for collaboration among security professionals [19].

Rowe [12] developed a taxonomy of offensive and defensive deceptions using a new approach derived from semantics in linguistics and rated the appropriateness of each of the deceptions for offense and defense in cyber war. The intent of the taxonomy was for military planning, but this taxonomy has received little attention within research circles or from industry.

In Rowe [13], the author observed that deception is a longstanding practice in warfare, and it would be natural to extend the practice to warfare within cyberspace. He surveyed the use of deception as a means of cyber defense and found that honeypots have been the most popular form of deception. Other uses identified by Rowe include fake information, false delays, false error messages, and identity deception. Rowe also proposed the use of strategic deceptions, such as advertising technical weaknesses in one's systems in the hope of inducing an attack that one knows could be handled. Rowe concludes that such deceptions can be difficult to implement since they tend to require coordination of large numbers of people and data.

Rowe et al. [14] developed a testbed for conducting defensive deception experiments with the normal random background of attacks from the Internet. The objective of the testbed was not to experiment in deceptive cyber defense, but to use deception as a means for investigating cyber attacks. The testbed was built on top of a honeypot modified to use various deception methods to fool an attacker. Their testbed permitted full interaction by an attacker with the system, thus enabling a wide range of deceptions. Their experiments indicate a number of conclusions: (1) attacks are less likely to occur on Friday, Saturday, and Sunday; (2) attacks on a newly used IP address are high at first and then decrease significantly over a few months; this suggests that a good way to reduce attacks on a new computer is to reuse an existing IP address when possible; (3) a few common vulnerabilities are repeatedly attacked. By maintaining up-to-date patches on these vulnerabilities, attack traffic will over time reduce, making it easier to handle attacks associated with less common vulnerabilities; (4) taking a system offline increases alerts due to ICMP traffic; since there are few relatively ICMP vulnerabilities and most are straightforward to fix, taking a system offline irregularly may encourage attackers to waste more time fruitlessly attacking a secure system instead of searching for a more susceptible target; (5) many of the commonly attacked vulnerabilities refer to features of the configuration of a site; by setting these parameters to uninteresting values or turning off the services may discourage interest by attackers; (6) the last packet received from a particular IP address shows a limited range of flags, and it may be possible to exploit this in defensive deception.

Neagoe and Bishop [9] analyzed the use of deception for cyber defense and noted that while in the past, defenders of systems used deception haphazardly, recent research has employed systematic deception methods. While this research has emphasized the notion of internal consistency, Neagoe and Bishop challenged this notion and explored possible uses of inconsistency in deception as a defense. They found that inconsistency can be as effective systematically consistent approaches, and has the added advantage of being easier to implement than the consistency.

Borders et al. [2] developed a system called OpenFire which used deception to interfere with the cyber reconnaissance activity. Unlike firewalls which block unwanted traffic, OpenFire accepted all traffic and forwarded unwanted messages to a cluster of decoy machines. To the outside, all ports and all IP addresses appeared open in an OpenFire network. They found that OpenFire reduced the number of attacks on real services by 65 % as compared to an unprotected system and by 46 % as compared to a honeypot-protected system.

Bowen et al. [3] proposed the use of trap-based defense mechanisms and a deployment platform for addressing the problem of insiders attempting to exfiltrate and use sensitive information. Their goal was to confuse and confound an adversary requiring more effort to identify real information from bogus information and provide a means of detecting when an attempt to exploit sensitive information has occurred. Under their scheme, "decoy documents" would be automatically generated and stored on a file system by the system with the aim of enticing a malicious user. They also embedded "stealthy beacons" within the decoys that would cause a signal to be emitted to a server indicating when and where the particular decoy was opened.

Ryu et al. [15] examined an information security system involving the deception system from an economic perspective. They developed a deception system model for enticing unauthorized users and restricting their access to the real system. Their model represents a system designer's defensive actions against intruders in a way that maximizes the difference between the intruders' cost and the system designer's cost of system protection. They found that when the unique information content of the deception system is high, the system designer should be cautious. Intruders may enter the deception system first and then, using the information gained from hacking the deception system, they may enter the real system. The optimal security strategy in this case is to raise the total protection level for the deception system as early as possible. High unique information content in the real system helps intruders to directly penetrate the real system during both periods. Therefore, the system designer needs to raise the protection level for the real system during both time points. Finally, when the common information content is high, the optimal security strategy is to increase the protection level for both systems during the earlier time period. The proposed model shows that intruders differ in behavior depending on the system's vulnerability at the time of intrusion as well as depending on their own economic incentives. The optimal results of the proposed model provide the system designer with insights on how to configure the level of protection for the two systems.

Bowen et al. [4] developed a prototype system for automatically injecting decoys used to detect attackers who are eavesdropping on network traffic. Key elements of the system included an automated decoy traffic generator, a decoy broadcaster, and trap-based decoys that were configured to expose the attacker when the attacker attempted to use them. They used human subjects and automated tools to evaluate the believability of the decoys and found that experienced judges could not distinguish them from authentic traffic. They also tested the system on a wireless network and successfully detected eavesdropping and exploitation attempts.

While the use of deception for defending information assets is not commonplace, clearly the research shows that it has its place within the information operations toolset. And yet none of the research reviewed actually puts into practice the vision introduced by Tirenin and Faatz [17]; that is, to be effective, the deception strategy must be dynamic, presenting a continually-changing situational picture to the attacker, preventing the attacker from understanding the targeted systems and networks.

3 Active Deception

We propose the broad categorization of deceptive techniques according to how the deception engages attackers. These are *active* and *passive* deception. It is our postulation that both types of deception are useful when engaging adversaries and that neither should be neglected when establishing a layered defense system for information systems. We discuss an approach to both that we have researched, developed, and partially implemented.

For active deception, we designed and developed interactive deception concepts for end node protection using illusionist techniques. We refer to the concept as *Legerdemain*, a simplification of the French words *léger de main*, referencing "sleight of hand" techniques used by magicians to secretly manipulate objects.[1] Illusionists use sleight of hand techniques to manipulate objects for entertainment purposes. Objects may be held in ways that are not visible to an audience, objects may be stowed and retrieved to and from pockets, and audience attention may be directed away from necessary object manipulations.

When creating a routine, an illusionist often combines multiple sleight of hand techniques for effect.[2] Depending upon the objectives of the illusionist, different

---

[1] See the Wikipedia entry for "sleight of hand" at http://en.wikipedia.org/wiki/Sleight_of_hand.

[2] For an fun explanation of sleight of hand techniques, see the video by illusionists Penn and Teller at http://www.youtube.com/watch?v=oXGr76CfoCs.

combinations may be employed as a series of manipulations that bewilders and delights an audience. All the while, a very intentional series of operations is performed without being discovered.

The Legerdemain project developed an end node protection concept based on illusionist techniques. Just as the illusionist manipulates objects to deceive the audience, critical program information (CPI) often exists in the form of software "objects" (e.g., algorithms, databases, decision criteria, authentication credentials, etc.). Such objects can be secretly manipulated for the purpose of confusing reverse engineers, attackers, and malicious insiders; hiding important items; and trapping unsuspecting adversaries. Likewise, CND protections may also be viewed as software objects that may be covertly deployed, replaced, moved, or otherwise manipulated. The Legerdemain approach applies illusionist techniques to manipulate both the items to be protected *and the protection mechanisms* in order to secure CPI against malicious users.

The possible manipulations of the deception system may be assembled into *strategies* which correspond to illusionist routines. The result is an ability to construct a catalog of end node protection mechanisms which can be reused. The catalog includes actions, end node architectural component locations, and attack models. The domain model was specified as an ontology for deceptive software protection techniques. The ontology maps concepts used to describe classes of software objects and for representing the manipulation of those software objects. With an ontology of locations, actions, attacks, and objects, sequences of actions are captured as operational strategies. The strategies are used and combined to deceive malicious users of protected systems.

Legerdemain actions and strategies were specified as design patterns which can be reused in multiple end node protection implementations. The design patterns were categorized into two groups: action patterns, which describe atomic operations involving protected items, and strategy patterns, which are composed of a sequence of actions. An *action pattern* describes an operation involving an object, a source location, and a destination location, as specified by the software protection ontology. The *hide* action pattern transfers an object from an in-band location to an out-of-band location. In-band and out-of-band locations are those visible and invisible, respectively, from the perspective of user space. The *obtain* pattern transfers an object from an out-of-band location to an in-band location (this is the opposite of *hide*). The *move* pattern transfers an object from one in-band location to another in-band location, or from one out-of-band location to another out-of-band location. These restrictions on the source and destination locations distinguish the move pattern and constrain its use within strategies. Lastly, the *dispose* pattern deletes an object altogether. This is somewhat of a special case in the sense that the destination location is non-existent. This pattern may be useful to an end node when there is no other recourse to protecting the CPI object.

A *strategy* pattern describes how to protect a *protected item* against a specified *attack vector* via a sequence of *action* patterns. In other words, a strategy is a sequence of actions intended to protect some software object from a specific attack. Two important aspects of a strategy pattern are therefore (1) its intended use, and (2) the type of attack it is intended to protect against.

These patterns may be applied to protect a number of different types of software object, including data files and configuration files which specify CPI. The strategy patterns may monitor software objects such as a keyboard interrupt handler, a system management mode (SMM) handler, or a password file. That is, the same design pattern may be applied to different kinds of software objects.

Once malicious activity has been detected, several forms of countermeasures may proceed. First, monitoring with no action may continue. This may be desirable for learning more about an adversary (as with a research honeypot). Second, malicious activity may be prevented overtly. This denies an attack without regard for the adversary's awareness of the denial. Lastly, deception may be used to delay or counteract the adversary.

Like the countermeasure patterns, deception strategies may involve continued monitoring or overt prevention. Two other deception strategies may also be used against an adversary: (1) prevention while providing no feedback to the adversary, and (2) prevention while providing false feedback to the adversary. Both deception approaches prevent an adversary action without making it known to the adversary. The former deception may be confusing and the latter deception may hide the protection actions for an extended period of time. All of these strategies may be effective in confusing an adversary, especially if applied inconsistently.

3.1   Game Modeling with Deception

Once a critical mass of action and deception design patterns has been implemented, the implementations may be used in an active cyber deception system. We developed a game theory model for actively deceiving targeted users. The formal "cyber deception game" is based on long-standing concepts from economics in which two or more players in a game execute strategies to their respective benefit. The game model consists of *two players*: the defender and the attacker. The software protection system is the defender and the intruding adversary is the attacker. The game model will be *asymmetric* in that the two players will not have access to the same information about the end node. The defender will have much more information about the end node state, but both players will have *incomplete* knowledge.[3] The asymmetry arises from the defender being aware of the attacker and the attacker's moves. The attacker will have little or no awareness of the defender moves since most of the defender moves may be executed in ways hidden from the attacker.

---

[3]For the defender to have complete knowledge requires an impractical number and placement of sensors and actuators spread throughout the end node.

The deception game design is modeled as a *dynamic*, or extensive, game in which there are many stages of the game. Moves are not strictly by turn, but may be simultaneous and sequential by each player. Moves by both players will be *repeated* as long as the defender chooses to continue the deception. To conclude the game, the defender may choose at any time to remove the attacker from the end node through deletion of all attacker network sessions, processes, user accounts, and files installed on the end node. The "solution" to the deception game, therefore, is a *mixed strategy* that is always under the control of the defender.

The *goals* of each player are included in the game design. For the defender, the anticipated goals are to continue the deception game as long as necessary and to collect information about the attacker and attack methods. For the attacker, the goal may be to establish an APT. Examples of intermediate attacker goals to establish an APT include the aforementioned reconnaissance, escalation of privilege, and malware installation. These goals will be associated with the attack vectors specified in the ontology.

For each player goal, we quantitatively modeled the *expected payoff* as a utility function. The utility functions add to or deduct from a cumulative score according to the objectives and overall payoff of achieving the goal. Thus, an important part of the game design was to determine the utility functions used by the defender and the attacker. For this, we developed attack graphs, which may or may not include loops. Without loops, the attack graphs may be viewed as attack trees for achieving the overall goals of the attacker. The root of the tree is the overall goal of the attacker, which is a successful breach of the end node security. In this way, attacker moves can be anticipated and deception used to prevent achievement of the overall goal. Ideally, the defender remains at least one step ahead of the attacker for all potential attacker moves.

Decisions by the defender use the game model, the defender and attacker utility functions, and a Competitive Markov Decision Process (CMDP). CMDP is a classic framework for making decisions in an asymmetric game as designed. Via CMDP, we analyzed the attacker's potential strategies and moves. The most likely attacker moves were prioritized and decisions with predecessor moves were made using the catalog of moves for the defender and the attack vectors for the attacker.

We implemented a deception game to secure and defend end nodes. The end nodes considered were Linux based and four candidate attacks were developed for analysis:

1. Data exfiltration,
2. Creating a persistent presence,
3. Remote control exploit, and
4. Denial of service.

The remote control exploit, where an end node is placed under the control of a user from across the network, was selected for further consideration because it can be a superset of the other three candidate attacks. Developing and solving a model for the remote control exploit game provided enough information to model the other attacks.

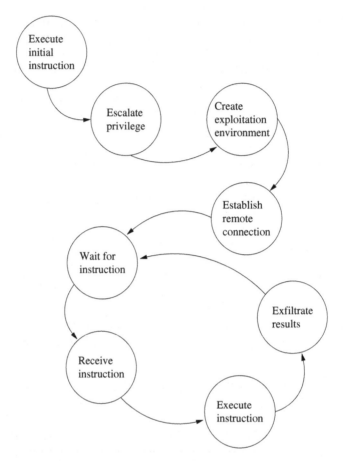

Fig. 1 Remote control attack graph

An attack graph, modeling each step of a remote control exploit and shown in Fig. 1, was created from root-kit reference literature. The purpose of the attack is to allow a remote user to take control of an end node by enabling the user to remotely execute instructions on that node. Once a connection is established on a victim node, an attacker can continually execute the process of sending and executing instructions.

A number of exploit techniques can be used at each step in the graph. The Mitre Common Vulnerabilities Enumeration (CVE) database was used to accumulate a list of vulnerabilities for each technique. These vulnerabilities and exploits for the vulnerabilities are listed in Table 1. Exploits were found using Metasploit and root-kit references. These resources provided tested and verified techniques for exploits which accelerated prototype implementation.

Table 1 Linux vulnerabilities, exploits, and scoring

Remote control execution	Vulnerabilities			Exploit	CVSS rating	
Execute initial instructions						
TCP/IP port accepting connections	CVE-2007-5244	CVE-2016-0273		Open marker file	9.3	7.5
USB device	CVE-2011-0640		CVE-2006-6881		6.9	7.5
Wireless connectors/drivers	CVE-2006-6332	CVE-2008-4395	CVE-2008-5134	Madwifi. overflow	7.5	8.3	10.0
Buffet overflow	CVE-2009-0065	CVE-2006-3705	CVE-2002-1522	SCTP exploit	10.0	7.5	5.0
Stack	CVE-2011-0404	CVE-2005-1099	CVE-2007-5243	CVE-2011-0495	Net support GLD buffer overflow ***INET_connect	7.5	10.0	9.3	6.0
Heap	CVE-2011-4913	CVE-2011-2497	CVE-2011-1493		7.8	8.3	7.5
Function pointer	CVE-2008-0009				2.1	
Return oriented programming		CVE-2005-1768	CVE-2004-1070		3.7	7.2
Printf	CVE-2009-0689	CVE-2008-1391	CVE-2003-0969		6.8	7.5	7.5
Injection attacks	CVE-2011-0923	CVE-2009-2946	CVE-2007-1974	HP data protector EXEC_CMD	10.0	9.3	7.5
Network session hijacking						
Man-in-the-middle	CVE-2011-3188	CVE-2002-1976			7.8	2.1
Network spoofing	CVE-2007-3843	CVE-2010-4648			4.3	3.3

(continued)

Table 1 (continued)

Remote control execution	Vulnerabilities				Exploit	CVSS rating			
Virus, malware	CVE-2000-0917	CVE-2006-4326	CVE-2006-6408		Format string	10.0	70.5	5.0	
Phishing, social engineering	CVE-2012-3976	CVE-2010-2654	CVE-2007-1970	CVE-2007-1796		5.8	5.8	5.0	5.0
Escalate privileges	CVE-2010-3081	CVE-2008-4258				7.2	6.2		
Modifying of exploiting uninitialized	CVE-2009-2692	CVE-2008-0600			Socket send page	7.2	7.2		
Kernel stack	CVE-2011-2319	CVE-2011-4062	CVE-2011-2517			4.3	7.2	7.2	
Kernel heap	CVE-2011-1759	CVE-2011-1477	CVE-2011-1017			6.2	4.6		7.2
Interprocess communications pathways	CVE-2009-1185	CVE-2011-2517	CVE-2009-2698		Udev netlink	7.2	7.2	7.2	
Drive vulnerabilities	CVE-2008-2812	CVE-2011-2211	CVE-2009-3043			7.2	7.2	4.9	
Trojan	CVE-2008-2040	CVE-2010-3182				7.2	6.9		
Establish connection to remote									
Fork process on port, create pipe to shell					Interact connection port inline				
Open network connection	CVE-2003-0019				Reverse TCP blind TCP	7.2			
Now user account	CVE-2008-4210				Add user UIDO	4.6			

4 Attacker Model

If an attacker assumes an exploit will not be detected and there will be no defensive presence on the victim node, then a Markov decision process (MDP) can be used to model the attacker's states and actions. Solving the MDP will produce an optimal strategy that allows the attacker to maximize their benefits. To solve for an attacker's strategy, this assumption and technique was applied.

5 The Attacker Game

The attacker is limited to two exploit options for each step prior to the execute and exfiltrate loop in Fig. 2: execute initial instructions, escalate privileges, create environment, and establish connection. The attack game diagram is so large that it cannot be inserted and legible; it is included here solely for illustration purposes. However, Table 2 is a tabular description of the diagram in Fig. 1. The exploits used in the model were selected from the exploits that have readily available code in Metasploit and are considered more reliable based on the Common Vulnerability Scoring System (CVSS) score. At each state, we compute the attacker moves from a number of possible actions and the selected action determines the state transition probabilities. All states before S7 and S8, connection established, and S16, instruction executed, have three actions to select from. Both S7 and S8 only have two state transitions because the only options at these states are to move to the wait instruction state, S9, or to be detected.

State 9 has seven actions to select between that represent the type of instruction sent. At state 17, the attacker must choose whether or not to continue executing and exfiltrating, or to end the exploit, S18. When the exploit is stopped, S18, an attacker must decide whether or not to start a new exploit on the node which will take the state machine back to S0, the initial state.

Most action has three possible outcomes: failure, success, and detection. If a chosen action fails, the attacker remains in the current state. If the action succeeds, the attacker moves to the next exploit state. If the attacker is detected, they return to the initial state, S0. Failure is defined as an exploit that does not complete successfully but does not cause unexpected results, e.g., crashing the machine. On the other hand, detection occurs when an unexpected event is triggered that indicates attacker presence. Detection therefore includes the probability of a defender becoming aware of attacker's presence due to evidence left behind from an exploit.

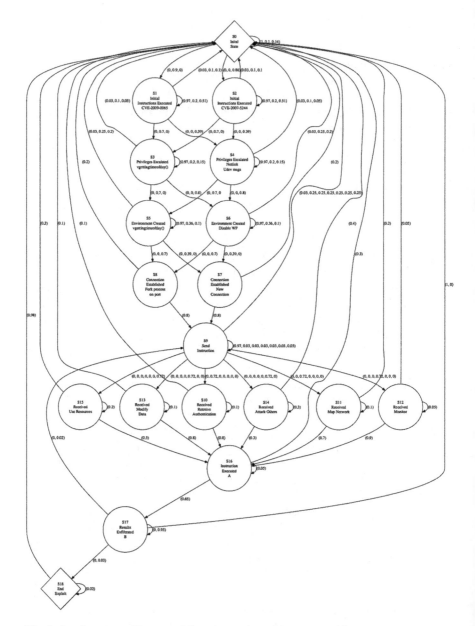

Fig. 2 Attacker game with two exploit options and transition opportunities

6 Attacker Actions

State transition probabilities depend on the action selected in a current state. For most states, the actions considered are no action, selection of first exploit, or selection of the second exploit. For example, if the first exploit is selected, the state transition probability for transition to 'second exploit completed' state equals zero and is not possible. In Fig. 3, the state transitions are the probabilities of that transition for the first action that can be taken in the current state, second action, etc., listed in that order in parenthesis. Table 2 details the attacker actions available for each state.

7 Threats

An attacker's payoff should be dependent on what they are attempting to do on a remote node. To create a model that can facilitate this concept, a variety of advanced threats were considered. The threats were categorized into five types:

1. Retrieving authentication files,
2. Mapping network,
3. Observing, monitoring state of machine,
4. Initiating new attacks, and
5. Using resources.

Each type of threat is used as an instruction that an attacker can send. The probability of success, failure, and detection is dependent on which type of threat they attempt to execute and eventually the payoff will also depend upon this.

8 Exploits and Metrics

The National Vulnerability Database (NVD) uses a standard metric to quantify the impact of all CVE vulnerabilities. This metric is known as the Common Vulnerability Scoring System, or CVSS. The scoring system provides a base score which quantifies the aspects that do not change with time or user environments of a vulnerability. It is a combination of an exploit ability and impact score. The range of this score is 0–10, inclusive.

The exploitability sub-score considers how a vulnerability can be accessed, i.e. locally or remotely, the complexity of the exploit, and the number of times authentication is needed. The score range is also between 0 and 10. The base score is used as a metric to compare vulnerabilities and the exploitability sub-score is used to estimate the probability of an exploit succeeding.

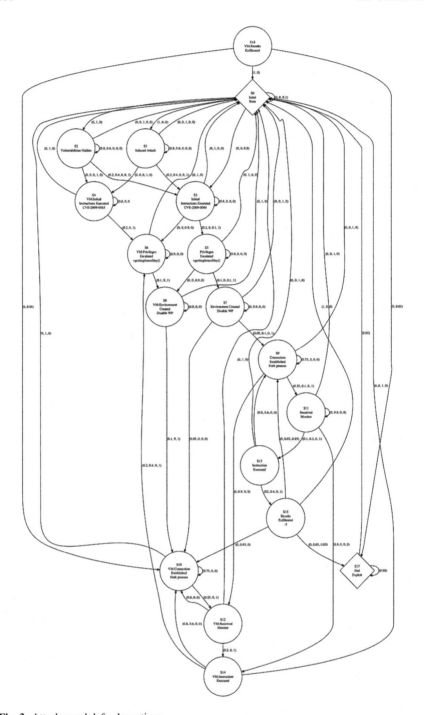

Fig. 3 Attacker and defender actions

Table 2 Attacker model: states, actions, and transition probabilities

State	State description	Actions	Action description	Next states	Prob. next state
S0	Initial stale	A(0) = 1	NO Action	S0, S1, S2	(1, 0, 0)
		A(0) = 2	CVE-2009-0065		(0.1, 0.9, 0)
		A(0) = 3	CVE-2007-5244		(0.14, 0, 0.86)
S1	Initial instructions executed CVE-2009-0065	A(1) = 1	No action	S0, S1, S3, S4	(0.03, 0.97, 0, 0)
		A(1) = 2	Vsyscall, gettimeofday()		(0.1, 0.2, 0.7, 0)
		A(1) = 3	CVE-2009-1185		(0.1, 0.51, 0, 0.39)
S2	Initial instructions executed CVE-2007-5244	A(2) = 1	No action	S0, S2, S3, S4	(0.03, 0.97, 0, 0)
		A(2) = 2	Vsyscall, gettimeofday()		(0.1, 0.2, 0.7, 0)
		A(2) = 3	CVE-2009-1185		(0.1, 0.51, 0, 0.39)
S3	Privileges escalated vsyscall	A(3) = 1	No action	S0, S3, S5, S6	(0.03, 0.97, 0, 0)
		A(3) = 2	Vsyscall, gettimeofday()		(0.1, 0.2, 0.7, 0)
		A(3) = 3	Disable WP		(0.05, 0.15, 0, 0.8)
S4	Privileges escalated netlink udev msgs	A(4) = 1	No action	S0, S4, S5, S6	(0.03, 0.97, 0, 0)
		A(4) = 2	Vsyscall, gettimeofday()		(0.1, 0.2, 0.7, 0)
		A(4) = 3	Disable WP		(0.05, 0.15, 0, 0.8)
S5	Environment created vsyscall	A(5) = 1	No action	S0, S5, S7, S8	(0.03, 0.97, 0, 0)
		A(5) = 2	CVE-2003-0019		(0.25, 0.36, 0.39, 0)
		A(5) = 3	Interact connect, port inline metasploit		(0.20, 0.1, 0, 0.7)
S6	Environment created. WP flag disabled	A(6) = 1	No action	S0, S6, S7, S8	(0.03, 0.97, 0, 0)
		A(6) = 2	CVE-2003-0019; New network connection Fork process on port, metasploit		(0.25, 0.36, 0.39, 0)
		A(6) = 3			(0.20, 0.1, 0, 0.7)
S7	Connection established, new connection	A(7) = 1	Move to send instruction		(0.2, 0.8)

(continued)

Table 2 (continued)

State	State description	Actions	Action description	Next states	Prob. next state
S8	Connection established, fork on port	A(8) = 1	Move to send instruction		(0.2, 0.8)
S9	Send instruction	A(9) = 1	No action	S0, S9, S10, S11, S12, S13, S14, S15	(0.03, 0.97, 0, 0, 0, 0, 0, 0)
		A(9) = 2	Retrieve authentication info		(0.25, 0.03, 0.72, 0, 0, 0, 0, 0)
		A(9) = 3	Map network		(0.25, 0.03, 0, 0.72, 0, 0, 0, 0)
		A(9) = 4	Monitoring/observation		(0.25, 0.03, 0, 0, 0.72, 0, 0, 0)
		A(9) = 5	Modify data		(0.25, 0.03, 0, 0, 0, 0.72, 0, 0)
		A(9) = 6	Attach oliver machines		(0.25, 0.03, 0, 0, 0, 0, 0.72, 0)
		A(9) = 7	Use or resources		(0.25, 0.03, 0, 0, 0, 0, 0, 0.72)
S10	Received retrieve authentication	A(10) = 1	Execute instruction	S0, S10, S16	(0.1, 0.1, 0.8)
S11	Received map network	A(11) = 1	Execute instruction	S0, S11, S16	(0.2, 0.1, 0.7)
S12	Received monitor	A(12) = 1	Execute instruction	S0, S12, S16	(0.05, 0.05, 0.9)
S13	Received modify data	A(13) = 1	Execute instruction	S0, S13, S16	(0.1, 0.1, 0.8)
S14	Received attack others	A(14) = 1	Execute instruction	S0, S14, S16	(0.3, 0.3, 0.4)
S15	Received use resources	A(15) = 1	Execute instruction	S0, S15, S16	(0.3, 0.2, 0.5)
S16	Instruction executed	A(16) = 1	Exfiltrate results	S0, S16, S17	(0.3, 0.05, 0.65)
S17	Results exfiltrated	A(17) = 1	No action	S0, S9, S17, S18	(1, 0, 0, 0)
		A(17) = 2	Continue, stop exploit		(0, 0.95, 0.03, 0.02)
S18	End of exploit	A(18) = 1	Start now attack	S0, S18	(0.98, 0.02)

9 Transition Probabilities

The attacker's model includes state transitions that are dependent on the probability of success, failure, and detection of the action. Reasonable state transition probabilities were determined using CVSS scores, subject matter expertise, root-kit reference literature, and rules of probability.

For state transitions dependent on an action that exploits a CVE vulnerability, the CVSS exploitability sub-score was used to calculate the probability of success. Detection probability was estimated by weighing multiple metrics from the CVSS scores and analysis of the imprint left on a system by an exploit. For example, if the exploit creates multiple processes, open sorts, files and logs, etc., it is more likely to be detected compared to an exploit that creates or uses only one of these items. Also, if a step in the exploit produces high network traffic or accesses protected files, it is more likely to be detected. Analysis of these events was used to determine detection probability of specific exploits and instructions. Additionally, the probability of detection is always greater than zero after successful completion of the first step of the exploit. After the first step, the machine is altered from its original state and there is always a probability of this change being detected.

Probability of success for exploits and instructions was estimated using the level of complexity of the task. Lower complexity tasks have a higher chance of success. For example, monitoring status is less complex than mapping the network, therefore monitoring has a higher probability of success than mapping. The probability of continuing the remote control exploit loop and beginning a new attack is high. It is assumed that an attacker continues the exploit while remaining undetected and will try to attack again after one successful exploit.

Root-kit sources, Perla and Oldani [11], were used to estimate the likelihood of success of exploits that use Vsyscall. Based on information from this source, Vsyscall is stored in the same memory location and includes several machine instructions that can invoke and execute system calls with attacker parameters. These exploits are considered to have a high success rate. The same type of information was considered for determining the success of the write protect (WP) flag exploits. All transition probabilities are included in Table 2 above.

10 Scoring

A remote control exploit has two states where an attacker can gain rewards: execute instructions, S16, and exfiltrate results, S17. These rewards are modeled for S16 and S17 in Table 2. There can be a relation between these two rewards because an attacker cannot exfiltrate results without first executing an instruction. An equation reflecting this relationship may be used to further define the relationship. Solving for an optimal policy requires numeric definition of rewards. The rewards are set to 1 and 5 for S16, α, and S17, β, respectively.

11 Attacker Optimal Solution

Finding an optimal attacker strategy involves solving for the long-run average performance, limiting the average reward, of the MDP. Algebraic equations can be used to construct a linear program that can solve for x_{sa}, the frequency of being in state, s, and choosing action, a, for the long-run frequency space, Filar and Vrieze [5]. Such a linear program includes an objective function, constraint equations, and a lower bound which are input to MatLab. The function solves for all x_{sa} that minimize the objective function within the defined constraints. Since an attacker is concerned with maximizing their payoff, the rewards are negated, -1 for $x_{16,2}$ and -5 for $x_{17,2}$ and $x_{17,2}$. The optimal minimizing solution for negative rewards is equivalent to the maximizing solution for positive reward. This problem produces the same frequencies solution for x_{sa}, but a negative optimal policy reward that must be negated to give the correct solution for positive rewards. The optimal x_{sa} frequencies for maximizing this model and the probability of an action given the state, $p(a|s)$ are indicated in Table 3. The latter is calculated by normalizing the frequencies of actions for a specific state and is the probability of choosing an action when in the current state.

The calculated solution is a pure strategy that always selects the action with highest probability of success for transition to a future state. According to the optimal solution, the maximum value of the objective function for this specific policy is 0.3945.

12 Defender Model

The goal of the defender is to apply deception strategies that minimize or eliminate the reward to be gained by the attacker. In other words, the defender attempts to stop an attacker using deception. If successful, the defender will collect information about the attacker and the attacker's methods which will be treated as additional payoffs for the defender. The deception model considers only the defender's actions given that the attacker will apply a pure strategy introduced previously.

13 Deception Model and Actions

The deception model was constructed under the assumption that the attacker will follow a pure strategy. In this situation, the defender knows what the attacker is attempting to do and the defender actions include different methods for deception against those actions. Referring to the state column in Table 4 the defender's actions for the initial state include inducing an attack by making the system appear vulnerable, preventing an attack by hiding vulnerabilities, and taking no action.

Table 3 Optimal solution of
attacker model: frequency
of state, action

Xsa (state_action)	Frequency	p(a|s)
X0_1	0	0
X0_2	0.1405	1
X0_3	0	0
X1_1	0	0
X1_2	0.158	1
XI_3	0	0
X2_1	0	0
X2_2	0	0
X2_3	0	0
X3_1	0	0
X3_2	0	0
X3_3	0.1301	1
X4_1	0	0
X4_2	0	0
X4_3	0	0
X5_1	0	0
X5_2	0	0
X5_3	0	0
X6_1	0	0
X6_2	0	0
X6_3	0.1157	1
X7_1	0	0
X8_1	0.081	1
X9_1	0	0
X9_2	0	0
X9_3	0	0
X9_4	0.1263	1
X9_5	0	0
X9_6	0	0
X9_7	0	0
X10_1	0	0
X11_1	0	0
X12_1	0.0957	1
X13_1	0	0
X14_1	0	0
X15_1	0	0
XI6_1	0.0907	1
X17_1	0	0
XI7_2	0.0608	1
X18_1	0.0012	1

The next two states, S1 and S2, represent that state of the node if configured to induce an attack, S1, and to prevent an attack, S2. Each state after this represents a state that can be achieved if the attacker continues executing each step of their exploit. Each step of the exploit includes two different states: exploit being executed on the target machine and exploit being executed in a virtual machine.

For all states following S2, the defender actions for a given are options for how to deceive what the attacker currently would be executing. For example, if in S5, the attacker's privileges have already been escalated on the target machine and at this point it is assumed the attacker is attempting to create the remote control exploit by disabling the WP flag. Having the a priori knowledge of the attacker's method, the actions for the defender include enabling the WP flag again, closing the network connection to end the attacker's exploit, moving the attacker's exploit into a virtual machine, or taking no action which could be an intentional choice or a result from not detecting the attacker. Similar actions are available at each state and include one or two options for deception, closing the network, and no action.

Selecting the shutdown connection action results in a transition back to the initial state while no action allows transition to the next future state. The deception actions have probabilities for transitioning to multiple states. The exfiltrate states, S15 and S16, take into consideration the probability of the attacker continuing the execute and exfiltrate loop or ending the current exploit. These states include a transition back to the connection established states, S9 and S10, respectively. Finally, S17 will remain in the same state if the attacker stops the exploit or transitions back to S0, initial state, if the attacker launches a new attack. All the defender's states, actions, and transition probabilities can be found in Table 4.

14   Transition Probabilities

Beginning at the initial state, the defender's actions include inducing an attack by making the system appear vulnerable, preventing an attack by hiding vulnerabilities, and taking no action. The first two actions have deterministic probabilities; the choice determines the next state with probability of transition equal to 1. The third action, no action, assumes that the attacker will begin an exploit and thus state transition probabilities are equal to the probabilities used in the attacker model if the attacker chooses that exploit action. Once the exploit has begun, at each state the defender's actions include deception, shutdown connection, and no action. Shutdown connection and no action have deterministic probabilities. Shutdown connection always results in a transition back to the initial state, while no action always transition to the next future state. The probabilities for no action should be modified to not be deterministic and to represent the probability of success and failure for the specific attacker exploit.

The deception action can either succeed of fail, which produces two state transition probabilities. The probability for success is estimated based on the complexity of the deception method and is always greater than the probability of

Table 4 Defender model: states, actions, and state transition probabilities

State	State description	Attacker action	Actions	Action description	Next state	P(s'|a)
S0	Initial state		A(0) = 1	Induce attack	S0, S1, S3, S4	(0, 1, 0, 0)
			A(0) = 2	Hide vulnerabilities, prevent		(0, 0, 1, 0)
			A(0) = 3	No action		(0.1, 0, 0, 0.9)
S1	Deception, induced attack	Execute CVE-2009-0065	A(1) = 1	Block messages	S0, S1, S3, S4	(0, 0.8, 0.2, 0)
			A(1) = 2	Free old slabs		(0, 0.6, 0.4, 0)
			A(1) = 3	Shutdown connection		(1, 0, 0, 0)
			A(1) = 4	Forward VM		(0, 0, 0, 1)
			A(1) = 5	No action		(0, 0, 1, 0)
S2	Vulnerablitites hidden	Execute CVE-2009-0065	A(2) = 1	Block messages	S0, S2, S3, S4	(0, 0.8, 0.2, 0)
			A(2) = 2	Free old slabs		(0, 0.6, 0.4, 0)
			A(2) = 3	Shutdown connection		(1, 0, 0, 0)
			A(2) = 4	Forward VM		(0, 0, 0, 1)
			A(2) = 5	No action		(0, 0, 1, 0)
S3	Initial instructions executed CVE-2009-0065	Execute syscal vgettingtimeofday()	A(3) = 1	Replace mooted vsyscalls with original	S0, S3, S5, S6	(0, 0.8, 0.2, 0)
			A(3) = 2	Close network connection		(1, 0, 0, 0)
			A(3) = 3	Use VM call vgettingtimeofday()		(0, 0, 0.1, 0.9)
			A(3) = 4	No action		(0, 0, 1, 0)
S4	VM, executed CVE-2009-0065	Execute syscal vgettingtimeofday()	A(4) = 1	Replace modified vsyscalls with orginal	S0, S4, S6	(0, 0.8, 0.2)
			A(4) = 2	Close network connection		(1, 0, 0)
			A(4) = 3	No action		(0, 0, 1)

(continued)

Table 4 (continued)

State	State description	Attacker action	Actions	Action description	Next state	P(s'|a)
S5	Privileges escalated, vgettingtimeofday()	Create environment disable WP flag	A(5) = 1	Enable WP	S0, S5, S7, S8	(0, 0.9, 0.1, 0)
			A(5) = 2	Close network connection		(1, 0, 0, 0)
			A(5) = 3	Use VM and disable WP		(0, 0, 0.1, 0.9)
			A(5) = 4	No action		(0, 0, 1, 0)
S6	VM, Executed vgettingtimeofday()	Create environment disable WP flag	A(6) = 1	Enable WP	S0, S6, S8	(0, 0.9, 0 1)
			A(6) = 2	Close network connection		(1, 0, 0)
			A(6) = 3	No action		(0, 0, 1)
S7	Environment created. WP disabled	Establish connection, fork process	A(7) = 1	Setup tunnel to VM	S0, S7, S9, S10	(0, 0, 0.05, 0.95)
			A(7) = 2	Setup connection on different machine		(0, 0.9, 0.1, 0)
			A(7) = 3	Close, block new network connection		(1, 0, 0, 0)
			A(7) = 4	No action		(0, 0, 1, 0)
S8	VM, Environment created WP disabled	Establish connection, fork process	A(8) = 1	Setup connection on different machine	S0, S8, S10	(0, 0.9, 0.1)
			A(8) = 2	Close now network connection		(1, 0, 0)
			A(8) = 3	No action		(0, 0, 1)

(continued)

Table 4 (continued)

State	State description	Attacker action	Actions	Action description	Next state	P(s'|a)
S9	Connection established, process forked on port	Send instruction, monitor system	A(9) = 1	Intercept instruction	S0, S9, S11, S12	(0, 0.75, 0.25, 0)
			A(9) = 2	Send instruction to VM		(0, 0, 0.1, 0.9)
			A(9) = 3	Close new network connection		(1, 0, 0, 0)
			A(9) = 4	No action		(0, 0, 1, 0)
S10	VM, connection established, process forked on port	Send instruction, monitor system	A(10) = 1	Intercept instruction	S0, S10, S12	(0, 0.75, 0.25)
			A(10) = 2	Close now network connection		(1, 0, 0)
			A(10) = 3	No action		(0, 0, 1)
			A(10) = 4			
S11	Received monitor instruction	Execute instruction, monitor system	A(11) = 1	Allow execution VM	S0, S11, S13, S14	(0, 0, 0.1, 0.9)
			A(11) = 2	Never execute, and make it appear did		(0, 0.8, 0.2, 0)
			A(11) = 3	Close network connection		(1, 0, 0, 0)
			A(11) = 4	No action		(0, 0, 1, 0)
S12	VM, received monitor instruction	Execute instruction, monitor system	A(12) = 1	Never execute, and make it appear did	S0, S10. S14	(0, 0.8, 0.2)
			A(12) = 2	Close network connection		(1, 0, 0)
			A(12) = 3	No action		(0, 0, 1)

(continued)

Table 4 (continued)

State	State description	Attacker action	Actions	Action description	Next state	P(s'|a)
S13	Instruction executed	Exfiltrate results	A(13) = 1	Send false info	S0, S9, S15	(0, 0.8, 0.2)
			A(13) = 2	Send real info, and modify system		(0, 0.6, 0.4)
			A(13) = 3	Close network connection		(1, 0, 0)
			A(13) = 4	No action		(0, 0, 1)
S14	VM, instruction routed	Exfiltrate results	A(14) = 1	Send fates into	S0, S10. 516	(0, 0.8, 0.2)
			A(14) = 2	Send real into, and modify system		(0, 0.6, 0.4)
			A(14) = 3	Close network connection		(1, 0, 0)
			At (14) = 4	No action		(0, 0, 1)
S15	Results exfiltrated	Continue attack or stop	A(15) = 1	Close network connection	S0, S9, S10, S17	(1, 0, 0, 0)
			A(15) = 2	Move connection to VM		(0, 0.02, 0.93, 0.05)
			A(1S) = 3	No action		(0, 0.95, 0, 0.05)
S16	VM, results exfiltrated	Continue attack or stop	A(16) = 1	Close network connection	S0, S10, 517	(1, 0, 0)
			AI(16) = 2	No action		(0, 0.95, 0.05)
S17	End exploit	New attack or stop	A(17) = 1		S0, S17	(0.98, 0.02)

failure. This is because the defender is assumed to have full observability and root access. The probability for success is always 0.9 or higher for the reasons stated previously. In Table 4, as before, the numbers indicated on state transitions are the probability of making that transition if the first action of the current state is selected, if the second action is selected, etc., in that order, in parenthesis.

15   Scoring

Determining how to score the defender's decisions is more complicated than the attacker's. Each successful deception action results in information about the attacker which could be considered a reward. Another option would be to set payoffs only at the states at which the attacker can potentially receive a payoff, execute, and exfiltrate states. There are many remaining factors that need to be considered before the final scoring system is decided.

An optimal solution can be found when a scoring method is selected. The linear program can be solved as soon as the objective function, constructed from payoffs, is defined. All of the equations and constraints have been defined based on the information presented in Table 4.

16   The Deception Game

A deception game between an attacker and a defender can be formulated as a pair of coupled partially observable Markov decision processes (POMDP), also known as a partially observable stochastic game (POSG). Linear programming used for solving the MDPs can be expanded to solve for the POSG, resulting in a pair of finite-state controllers that describe optimal strategies for executing the game. This problem can be modeled with the defender having full observability and the attacker having partial observability. To solve for this problem, all defender and attacker actions for each state must be defined and every combination of these actions must be considered. Additionally, the attacker's observation at each state must be defined and factored into the state transition probabilities. Figure 3 models the defender's actions given a fixed exploit action of the attacker. In this model, circles represent attacker actions and rectangles are defender actions.

17   Observations

The defender model assumes that the attacker will never be aware of another player. This simplifies solving the game but is not realistic. Due to delay times of virtual machines and unexpected behavior of the target machine, it is likely that the attacker may realize that deceit is occurring. These two factors are what will be considered as observations that the attacker can make and changes the aspects of the game. Other observables for an attacker include network traffic messages and logs, processes once the attacker has escalated privileges. A full list of observables must be constructed to accurately solve the POSG.

18   Passive Deception

Passive deception is an approach to counteracting a malicious user who is performing reconnaissance. The use of deception has been studied, but only accomplished on a limited basis and mainly in the form of honeypots and honeynets. In these cases, however, the primary application of honeypots and honeynets has been to redirect a suspicious user to a segregated area for the purpose of either (1) isolating the user or (2) studying the malicious user's methods and behaviors. In both types of deception, a malicious user may be trapped, studied, and removed from the protected system. With an APT, however, the malicious user may easily attack again with awareness of the deception and some understanding of its implementation. With each successive attack, the APT knowledge is greater.

We propose a scalable cyber deception concept which uses passive deception on a wide scale and over multiple levels as a sensing system. The passive deception essentially serves as an early warning system for malicious use. Since the trickery is expanded beyond prior work in both breadth and depth of deployment, the concept is referred to as *Polydeception*. Polydeception is intended to provoke a malicious user into exposing itself as it attempts to move about the protected network system. The Polydeception concept may use traditional honeypots and honeynets to isolate or study a malicious user, but specifically it aims to quickly sense, detect, and expose malicious users.

18.1   Polydeception

The Polydeception concept is illustrated in Fig. 4. It is assumed that there is a network defense perimeter, but that within the boundary is a blend of legitimate systems and services and of honeynets. In an attack scenario, some APT resides inside the boundary (hiding under the keyboard). Within the protected network system, there are instantiations of virtual machines (VM)—perhaps via a cloud infrastructure—which provide the legitimate services of the mission critical system (the darker nodes). Normal services are provided by an anonymous peer-to-peer (P2P) network in which each service node is only aware of—and communicates with—the next service node in the service path. Other nodes in the service path are unknown to each node. Each service node may potentially communicate with nodes not on the service path, but are aware only of which node is next on

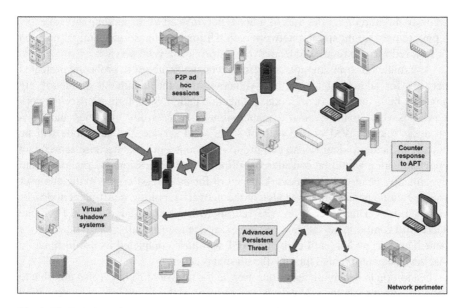

Fig. 4 The Polydeception concept

the true path. The client-server path of virtual machine service nodes (the wide arrows) is anonymous, dynamic, and random for each session, forming an ad hoc communication path for each invocation of the legitimate service.

Other network nodes are instantiated as virtual machines as well. These are decoy nodes which do not provide services as part of the protected system. The instantiation of these nodes is also very fluid and dynamic, with decoy nodes being created and destroyed continuously. Each node presents different information to all users (i.e., malicious or not), most of which will be for the purpose of confusing or withholding information from users seeking information about the protected network system.

Each decoy node deceives malicious users in multiple ways. A decoy node may spoof configuration information, such as its operating system type, OS version number, OS patch or update number, available network services, application-layer services, open network ports, etc. The decoy node configuration may have different combinations of configuration characteristics with each instantiation of a VM.

Furthermore, the responses to queries for this information may consist of different dynamically-selected options, including error conditions, security or authorization responses, or no response. An example error condition is HTTP error 404 when a non-existing web page is requested. An example security response is that a gateway or host destination address is prohibited. Lastly, an example of no response occurs when a ping service is unsupported. Deception can be provided by responding in false ways rather than always providing the appropriate response. So, for example, instead of returning HTTP error 404, the decoy HTTP service might provide a false web page. Instead of returning a security response, the prohibited gateway might redirect the user to a honeypot. Instead of responding with silence, a ping request might spoof a network path. Ultimately, the services might respond falsely, redirect the request to a honeypot, or provide a valid service.

As a malicious user attempts to perform reconnaissance or to probe the protected network for additional targets and vulnerabilities, the likelihood of identifying anything but a decoy network node, connection path, OS configuration, or network service is very small. Even if a malicious user successfully identifies and compromises a single VM on the path of some valid service, the next node will be equally difficult to identify and compromise without a mistake because subsequent communication would be considered a different service session. By randomly and dynamically changing the network structure for each session, the malicious user would gain no persistent knowledge of the network topology, constituent nodes, or valid services. Thus, even if an APT returns with the knowledge gained from one successful compromise, the next session or service request will create a new service path. The path node identified by the APT will have completed its useful function and have been terminated through the normal process.

In addition to deceiving malicious users at the network level, it is also possible to deceive within each node. The deception concepts so far have focused on deception via the node external interfaces. However, the possibility of deception applies from within an end node. On a legitimate end node VM, for example, decoy processes and files may be provided and monitored while obfuscating the presence and location of true processes and files. If an adversary reads, copies, or corrupts a decoy, the intent is clearly exposed. Furthermore, if the presence and location of processes and files is also dynamic, confusion for the adversary is magnified. A malicious user no longer knows what is legitimate and what is a decoy at any level of the system. Any surreptitious activity yields the risk of being exposed and defeated.

The innovation in the approach is that the technique of deception becomes ubiquitous and dynamic on the network and within critical end nodes. Deception is employed on a widespread basis and is constantly changing. Every step a malicious user takes is very likely to raise alarms and expose the intrusion and malicious intent. The concept uses passive deception to act as sensing system for CND C2 using active deception.

Another benefit is that a perimeter defense is no longer the primary protection mechanism. While still necessary and valuable, even adversaries who gain admission to the protected network can be detected and defeated. Additionally, the Polydeception concept provides protection against trusted insiders who behave with malicious intent. The deception approach offers the possibility that adversary success will be exponentially more difficult and much more risky.

Technology development is necessary to prove out the Polydeception concept. Furthermore, these technology developments must be integrated into a seamless system of deception. These technologies include:

- *The application of cloud computing platforms.* Cloud computing technology offers much of the functionality needed to instantiate and terminate VMs. However, since many of the VMs in the cloud would be topologically interconnected, some modifications would need to be developed. Also, hardening of the distributed hardware platforms would be crucial to the security of the overall Polydeception approach.

- *The use of anonymous P2P.* An opennet version of an anonymous peer-to-peer network would need to be developed and implemented as part of Polydeception. The anonymous P2P would need to be integrated with the cloud computing environment to establish legitimate services for normal users of the mission critical system. The Onion Router (TOR) provides much of the dynamic communication path capability for Polydeception, but would need to be extended to detect and report reconnaissance.
- *Distributed autonomous control of spoofing.* The Polydeception concept includes far more obfuscation than the communication path provided by the P2P network. Spoofing of the OS, hardware platform, network services, etc., is random and dynamic. Centralized selection and instantiation of the VMs would present a high-value target for attackers. Therefore, a distributed but coordinated mechanism provides greater security, scalability, and flexibility.
- *Spoofing of end node software systems.* Each decoy VM may have random configuration characteristics that can be spoofed. The technology to randomly select and instantiate different software systems and services on the VMs must be developed. These systems and services include:

  - Node identification information. This includes the operating system type, version number, and patch or update level.
  - Network systems and services. This includes network services such as ping, traceroute, FTP, SMTP, and HTTP.
  - Open communication ports. Spoofing ports by randomly and perhaps dynamically responding (or not) to port probes.
  - Communication topology. Each decoy end node will be connected topologically to other decoy nodes (and possibly to legitimate service nodes). These connections must be managed and spoofed as needed.

- *Types of deception.* Methods for deceiving users—such as lying, responding with silence, redirecting to a honeypot, etc.—must be developed.

These are all component technologies which contribute to Polydeception. Much of the component technology for Polydeception currently exists, but requires adaptation and enhancement to fully provide ubiquitous deception. Significant integration is required to apply the component technologies and achieve the goals of Polydeception.

19 Summary

Effective cyber deception calls for an amalgam of techniques. We argue that both active and passive cyber deception can be employed to achieve this effectiveness. As presented here, we submit two, high-level forms of cyber deception corresponding to each of the categories of cyber deception: active and passive. Respectively, there are referred to as Legerdemain and Polydeception.

Acknowledgments The author would like acknowledge the U.S. Air Force Research Laboratory Small Business Innovation Research program for partially funding this work under contracts AFRL-RY-WP-TR-2010-1170 and AFRL-RY-WP-TR-2013-0014. Additional research and development has been funded by Sentar, Inc.

References

1. Abbasi, F. H., & Harris, R. J. (2009). Experiences with a generation III virtual honeynet. In *Telecommunication Networks and Applications Conference (ATNAC)* (pp. 1–6).
2. Borders, K., Falk, L., & Prakash, A. (2007). *OpenFire: Using deception to reduce network attacks.* Paper presented at the Third International Conference on Security and Privacy in Communications Networks and the Workshops (SecureComm 2007)
3. Bowen, B. M., Hershkop, S., Keromytis, A. D., & Stolfo, S. J. (2009). Baiting inside attackers using decoy documents In *Security and privacy in communication networks* (pp. 51–70). Berlin: Springer.
4. Bowen, B. M., Kemerlis, V. P., Prabhu, P., Keromytis, A. D., & Stolfo, S. J. (2010). *Automating the injection of believable decoys to detect snooping.* Paper presented at the Proceedings of the third ACM conference on Wireless network security.
5. Filar, J., and Vrieze, K. (1997). *Competitive Markov Decision Processes*, Springer, NY.
6. Gerwehr, S., & Anderson, R. H. (2000). *Employing deception in INFOSEC.* Paper presented at the Information Survivability Workshop (isw2000).
7. Levine, J., LaBella, R., Owen, H., Contis, D., & Culver, B. (2003). *The use of Honeynets to detect exploited systems across large enterprise networks.* Paper presented at the Information Assurance Workshop, 2003. IEEE Systems, Man and Cybernetics Society.
8. Michael, J. B., & Wingfield, T. C. (2003). Lawful cyber decoy policy. In D. Gritzalis, S. C. Vimercati, P. Samarati & K. Sokratis (Eds.), *Security and privacy in the age of uncertainty* (pp. 483-488). Boston, MA: Kluwer.
9. Neagoe, V., & Bishop, M. (2007). Inconsistency in deception for defense. In *Proceedings of the 2006 Workshop on New Security Paradigms* (pp. 31-38). New York: ACM Press.
10. Niels, P., & Thorsten, H. (2007). *Virtual honeypots: from botnet tracking to intrusion detection:* Addison-Wesley Professional.
11. Perla, E., and Oldani, M. (2011). PART III, Remote Kernel Exploitation in: *A Guide to Kernel Exploitation: Attacking the Core*, by Syngress, Amsterdam, NL.
12. Rowe, N. C. (2006). A Taxonomy of deception in cyberspace. In *International Conference on Information Warfare and Security*. Princess Anne, MD.
13. Rowe, N. C. (2007). Deception in defense of computer systems from cyber-attack. In A. Colarik & L. Janczewski (Eds.), *Encyclopedia of Cyber War and Cyber Terrorism.* Hershey, PA: The Idea Group.
14. Rowe, N. C., Goh, H., Lim, S., & Duong, B. (2007). Experiments with a testbed for automated defensive deception planning for cyber-attacks. In L. Armistead (Ed.), *2nd International Conference on i-Warfare and Security (ICIW 2007)* (pp. 185-194). Monterey, California, USA.
15. Ryu, C., Sharman, R., Rao, H. R., & Upadhyaya, S. (2009). Security protection design for deception and real system regimes: A model and analysis. *European Journal of Operational Research, 201*(2), 545–556.
16. Spitzner, L. (2003). *Honeypots: Tracking hackers.* Boston, MA: Pearson Education.
17. Tirenin, W., & Faatz, D. (1999). A concept for strategic cyber defense. In *Military Communications Conference Proceedings* (Vol. 1, pp. 458-463). Atlantic City, NJ: IEEE.

18. Yuill, J. J. (2006). *Defensive computer-security deception operations: Processes, principles and techniques.* Unpublished dissertation, North Carolina State University.
19. Yuill, J. J., Denning, D., & Feer, F. (2006). Using deception to hide things from hackers: Processes, principles, and techniques. *Journal of Information Warfare, 5*(3), 26–40.

Cyber-Deception and Attribution in Capture-the-Flag Exercises

Eric Nunes, Nimish Kulkarni, Paulo Shakarian, Andrew Ruef, and Jay Little

Abstract Attributing the culprit of a cyber-attack is widely considered one of the major technical and policy challenges of cyber-security. The lack of ground truth for an individual responsible for a given attack has limited previous studies. Here, we overcome this limitation by leveraging DEFCON capture-the-flag (CTF) exercise data where the actual ground-truth is know. In this work, we use various classification techniques to identify the culprit in a cyberattack and find that deceptive activities account for the majority of misclassified samples. We also explore several heuristics to alleviate some of the misclassification caused by deception.

1 Introduction

Attributing the culprit of a cyber-attack is widely considered one of the major technical and policy challenges of cyber-security. The lack of ground truth for an individual responsible for a given attack has limited previous studies. In this study, we take an important first step toward developing computational techniques toward attributing the actual culprit (here hacking group) responsible for a given cyber-attack. We leverage DEFCON capture-the-flag (CTF) exercise data which we have processed to be amenable to various machine learning approaches. Here, we use various classification techniques to identify the culprit in a cyber-attack and find that deceptive activities account for the majority of misclassified samples. We also explore several heuristics to alleviate some of the misclassification caused by deception. Our specific contributions are as follows:

- We assemble a dataset of cyber-attacks with ground truth derived from the traffic of the CTF held at DEFCON 21 in 2013.

E. Nunes (✉) • N. Kulkarni • P. Shakarian
School of Computing, Informatics and Decision Systems Engineering, Arizona State University, Tempe, AZ 85281, USA
e-mail: enunes1@asu.edu; nimish.kulkarni@asu.edu; shak@asu.edu

A. Ruef • J. Little
Trail of Bits, Inc., New York, NY 10003, USA
e-mail: andrew@trailofbits.com; jay@trailofbits.com

© Springer International Publishing Switzerland 2016
S. Jajodia et al. (eds.), *Cyber Deception*, DOI 10.1007/978-3-319-32699-3_7

- We analyze this dataset to identify cyber-attacks where deception occurred.
- We frame cyber-attribution as a multi-class classification problem and leverage several machine learning approaches. We find that deceptive incidents account for the vast majority of misclassified samples.
- We introduce several pruning techniques and show that they can reduce the effect of deception as well as provide insight into the conditions in which deception was employed by the participants of the CTF.

The remainder of the chapter is organized as follows: we review related work in Sect. 2, describe out dataset in Sect. 3 in addition to analysis regarding the presence of deceptive and duplicate attacks. We then briefly describe the machine learning approaches used to evaluate the dataset and also present experimental results in Sect. 4. We then propose different pruning techniques to improve the baseline performance in Sect. 5. Finally conclusion and future work are discussed in Sects. 6 and 7 respectively.

2 Related Work

In our text on cyber-warfare [1], we discuss the difficulties of cyber-attribution and how an intelligence analyst must also explore the deception hypothesis in a cyber-warfare scenario. When compared to other domains of warfare, there is a much greater potential for evidence found in the aftermath of cyber-attack to be planted by the adversary for purposes of deception. The policy implications of cyber-attribution have also been discussed in [2] where the authors point out that anonymity, ability to launch multi-stage attacks, and attack speed pose significant challenges to cyber attribution.

In an early survey on cyber-attribution [3], the authors point out that technical attribution will generally identify machines, as opposed to a given hacker and his/her affiliations. While we will use technical information in our approach, we have ground truth data on the group involved by the nature of the capture-the-flag data. This will allow our approach to profile the tactics, techniques, and procedures of a given group as we have ground-truth information on a hacking group as opposed to machines. An example of such an approach is the WOMBAT attribution method [4] which attributes behavior to IP sources that are potentially linked to some root cause determined through a clustering technique. Similarly, other work [5] combines cluster analysis with a component for multi-criteria decision analysis and studied an implementation of this approach using honeypot data again, this approach lacks any ground truth of the actual hacker or hacking group. In other work on attribution [6], the authors study the problem of attributing abnormal internal behavior to a malicious insider over the course of an advanced persistent threat (APT) a different type of attribution problem than the one we propose to study. Outside of cyber-security, attribution has also been studied in other contexts. Of particular note is the work of Walls [7]. Here, the author look at attribution based on forensic information in a much different problem. They consider diverse sources, but do not seek to overcome inconsistency caused by intentional deception nor do they apply their methods to cyber-attacks. More recently the Q model has been proposed [8]. The framework of the Q model works by letting the analyst ask a range of relevant questions both technical and non-technical to aid in his process of attributing the attack. It provides a theoretical map towards cyber-attribution but does not address the issues of deception and does not evaluate the model on a relevant dataset.

Concurrently, we have devised a formal logical framework for reasoning about cyber-attribution [9, 10]. However, we have not studied how this framework can be instantiated on a real world dataset and, to date, we have not reported on an implementation or experiments in the literature.

We note that none of the previous work on cyber-attribution leverages a data set with ground truth information of actual hacker groups—which is the main novelty of this chapter.

3 Dataset

3.1 DEFCON CTF

The DEFCON security conference sponsors and hosts a capture the flag (CTF) competition every year, held on site with the conference in Las Vegas, Nevada. DEFCON CTF is one of the oldest and best known competitions. The ctftime.org system provides a ranking for CTF teams and CTF competitions, and in this system DEFCON CTF has the highest average weight of all other CTF competitions.

CTF competitions can be categorized by what role the competitors play in the competition: either red team, blue team, or a combination. In a blue team focused CTF the competitors harden their systems against a red team played by the organizers of the CTF. In a combined red/blue team CTF every team plays both blue and red team simultaneously. The NCCDC and CDX competitions are examples of a blue team CTF, while DEFCON CTF is a combined red/blue team. Each team is simultaneously responsible for hardening and defending their systems as well as identifying vulnerabilities and exploiting them in other teams systems.

The game environment is created primarily by the DEFCON CTF organizers. The game focuses around programs (known in the game as services) written by the organizers. These services are engineered to contain specific vulnerabilities. The binary image of the service is made available to each team at the start of the game, but no other information about the service is released. Part of the challenge of the game is identifying the purpose of each service as well as the vulnerabilities present in the service. Identification of vulnerabilities serves both a defensive and offensive goal. Once a vulnerability has been identified, a team may patch this vulnerability in the binary program. Additionally, the teams may create exploits for that vulnerability and use them to attack other teams and capture digital flags from those teams systems.

Each team is also provided with a server running the services. This server contains the digital flags to be defended. To deter defensive actions such as powering off the server or stopping the services, the white team conducts periodic availability tests of the services running on each teams server. A teams score is the sum of the value of the flags they have captured, minus the sum of the flags that have been captured from that team, multiplied by an availability score determined by how often the white team was able to test that teams services. This scoring model incentivizes teams to keep their server online, identify the vulnerabilities in services and patch them quickly, and exploit other teams services to capture their flags. It disincentivizes teams from performing host-level blocking and shutting down services, as this massively impacts their final score.

This game environment can be viewed as a microcosm of the global Internet and the careful game of cat and mouse between hacking groups and companies. Teams are free to use different technical means to discover vulnerabilities. They may use fuzzing and reverse engineering on their own programs, or, they may monitor the network data sent to their services and dynamically study the effects that network data has on unpatched services. If a team discovers a vulnerability and uses it against another team, the first team may discover that their exploit is re-purposed and used against them within minutes.

3.2 DEFCON CTF Data

The organizers of DEFCON CTF capture all of the network traffic sent and received by each team, and publish this traffic at the end of the competition [15]. This includes IP addresses for source and destination, as well as the full data sent and received and the time the data was sent or received. This data is not available to contestants, depending on the organizers choice from year to year the contestants either have a real time feed but with the IP address obscured, or, a full feed delivered on a time delay of minutes to hours.

In addition to the traffic captures, copies of the vulnerable services are distributed by the organizers. The organizers usually do not disclose the vulnerabilities they engineered into each service, however, competitors frequently disclose this information publicly after the game is finished as technical write-ups.

The full interaction of all teams in the game environment are captured by this data. We cannot build a total picture of the game at any point in time, since there is state information from the servers that is not captured, but any exploit attempt would have to travel over the network and that would be observed in the data set.

3.3 Analysis

The CTF data set is very large, about 170 GB compressed. We used multiple systems with distributed and coordinated processing to analyze the entire data set. Fortunately, analyzing individual streams is an embarrassingly parallel task. To analyze this data, we identified the TCP ports associated with each vulnerable service. From this information, we used the open source tool tcpflow[1] to process the network captures into a set of files, with each file representing data sent or received on a particular connection.

This produced a corpus of data that could be searched and processed with standard UNIX tools, like grep. Further analysis of the game environment provided an indicator of when a data file contained an exploit. The game stored keys for services in a standard, hard-coded location on each competitors server. By searching for the text of this location in the data, we identified data files that contained exploits for services.

With these data files identified, we analyzed some of them by hand using the Interactive Disassembler (IDA) to determine if the data contained shell-code, and they did. We used an automated tool to produce a summary of each data file as a JSON encoded element. Included in this summary was a hash of the contents of the file and a histogram of the processor instructions contained in the file. These JSON files were the final output of the low level analysis, transforming hundreds of gigabytes of network traffic into a manageable set of facts about exploit traffic in the data. Each JSON file is a list of tuples (time-stamp, hash, byte-histogram, instruction-histogram). These individual fields of the tuple are listed in Table 1.

From this pre-processing of the network data (packets) we have around ten million network attacks. There are 20 teams in the CTF competition. In order to attribute an attack to a particular team, apart from analyzing the payloads used by the team, we also need to analyze the behavior of the attacking team towards his adversary. For this purpose we separate the network attacks according to the team

Table 1 Fields in a instance of network attack

Field	Intuition
byte_hist	Histogram of byte sequences in the payload
inst_hist	Histogram of instructions used in the payload
from_team	The team where the payload originates (attacking team)
to_team	The team being attacked by the exploit
svc	The service that the payload is running
payload_hash	Indicates the payload used in the attack (md5)
time	Indicates the date and time of the attack

---

[1] https://github.com/simsong/tcpflow.

Table 2 Example event from the dataset

Field	Value
byte_hist	0×43:245, 0×69:8, 0×3a:9, 0×5d:1, ...
inst_hist	cmp:12, svcmi:2, subs:8, movtmi:60 . . . . . .
from_team	Men in black hats
to_team	Robot Mafia
svc	02345
payload_hash	2cc03b4e0053cde24400bbd80890446c
time	2013-08-03T23:45:17

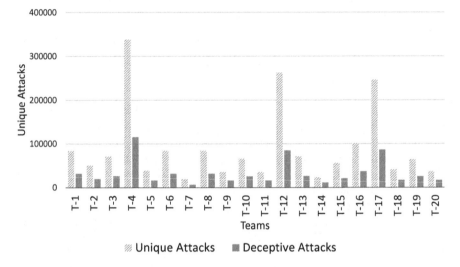

Fig. 1 Unique deceptive attacks directed towards each team

being targeted. Thus we have 20 such subsets and we list them by team name in Table 3. An example of an event in the dataset is shown in Table 2.

We now discuss two important observations from the dataset, that makes the task of attributing a observed network attack to a team difficult.

Deception In the context of this paper we define an attack to be deceptive when multiple adversaries get mapped to a single attack pattern. In the current setting we define deception as the scenario when the same exploit is used by multiple teams to target the same team. Figure 1 shows the distribution of unique deception attacks with respect to the total unique attacks in the dataset based on the target team. These unique deceptive attacks amount to just under 35 % of the total unique attacks.

Duplicate Attacks A duplicate attack occurs when the same team uses the same payload to attack a team at different time instances. Duplicate attacks can be attributed to two reasons. First when a team is trying to compromise other system, it just does not launch a single attack but a wave of attacks with very little time difference between consecutive attacks. Second, once a successful payload is

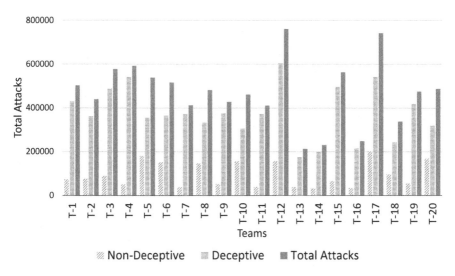

Fig. 2 Total attacks and duplicate attacks (Deceptive and Non-deceptive) directed towards each team

created which can penetrate the defense of other systems, it is used more by the original attacker as well as the deceptive one as compared to other payloads. We group duplicates as being non-deceptive and deceptive. Non-deceptive duplicate are the duplicates of the team that first initiated the use of a particular payload. On the other hand deceptive duplicates are all the attacks from the teams that are being deceptive. Deceptive duplicates form a large portion of the dataset as seen in Fig. 2.

Analyzing the number of teams that use a particular payload, gives us insights into the deceptive behavior of teams. We plot the usage of unique payloads with respect to the number of teams using them in their attacks. We use 4 different categories namely payloads used by a single team, payloads used by two teams, payloads used by three teams and payloads used by more than three teams. Figure 3 shows the plot for each target team. A large fraction of unique payloads fall in the first two categories (one team and two teams).

4   Baseline Approaches

From the dataset we have the ground truth available for all the samples. Hence we use supervised machine learning approaches to predict the attacking team. The ground truth corresponds to a team mentioned in Table 3.

Decision Tree (DT) For baseline comparisons we first implemented a decision tree classifier. This hierarchical recursive partitioning algorithm is widely used for classification problems. We built the decision tree by finding the attribute that

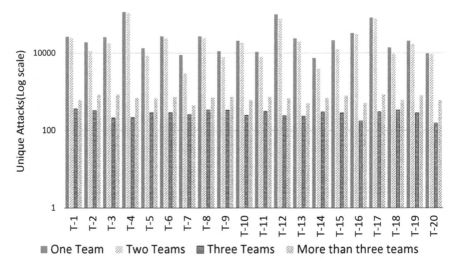

Fig. 3 Attacks on each target team by one team, two teams, three teams and more than three teams

Table 3 Teams in the CTF competition

Notation	Team
T-1	9447
T-2	APT8
T-3	Alternatives
T-4	PPP
T-5	Robot Mafia
T-6	Samurai
T-7	The European Nopsled Team
T-8	WOWHacker-BIOS
T-9	[Technopandas]
T-10	Blue lotus
T-11	Clgt
T-12	Men in black hats
T-13	More smoked leet chicken
T-14	Pwnies
T-15	Pwningyeti
T-16	Routards
T-17	raon_ASRT (whois)
T-18	Shell corp
T-19	Shellphish
T-20	Sutegoma2

maximizes information gain at each split. This attribute is termed as the best split attribute and is used to split the node. Higher the information gain, the more pure the nodes that are split will be. During the testing phase, we check the test sample for the presence or absence of the best split attribute at each node till we reach the leaf node. The team that has majority samples at the leaf node, is predicted as the attack team for the test sample. In order to avoid over-fitting we terminate the tree, when the number of samples in the node are less than 0.1 % of the training data.

Random Forest (RF) Random forest is an ensemble classification method proposed by Breiman [11]. It is based on the idea of generating multiple predictors which are then used in combination to classify unseen samples. The strength of random forest lies in injecting randomness to build each classifier and using random low dimensional subspaces to split the data at each node in a classifier. We use a random forest which combines bagging [11] for each tree with random feature selection [12] at each node to split the data thus generating multiple decision tree classifiers. To split the data at each node we use information gain with random subspace projection. The information gain indicates the amount of purity in the node with respect to class labels. More pure nodes result in higher information gain. Hence we try to find the splits that maximize the information gain. The advantage of using random forest over a single decision tree is low variance and the notion that weak learners when combined together have a strong predictive power. During the test phase, each test sample gets a prediction from each individual decision tree (weak learner) giving its own opinion on test sample. The final decision is made by a majority vote among those trees.

Support Vector Machine (SVM) Support vector machines is a popular supervised classification technique proposed by Vapnik [13]. SVM's works by finding a separating margin that maximizes the geometric distance between classes. The separating margin is termed as a hyperplane. We use the popular LibSVM implementation [14] which is publicly available. SVM is inherently a binary classifier, and it deals with multi-class classification problems by implementing several 1-vs-1 or 1-vs-all binary classifiers which adds to the complexity as the number of classes increases.

Logistic Regression (LOG-REG) Logistic regression classifies samples by computing the odds ratio. The odds ratio gives the strength of association between the features and the class. As opposed to linear regression, the output of logistic regression is the class probability of the sample belonging to that class. We implement the multinomial logistic regression which handles multi-class classification.

4.1 Experimental Results

For our baseline experiments, we separate the attacks based on the team being targeted. Thus we have 20 attack datasets. We then sort the attack according to time. We reserve the first 90 % of the attacks for training and the rest 10 % for testing. Attacker prediction accuracy is used as the performance measure for the experiment. Accuracy is defined as the fraction of correctly classified test samples. Figure 4

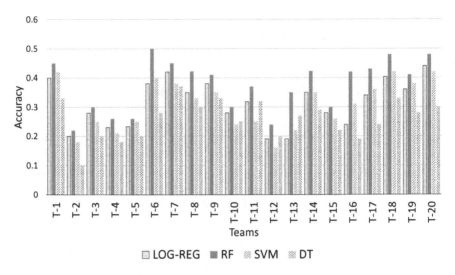

Fig. 4 Team prediction accuracy for LOG-REG, RF, SVM and DT

Table 4 Summary of prediction results averaged across all teams	Method	Average performance
	Decision tree (DT)	0.26
	Logistic regression (LOG-REG)	0.31
	Support vector machine (SVM)	0.30
	Random Forest (RF)	**0.37**

shows the accuracy for predicting the attacker for each target team. Machine learning techniques significantly outperform random guessing which would have an average accuracy of choosing 1 out of 19 teams attacking yielding an accuracy of 0.053. For this experiment random forest classifier performs better than logistic regression, support vector machine and decision tree for all the target teams with an average accuracy of 0.37. Table 4 below summarizes the average performance for each method.

4.2 Misclassified Samples

Misclassification can be attributed to the following sources,

- Non-deceptive duplicate attacks attributed to one of the deceptive teams.
- Deceptive duplicates attributed to some other deceptive team.
- Payloads that were not encountered during the training phase.

The first two sources of error make up the majority of misclassifications, since a given attack can be attributed to any of the 19 teams.

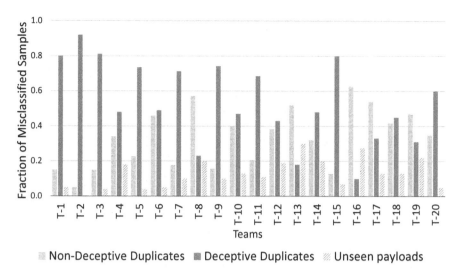

Non-Deceptive Duplicates   ■ Deceptive Duplicates   Unseen payloads

Fig. 5 Sources of error in the misclassified samples

Figure 5 shows the distribution of the above mentioned sources of misclassification for each team. Deceptive duplicates form the majority of misclassifications. This is not surprising given the fact that deceptive duplicates make up almost 90 % of the total attacks (see Fig. 2).

4.2.1 Average Prediction Probability

Figure 6 shows the average probability of correctly classified and misclassified samples. The reported average probabilities are from the random forest classifier which performs the best among the baseline approaches (see Table 4). To compute this average probability we look at the predicted probability for each test sample rather than the prediction. For random forest the predicted probability is the average of individual decision trees in the forest. It is clear that the classifier predicts the correct team with higher probability as opposed to misclassified samples which are predicted with less confidence (probability).

5 Pruning

We explore different pruning techniques to address misclassification issues with respect to deceptive and non-deceptive duplicates. The pruning techniques are only applied to the training data, while the test data is maintained at 10 % as mentioned in Sect. 4.1. We use the random forest classifier for all the pruning techniques.

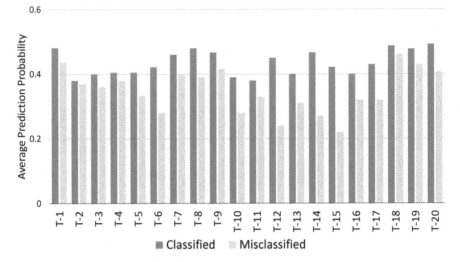

Fig. 6 Average prediction probability for correctly classified and misclassified samples

Table 5 Summary of prediction results averaged across all teams

Method	Average performance
Baseline approach (RF)	0.37
All-but-majority pruning (RF)	0.40
All-but-K-majority pruning (RF)	**0.42**
All-but-earliest pruning (RF)	0.34
All-but-most-recent pruning (RF)	0.36

These pruning techniques are briefly described as follows,

- *All-but-majority*: In this pruning we only consider the duplicates of the most attacking team given a payload and prune other duplicates.
- *All-but-K-majority*: Only consider the duplicates of the top K most frequent attacks given a payload and prunes the rest of the duplicates.
- *All-but-earliest-majority*: We only retain the duplicates of the team that initiates the attack given a payload, rest all duplicates are pruned.
- *All-but-most-recent-majority*: In this pruning we retain the duplicates of the team that last used the payload in the training data, rest all duplicates are pruned.

Table 5 gives the summary of the prediction results for all the pruning techniques in comparison with the random forest baseline approach. In the pruning techniques All-but-K-majority works best with an average accuracy of 0.42.

All-but-Majority (P-1) In this pruning technique, for each payload, we only retain duplicates of the most frequent attacking team and prune the duplicates of all other teams. This pruned set is then used to train the random forest classifier. Table 6 shows the classifier performance in comparison with the baseline method. All-but-majority pruning technique has better performance on the test set than the baseline

Table 6 Pruning technique performance comparison

Teams	RF	P-1(RF)	P-2(RF)	P-3(RF)	P-4(RF)
T-1	0.45	0.16	**0.46**	0.15	0.15
T-2	0.22	0.28	**0.30**	0.15	0.14
T-3	0.30	0.53	0.29	**0.57**	**0.57**
T-4	0.26	**0.33**	0.27	0.31	0.32
T-5	0.26	0.38	**0.45**	0.40	0.42
T-6	**0.50**	0.27	0.24	0.31	0.26
T-7	0.45	**0.59**	0.58	0.19	0.49
T-8	0.42	0.52	0.52	0.51	**0.55**
T-9	0.41	0.65	**0.68**	0.52	0.53
T-10	0.30	0.54	0.34	0.55	**0.57**
T-11	**0.37**	0.27	0.35	0.27	0.29
T-12	0.24	**0.37**	**0.37**	0.25	0.22
T-13	0.35	0.27	**0.37**	0.29	0.27
T-14	**0.42**	0.27	0.40	0.30	0.30
T-15	**0.30**	0.20	0.27	0.21	0.20
T-16	**0.42**	0.28	0.22	0.32	0.31
T-17	0.43	**0.45**	0.35	0.43	0.40
T-18	**0.48**	0.39	0.43	0.41	0.40
T-19	0.41	**0.65**	0.58	0.54	0.60
T-20	**0.48**	0.16	0.16	0.16	0.17

approach for 11 out of 20 teams. Using this pruning technique does benefit majority of the teams as the prediction accuracy improves for them, but for some teams the performance drops. The reason for the drop in performance for some teams is due to the fact that training set gets dominated by a single team which does not have majority in testing set. Since the majority team gets represented in most of the leaves of the random forest classifier, it gets predicted more often leading to high misclassifications.

All-but-K-Majority (P-2) In order to address the issue of one team dominating in the training set, we use the all-but-K-majority where we consider the K most frequent teams for a payload under consideration. After trying out different values of K we select K = 3, which gives the best performance. For higher values of K, the pruning behaves like the baseline approach and for lower values it behaves like All-but-majority. On average each team gains about 40 K samples in the training set as compared to all-but-majority pruning. Table 6 shows the classifier performance. In this case also pruning performs better than baseline in 11 out of 20 teams, but as compared to all-but-majority the performance for most teams is better.

All-but-Earliest (P-3) For this pruning we only retain the duplicates of the team that initiated the attack using a particular payload. This pruning technique retains all the non-deceptive duplicates while getting rid of the deceptive ones. Table 6 shows the classifier performance. This pruning technique performs better than the baseline

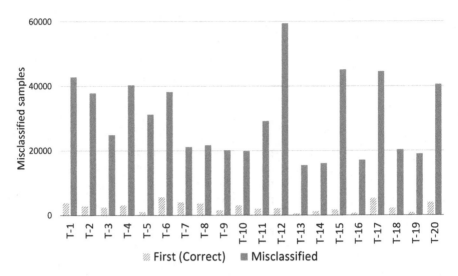

Fig. 7 Samples correctly classified by all-but-earliest and not by all-but-most-recent

approach for 8 out of 20 teams. Comparing this result to all-but-majority (including all-but-K-majority) pruning indicates that deceptive duplicates are informative in attributing an attack to a team and should not be ignored completely.

All-but-Most-Recent (P-4) In this pruning we repeat a similar procedure like All-but-earliest but instead of retaining the duplicates of the team that initiated an attack, we retain the duplicates of the team that used the payload last in the training set. Since the data is sorted according to time, the last attacker becomes the most recent attacker for the test set. Table 6 shows the classifier performance.

5.1 Discussion

On further analysis of the misclassified samples from all-but-earliest and all-but-most-recent provides an interesting observation. Though majority of the misclassified samples between the two pruning techniques are similar, there is a fraction of samples which were correctly classified by all-but-earliest but misclassified by all-but-most-recent and vice versa. Let first (correct) denote the number of samples that were correctly classified from the misclassified samples of all-but-most-recent majority pruning experiment. Similarly last (correct) be the samples that were correctly classified from the misclassified samples of all-but-earliest majority pruning technique. Figure 7 shows the number of samples that all-but-earliest pruning was able to classify correctly that were misclassified by all-but-most-recent. Figure 8 shows a similar result for the other case. For both cases the correctly

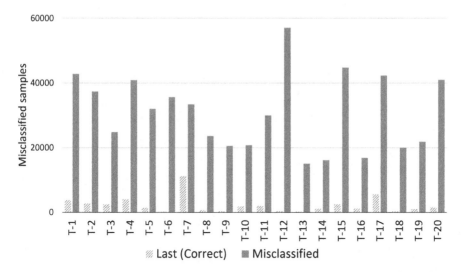

Fig. 8 Samples correctly classified by all-but-most-recent and not by all-but-earliest

classified samples make up around 5–10 % of the misclassified samples for each team. This result shows that using the two pruning techniques together to make attribution decision, would lead to higher performance as opposed to using only one.

5.2 Ensemble Classifier

As discussed earlier using pruning techniques together would improve the prediction accuracy as opposed to using the pruning techniques individually. We perform an experiment to demonstrate it. We generate one prediction from each of the three pruning techniques namely, all-but-K-majority (since this pruning technique performs better than all-but-majority), all-but-earliest and all-but-most-recent. We call these predictions candidates for being the most likely attackers. We then define conditions to predict the actual attacker as follows:

1. *Predict the team that satisfies atleast two pruning cases.*
2. *If all the attacking candidates are different, then predict the all-but-K-majority prediction.*

Table 7 shows the results of this ensemble framework in comparison with the baseline approach and the different pruning techniques. The results are averaged across all teams. This 3-candidate ensemble framework performs the best with an average accuracy of 0.48 better than all the approaches indicating that using the pruning techniques in conjunction with each other improves the prediction accuracy.

Table 7 Summary of prediction results averaged across all teams

Method	Average performance
Baseline approach (RF)	0.37
All-but-majority pruning (RF)	0.40
All-but-K-majority pruning (RF)	0.42
All-but-earliest pruning (RF)	0.34
All-but-most-recent pruning (RF)	0.36
3-Candidate ensemble model (RF)	**0.48**

6 Conclusion

In this paper, we study cyber-attribution by examining DEFCON CTF data—which provides us with ground-truth on the culprit responsible for each attack. We frame cyber-attribution as a classification problem and examine it using several machine learning approaches. We find that deceptive incidents account for the vast majority of misclassified samples and introduce heuristic pruning techniques that alleviate this problem somewhat. We also observe that using the pruning techniques together to make a classification decision, improves the accuracy as against using a single pruning technique.

7 Future Work

Moving forward, we look to employ a more principled approach to counter deception based on our previously established theoretical framework for reasoning about cyber-attribution [9, 10]. In particular we wish to employ temporal reasoning to tackle the problem of deceptive attacks. Studying the problem from an attacker's perspective would give us more insights in their behavior to commit deception and attacking a particular team in the competition. This opens up interesting research questions in particular identifying hacking group from a series of attacks over a period of time, differentiating between deceptive hacking groups in time series data. This is a knowledge engineering challenge which calls for development of efficient and scalable algorithms.

Acknowledgements Some of the authors of this work were supported by the U.S. Department of the Navy, Office of Naval Research, grant N00014-15-1-2742 as well as the Arizona State University Global Security Initiative (GSI). Any opinions, findings, and conclusions or recommendations expressed in this material are those of the author(s) and do not necessarily reflect the views of the Office of Naval Research.

References

1. Shakarian, Paulo, Jana Shakarian, and Andrew Ruef. Introduction to cyber-warfare: A multidisciplinary approach. Newnes, 2013.
2. Tsagourias, Nicholas. "Nicolas Politis: initiatives to outlaw war and define aggression, and the narrative of progress in international law." European Journal of International Law 23.1 (2012): 255–266.
3. Boebert, W. Earl. "A survey of challenges in attribution." Proceedings of a workshop on Deterring CyberAttacks. 2010.
4. Dacier, Marc, Van-Hau Pham, and Olivier Thonnard. "The WOMBAT Attack Attribution method: some results." Information Systems Security. Springer Berlin Heidelberg, 2009. 19–37.
5. Thonnard, Olivier, Wim Mees, and Marc Dacier. "On a multicriteria clustering approach for attack attribution." ACM SIGKDD Explorations Newsletter 12.1 (2010): 11–20.
6. Kalutarage, Harsha K., et al. "Sensing for suspicion at scale: A Bayesian approach for cyber conflict attribution and reasoning." Cyber conflict (CYCON), 2012 4th international conference on. IEEE, 2012.
7. Walls, Robert J. Inference-based Forensics for Extracting Information from Diverse Sources. Diss. University of Massachusetts Amherst, 2014.
8. Rid, Thomas, and Ben Buchanan. "Attributing Cyber Attacks." Journal of Strategic Studies 38.1–2 (2015): 4–37.
9. Jajodia, Sushil. "Advances in Information Security." (2015).
10. Shakarian, Paulo, Gerardo I. Simari, and Marcelo A. Falappa. "Belief revision in structured probabilistic argumentation." Foundations of Information and Knowledge Systems. Springer International Publishing, 2014. 324–343.
11. Breiman, Leo. "Random forests." Machine learning 45.1 (2001): 5–32.
12. Breiman, Leo. "Bagging predictors." Machine learning 24.2 (1996): 123–140.
13. Cortes, Corinna, and Vladimir Vapnik. "Support-vector networks." Machine learning 20.3 (1995): 273–297.
14. Chang, Chih-Chung, and Chih-Jen Lin. "LIBSVM: A library for support vector machines." ACM Transactions on Intelligent Systems and Technology (TIST) 2.3 (2011): 27.
15. DEFCON: https://media.defcon.org/ (2013).

Deceiving Attackers by Creating a Virtual Attack Surface

Massimiliano Albanese, Ermanno Battista, and Sushil Jajodia

Abstract Cyber attacks are typically preceded by a reconnaissance phase in which attackers aim at collecting valuable information about the target system, including network topology, service dependencies, operating systems, and unpatched vulnerabilities. Unfortunately, when system configurations are static, attackers will always be able, given enough time, to acquire accurate knowledge about the target system through a variety of tools—including operating system and service fingerprinting—and engineer effective exploits. To address this important problem, many techniques have been devised to dynamically change some aspects of a system's configuration in order to introduce uncertainty for the attacker. In this chapter, we present a graph-based approach for manipulating the attacker's view of a system's attack surface, which addresses several limitations of existing techniques. To achieve this objective, we formalize the notions of *system view* and *distance between views*. We then define a principled approach to manipulating responses to attacker's probes so as to induce an external view of the system that satisfies certain desirable properties. In particular, we propose efficient algorithmic solutions to two classes of problems, namely (1) inducing an external view that is at a minimum distance from the internal view, while minimizing the cost for the defender; (2) inducing an external view that maximizes the distance from the internal view, given an upper bound on the cost for the defender. In order to demonstrate practical applicability of the proposed approach, we present deception-based techniques for defeating an attacker's effort to fingerprint operating systems and services on the target system. These techniques consist in manipulating outgoing traffic so that it resembles traffic generated by a completely different system. Experimental results show that our approach can efficiently and effectively deceive an attacker.

M. Albanese (✉) • S. Jajodia
Center for Secure Information Systems, George Mason University, Fairfax, VA 22030, USA
e-mail: malbanes@gmu.edu; jajodia@gmu.edu

E. Battista
University of Naples, Naples, Italy
e-mail: ermanno.battista@unina.it

© Springer International Publishing Switzerland 2016   167
S. Jajodia et al. (eds.), *Cyber Deception*, DOI 10.1007/978-3-319-32699-3_8

1 Introduction

Today's approach to cyber defense is governed by slow and deliberative processes such as deployment of security patches, testing, episodic penetration exercises, and human-in-the-loop monitoring of security events. Adversaries can greatly benefit from this situation, and can continuously and systematically probe target networks with the confidence that those networks will change slowly if at all. In fact, cyber attacks are typically preceded by a reconnaissance phase in which adversaries aim at collecting valuable information about the target system, including network topology, service dependencies, operating systems and applications, and unpatched vulnerabilities. As most system configurations are static—hosts, networks, software, and services do not reconfigure, adapt, or regenerate except in deterministic ways to support maintenance and uptime requirements—it is only a matter of time for attackers to acquire accurate knowledge about the target system. A vast array of automated tools and techniques exist to facilitate this task, including OS fingerprinting and service fingerprinting tools.[1] Specifically, operating system fingerprinting aims at determining the operating system of a remote host in either a passive way, through sniffing and traffic analysis, or an active way, through probing. Similarly, service fingerprinting aims at determining what services are running on a remote host. The information collected during reconnaissance will eventually enable attackers to engineer reliable exploits and plan attacks.

In order to address this important problem, significant work has been done in the area Adaptive Cyber Defense (ACD), which includes concepts such as Moving Target Defense (MTD), artificial diversity, and bio-inspired defenses. Essentially, a number of techniques have been proposed to dynamically change a system's attack surface by periodically reconfiguring some aspects of the system. In [18], a system's attack surface has been defined as the *"subset of the system's resources (methods, channels, and data) that can be potentially used by an attacker to launch an attack"*. Intuitively, dynamically reconfiguring a system is expected to introduce uncertainty for the attacker and increase the cost of the reconnaissance effort. However, one of the major drawbacks of current approaches is that periodically reconfiguring a system may introduce a costly overhead for legitimate users, as well as the potential for denial of service conditions. Additionally, most of the existing techniques are purely proactive in nature or do not adequately consider the attacker's behavior.

The work we present in this chapter advances the state of the art in adaptive cyber defense by proposing a graph-based approach for manipulating the attacker's perception of a system's attack surface. To achieve this objective, we formalize the notion of *system view* as well as the notion of *distance between views*. We refer to the attacker's view of the system as the *external view* and to the defender's view as the *internal view*. A system's attack surface can then be thought of as the subset of the internal view that would be exposed to potential attackers if no deceptive

---

[1]We present a taxonomy of OS fingerprinting tools in Sect. 6.

strategy is adopted. Starting from these definitions, we develop a principled yet practical approach to manipulate outgoing traffic so as to induce an external view of the system that satisfies certain desirable properties. In particular, we propose efficient algorithmic solutions to different classes of problems, namely (1) inducing an external view that is at a minimum distance from the internal view, while minimizing the cost for the defender; (2) inducing an external view that maximizes the distance from the internal one, given an upper bound on the cost for the defender.

We go beyond simply introducing uncertainty for the attacker, and deceive potential intruders into making incorrect inferences about important system characteristics, including operating systems and active services. In order to demonstrate practical applicability of the proposed approach, we present deception-based techniques for defeating an attacker's effort to fingerprint operating systems and services on the target system. Differently from many existing techniques, we do so without changing the actual configuration of the system. In fact, our approach mainly consists in manipulating outgoing traffic such that, not only important details about operating systems and services are not revealed, but the traffic also resembles traffic generated by hosts and networks with different characteristics. Experiments conducted on a prototypal implementation show that the overhead introduced by the proposed approach is negligible, thus rendering this solution completely transparent to legitimate users. At the same time, our approach can effectively deceive the attackers, and steer them away from critical resources we wish to protect.

The remainder of the chapter is organized as follows. Section 2 discusses related work. Section 3 discusses the threat model whereas Sect. 4 presents a motivating example. Next, Sect. 5 provides a detailed description of our approach and presents the problem statement as well as the proposed algorithms. Then, Sect. 6 presents in detail the specific techniques we have designed to defeat OS and service fingerprinting. Finally, Sect. 7 reports the results of our experiments, and Sect. 8 gives concluding remarks.

2 Related Work

Moving Target Defense (MTD) defines mechanisms and strategies to increase complexity and cost for attackers [13]. MTD approaches aiming at selectively altering a system's attack surface usually involve reconfiguring the system in order to make attacks' preconditions unstable.

Dunlop et al. [12] propose a mechanism to dynamically hide addresses of IPv6 packets to achieve anonymity. This is done by adding virtual network interface controllers and sharing a secret among all the hosts in the network. In [11], Duan et al. present a proactive Random Route Mutation technique to randomly change the route of network flows to defend against eavesdropping and DoS attacks. In their implementation, they use OpenFlow Switches and a centralized controller to define the route of each flow. Jafarian et al. [15] use an IP virtualization scheme based on virtual DNS entries and Software Defined Networks. Their goal is to hide network assets from scanners. Using OpenFlow, each host is associated with a range of virtual IP addresses and mutates its IP address within its pool. A similar identity virtualization approach is presented in [3]. In Chap. 8 of [16], an approach based on diverse virtual servers is presented. Each server is configured with a set of software stacks, and a rotational scheme is employed for substituting different software stacks for any given request. This creates a dynamic and uncertain attack surface. Also, Casola et al. [8, 9] propose an MTD approach for protecting resource-constrained distributed devices through fine-grained reconfiguration at different architectural layers.

These solutions tend to reconfigure a system in order to modify its external attack surface. On the other hand, the external view of the system is usually inferred by attackers based on the results of probing and scanning tools. Starting from this observation, our approach consists in modifying system *responses* to probes in order to expose an *external view* of the system that is significantly different from the actual attack surface, without altering the system itself.

Reconnaissance tools, such as *nmap* or *Xprobe2*, can identify a service or an operating system by analyzing packets that can reveal implementation specific details about the host [17, 22]. Network protocol *fingerprinting* refers to the process of identifying specific *features* of a network protocol implementation by analyzing its input/output behavior [20]. These features may reveal specific information such as protocol version, vendor information, and configurations. Reconnaissance tools store known system's features and compare them against scan responses in order to match a fingerprint. Watson et al. [22] adopted protocol scrubbers in order to avoid revealing implementation-specific information and restrict an attacker's ability to determine the OS of a protected host. Moreover, some proof-of-concept software and kernel patches have been proposed to alter a system fingerprint [7], such as IP Personality and Stealth Patch. Amongst the various techniques that have been proposed to defeat fingerprinting [19], a very simple and intuitive one consists in modifying the default values of a TCP/IP stack implementation, such as the TTL, Window Size, or TCP Options. Other approaches to defeating fingerprinting [21] include altering public service banners and searching content files for 'incriminating' strings that can give away information about the OS. For instance, web pages may include automatically generated comments that identify the authoring tool.

Honeypots have been traditionally used to try to divert attackers away from critical resources. Although our approach and honeypots share a common goal, they are significantly different. Our approach does not alter the system while honeypot-based solutions introduce vulnerable machines in order to either capture the attacker [1] or collect information for forensic purposes [10]. Instead, we aim at deceiving attackers by manipulating their view of the target system and forcing them to plan attacks based on inaccurate knowledge, so that the attacks will likely fail. To the best of our knowledge, we are the first to propose an adaptive and comprehensive approach to changing the attacker's view of a system's attack surface without reconfiguring the system itself [4].

3   Threat Model

We assume that an external adversary is attempting to infer a detailed view of the target network using reconnaissance tools such as *nmap* [17] to discover active hosts in the network. The information an attacker may attempt to discover includes operating systems, exposed services and their version, network layout, and routing information. The attacker will then leverage this knowledge to plan and execute attacks aimed at exploiting exposed services. We also assume that the attacker will use an OS fingerprint technique based on sending valid and invalid IP packets and analyzing the respective responses. We do not explicitly address techniques based on timing and data analysis. Moreover, we limit service fingerprinting to the case of TCP probes, as it is the case for most common probing tools.

The attacker's strategy is illustrated by the flowchart of Fig. 1. The goal is to launch an attack against one of the hosts in the target network. Multiple stages of this attack strategy (marked with a red cross in the figure) can be defeated using our approach. For instance, we may expose services with no publicly available exploits. On the other hand, for a given service, we may expose exploitable vulnerabilities which do not correspond to the actual vulnerabilities of that service.

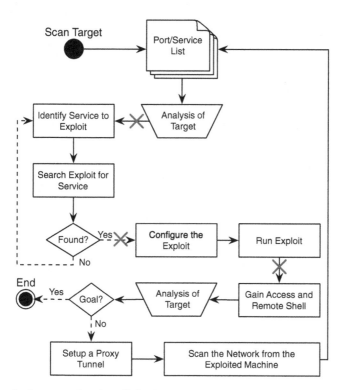

Fig. 1 Attacker's strategy flowchart (Color figure online)

4 Motivating Example

As a reference scenario, we consider the networked system of Fig. 2, modeling the IT infrastructure of an *e-commerce company*. Customers can access publicly available services through a website hosted in the DMZ. The business logic and critical services are deployed in the Intranet, and some of these services need to be accessible through the Internet in order to allow company branches to process orders and query the inventory.

Our goal is to modify the attacker's view of the system. In order to do so, we only modify system-dependent information exposed by system-specific protocol implementations. We adopt a graph-based strategy to generate different views of the system, such as the one in Fig. 3, by repeatedly applying view manipulation primitives, which are implemented by filtering and altering outgoing traffic.

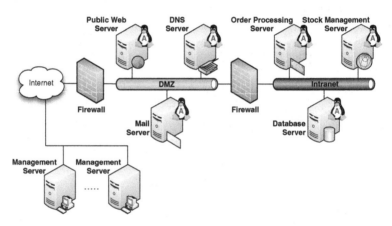

Fig. 2 Topology and configuration of the reference system

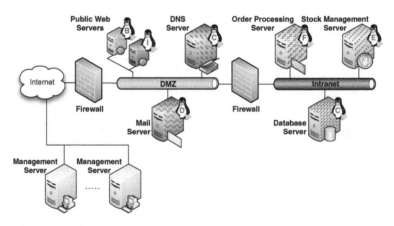

Fig. 3 Topology and configuration of the reference system as presented to the attacker

In Fig. 3, we depict *manipulated* services with a different color/texture—compared to Fig. 2—of the machine where they are deployed. A change in the operating system is represented by a different letter on the top right corner of each machine. In this example, applying several manipulation steps to the original view, we move from a scenario where all the servers have the same operating system to a scenario in which each operating system is different from the real one and the others. In the same way, all the services under our control are altered. For instance, we alter the *database server* fingerprint so as it will be recognized as an implementation from a different vendor. As for the *public web server*, we want it to act like two web servers in a load balancing configuration. To do so we mutate, with a certain frequency, both the OS and service fingerprints, and modify packet level parameters. In this way, we can force the attacker to believe that multiple servers need to be compromised in order to disrupt the service.

5 Our Approach

In order to achieve our goal of inducing an attacker's view of the system's attack surface that is measurably different from the internal view, we first need to formalize the notions of *view*, *manipulation primitive*, and *distance between views* (Sect. 5.1).

5.1 View Model

In the following, we assume a system is a set $S = \{s_1, s_2, \ldots, s_n\}$ of devices (e.g., hosts, firewalls), and use Ψ to denote the set of services that can be offered by hosts in S. The defender's and attacker's knowledge of the system is represented by views, as defined below.

Definition 1 (System's View). Given a system S, a *view* of S is a triple $V = (S_o, C, \nu)$, where $S_o \subseteq S$ is a set of observable devices, $C \subseteq S_o \times S_o$ represents connectivity between elements in S_o, and $\nu : S_o \to 2^{\Psi}$ is a function that maps each host in S_o to the set of services it offers.

Intuitively, a view represents knowledge of a subset of the system and includes information about the topology and about services offered by reachable hosts.[2]

Definition 2 (Manipulation Primitive). Given a system S and a set $\mathcal{V}$ of views of S, a *manipulation primitive* is a function $\pi : \mathcal{V} \to \mathcal{V}$ that transforms a view $V' \in \mathcal{V}$ into a view V" $\in \mathcal{V}$. Let Π denote a family of such functions. For each $\pi \in \Pi$, the following properties must hold.

---

[2]A more complete definition of view could incorporate information about service dependencies and vulnerabilities, similarly to what proposed in [2].

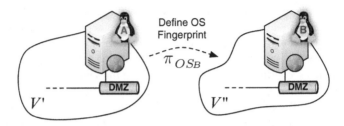

Fig. 4 Example of manipulation primitive

$$(\forall V \in \mathcal{V}) \; \pi(V) \neq V \; (\forall V \in \mathcal{V}) \; (\nexists \langle \pi_1, \pi_2, \ldots, \pi_m \rangle \in \Pi^m \; |$$

$$\pi_1(\pi_2(\ldots \pi_m(V)\ldots)) = \pi(V))$$

Intuitively, a manipulation primitive is an atomic transformation that can be applied to a view to obtain a different view.

Example 1. A possible manipulation primitive is $\pi_{OS_B}(\cdot)$, which transforms a view V' into a view V'' by changing the operating system fingerprint of a selected host.[3] Figure 4 illustrates the effect of this primitive on the system's view.

Definition 3 (View Manipulation Graph). Given a system S, a set $\mathcal{V}$ of views of S, and a family Π of manipulation primitives, a view manipulation graph for S is a directed graph $G = (\mathcal{V}', \mathcal{E}, \ell)$, where

- $\mathcal{V}' \subseteq \mathcal{V}$ is a set of views of S;
- $\mathcal{E} \subseteq \mathcal{V} \times \mathcal{V}$ is a set of edges;
- $\ell : \mathcal{E} \to \Pi$ is a function that associates with each edge $(V', V'') \in \mathcal{E}$ a manipulation primitive $\pi \in \Pi$ such that $V'' = \pi(V')$.

The node representing the internal view has no incoming edges. All other nodes represent possible external views.

Figure 5 shows an example of *view manipulation graph*. After applying any $\pi \in \Pi$, a new view is generated. By analyzing the graph, one can enumerate all possible ways to generate external views starting from the internal view V of a system S.

Definition 4 (Distance). Given a system S and a set $\mathcal{V}$ of views of S, a distance over $\mathcal{V}$ is a function $\delta : \mathcal{V} \times \mathcal{V} \to \mathbb{R}$ such that, $\forall \; V', V'', V''' \in \mathcal{V}$, the following properties hold:

$$\delta\left(V', V''\right) \geq 0$$

$$\delta\left(V', V''\right) = 0 \iff V' = V''$$

---

[3]Each primitive may have a set of specific parameters, which we omit to simplify the notation.

Fig. 5 Example of *view manipulation graph*

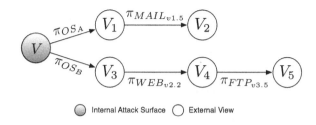

⬤ Internal Attack Surface ◯ External View

$$\delta \left(V', V'' \right) = \delta \left(V'', V' \right)$$

$$\delta \left(V', V''' \right) \leq \delta \left(V', V'' \right) + \delta \left(V'', V''' \right)$$

Example 2. In the simplest case, the distance can be measured by looking at the elements that change between views. To do so, we consider the difference in the number of hosts between two views. Then, for each host that is present in both views we add one if OS fingerprints differ and we add one if service fingerprints differs. More sophisticated distances can be defined, but this is beyond the scope of this chapter.

Definition 5 (Path Set). Given a view manipulation graph G, the path set P_G for G is the set of all possible paths (sequence of edges) in G. We denote the kth path, of length m_k, in P_G as $p_k = \langle \pi_{i_1}, \pi_{i_2} \ldots, \pi_{i_{m_k}} \rangle$.

Example 3. Consider the view manipulation graph G in Fig. 5. The set of all possible paths is $P_G = \{\langle \pi_{OS_A} \rangle, \langle \pi_{MAIL_{v1.5}} \rangle, \langle \pi_{OS_A}, \pi_{MAIL_{v1.5}} \rangle, \langle \pi_{OS_B} \rangle, \langle \pi_{WEB_{v2.2}} \rangle, \langle \pi_{FTP_{v3.5}} \rangle, \langle \pi_{OS_B}, \pi_{WEB_{v2.2}} \rangle, \langle \pi_{WEB_{v2.2}}, \pi_{FTP_{v3.5}} \rangle, \langle \pi_{OS_B}, \pi_{WEB_{v2.2}}, \pi_{FTP_{v3.5}} \rangle\}$.

We will use the notation $V_a \xrightarrow{p_k} V_b$ to refer to a path p_k originating from V_a and ending in V_b. For instance, in Fig. 5, the path which goes from V to V_5 is denoted as $V \xrightarrow{p_9} V_5 = \langle \pi_{OS_B}, \pi_{WEB_{v2.2}}, \pi_{FTP_{v3.5}} \rangle$.

Definition 6 (Cost Function). Given a path set P_G, a *cost function* is a function $f_c : P_G \to \mathbb{R}$ that associates a cost to each path in P_G. The following properties must hold.

$$f_c \left(\langle \pi_i \rangle \right) \geq 0 \tag{1}$$

$$f_c \left(\langle \pi_{i_j}, \pi_{i_{j+1}} \rangle \right) \geq min \left[f_c \left(\langle \pi_{i_j} \rangle \right), f_c \left(\langle \pi_{i_{j+1}} \rangle \right) \right] \tag{2}$$

$$f_c \left(\langle \pi_{i_j}, \pi_{i_{j+1}} \rangle \right) \leq f_c \left(\langle \pi_{i_j} \rangle \right) + f_c \left(\langle \pi_{i_{j+1}} \rangle \right) \tag{3}$$

If Eq. 3 holds strictly, then f_c is said to be *additive*.

5.2 Problem Statement

We can now formalize the two related problems we are addressing in this chapter.

Problem 1. Given a view manipulation graph G, its internal view V_i and a distance threshold $d \in \mathbb{R}$, find an external view V_d and a path $V_i \xrightarrow{p_d} V_d$ that minimizes $f_c(p_d)$ subject to $\delta(V_i, V_d) \geq d$.

Problem 2. Given a view manipulation graph G, its internal view V_i and a budget $b \in \mathbb{R}$, find an external view V_b and a path $V_i \xrightarrow{p_b} V_b$ that maximizes $\delta(V_i, V_b)$ subject to $f_c(p_b) \leq b$.

5.3 Algorithms

In this section, we present heuristic algorithms to solve the problems defined in Sect. 5.2. The algorithms start from the internal view and explore the state space by iteratively traversing the k most promising outgoing edges of each node traversed, until a termination condition is reached. To quantify the benefit of traversing a given edge, we define a *benefit score* as the ratio of the distance between the corresponding views to the cost for achieving that distance. For each node in the graph, we only traverse the k outgoing edges with the highest values of the benefit score.

5.3.1 Algorithm *TopKDistance*

To solve Problem 1, we first generate the view manipulation graph and then execute the heuristic *TopKDistance* algorithm. For efficiency purposes, we only generate a sub-graph G_d of the complete view manipulation graph G, such that generation along a given path stops when the distance from the internal view becomes equal to or larger than the minimum required distance d—in fact, any additional edge would increase the cost of the solution, thus it would not be included in the optimal path.

Algorithm 1 describes how we generate the sub-graph G_d. The algorithm uses a queue Q to store the vertices to be processed. At each iteration of the while loop (Line 3), a vertex v is popped from the queue and the maximum distance from the internal view O is updated (Line 4). The constant MAX_INDEGREE (Line 5) is used to test if a node has been fully processed. When the in-degree is equal to MAX_INDEGREE, the node has been linked to all the nodes that differ by only one element. In this case, there are no new vertices that can be generated starting from this node. Given the set $C(s)$ of all admissible configurations for a device $s \in S$, MAX_INDEGREE can be computed as MAX_INDEGREE $= \sum_{s \in S} (|C(s)| - 1)$.

On Line 6, the set of v's predecessors is retrieved. The function *getCombinations* (Line 7) generates all possible configurations that differ from v's configuration by only one element. Both v and *predecessors* are excluded from the returned array.

Algorithm 1 $GenerateGraph(V_o, C, d)$

Input: The internal view V_o, the set of admissible configurations C for each host, the minimum required distance d.
Output: A subgraph G_d of the complete view manipulation graph with vertices within distance d of V_o.
1: // Initialization: Q is the queue of vertices to be processed (initially empty); G_d is also initially empty.
2: $maxDistance \leftarrow 0$; $G_d.addVertex(V_o)$; $Q.push(V_o)$;
3: **while** $Q \neq \emptyset$ **do**
4:    $V \leftarrow Q.pop()$; $maxDistance \leftarrow \max(maxDistance, \delta(V_o, V))$;
5:    **if** $V.indegree <$ MAX_INDEGREE **then**
6:       $predecessors \leftarrow G_d.getPredecessors(V)$
7:       $newVertices \leftarrow getCombinations(V, predecessors, C)$
8:       **for all** $V_n \in newVertices$ **do**
9:          $G_d.addVertex(V_n)$
10:          $G_d.addDirectedEdge(V, V_n, cost(V, V_n), \delta(V, V_n))$; //Add an edge from V to V_n
11:          $verticesToLink \leftarrow getOneChangeVertices(G_d, V_n)$
12:          **for all** $V' \in verticesToLink$ **do**
13:             $G_d.addBidirectEdge(V, V_n, cost(V_n, V'), \delta(V_n, V'))$ //Add a bidirectional edge between V_n and V'
14:          **end for**
15:          **if** $maxDistance \leq d$ **then**
16:             $Q.push(V_n)$
17:          **end if**
18:       **end for**
19:    **end if**
20: **end while**

The function $getOneChangeVertices$ (Line 11) returns an array of all the vertices whose configuration differs just for one element from the vertex given as input. Those vertices need to be linked with the vertex v (Lines 12–14). Lines 15–17 check if the maximum distance d has been reached. If this is not the case, the newly generated vertex is pushed into the queue to be examined.

Once the subgraph has been generated, we can run the top-k analysis using the $TopKDistance$ algorithm (Algorithm 2), which recursively traverses the subgraph to find a solution. We use v to denote the vertex under evaluation in each recursive call. Line 1 creates an empty list to store all the paths discovered from v. Line 2 is one of the two termination conditions. It checks whether the current distance is greater than or equal to d or no other nodes can be reached. The second term in the termination condition takes into account both the case of a node with no outgoing edges and the case of a node whose successors are also its predecessors. We do not consider edges directed to predecessors in order to construct loop-free paths. If the termination condition is satisfied, a solution has been found and a path from v to v_o can be constructed by closing the recursion stack. Line 3 creates an empty path and adds v to it. Then, the path list is updated and returned. On Line 6, all the edges originating from v are sorted by decreasing benefit score. Then, Lines 7–13 perform the top-k analysis and $TopKDistance$ is recursively invoked for each of the best k destinations,. The result is a list of paths having vertex v as the origin (Lines 10–12).

Algorithm 2 $TopKDistance(G, V, k, d, cc, cd)$

Input: A graph G, a vertex V, an integer k, a minimum distance d, the current cost cc, and the current distance cd.
Output: List of paths $pathList$.
1: $pathList \leftarrow \emptyset$
2: **if** $cd \geq d \vee V.successors \setminus (V.successors \cap V.predecessors) = \emptyset$ **then**
3:     $p \leftarrow emptyPath$; $p.addVertex(V)$; $pathList \leftarrow \{p\}$;
4:     **return** $pathList$
5: **end if**
6: $sort(V.outgoingEdges)$; $numToEval \leftarrow min(k, |V.outgoingEdges|)$;
7: **for all** $i \in [1, \leq numToEval]$ **do**
8:     $V_n \leftarrow V.outgoingEdges[i].getDestination()$
9:     $eval \leftarrow TopKDistance(G, V_n, k, d, update(cc), update(cd))$
10:     **for all** $p \in eval$ **do**
11:         $p.addVertex(V)$; $p.addDirectedEdge(V, V_n, \delta(V, V_n), cost(V, V_n))$; //Add an edge from V to V_n
12:     **end for**
13: **end for**
14: **return** $pathList$

5.3.2 Algorithm *TopKBudget*

Algorithm *TopKBudget* (Algorithm 3) implements both graph generation and exploration in order to improve time efficiency in the resolution of Problem 2. Our approach is to generate the graph only in the k most promising directions in order to limit graph generation. The algorithm uses a queue to store examined paths that may represent a solution.

Line 5 retrieves a path p from the queue and its last vertex v. Lines 6–17 perform graph generation starting from v. The generation process is similar to the one described for Algorithm 1. On Line 18, all the successors of v (generated or linked at this stage) have been computed. All the successor nodes are used to compute the *importance* level of v. We sum v's benefit (distance/cost) and an importance estimation of v'successors. This estimation provides some knowledge about the solution we may discover by further exploring from v. It is done by the function *estimate* which returns the value of the maximum benefit of vertices reachable from v's successor v_n. A triple including the importance level, the successor v_n, and the path p under examination are then added to a list (Lines 20–21).

On Line 24, the list is sorted by decreasing importance level. Line 26 checks if there is no further exploration to perform. In this case, the path under examination is added to the list of solutions. Otherwise (Lines 29–32), new paths are generated from p. Each of the paths is p plus a new node that is in the top k successors of v. All the newly generated paths are then pushed into the queue for further examination. When the queue becomes empty, the complete list of solutions is returned.

6 Fingerprinting

Operating System (OS) fingerprinting is the practice of determining the operating system of a remote host on a network. This may be accomplished either *passively*— by sniffing and analyzing network packets traveling between hosts—or *actively*—by sending carefully crafted packets to the target host and analyzing the responses [21].

Algorithm 3 *TopKBudget*(G, V_o, C, b)

Input: A graph G, the internal view V_o, the set of admissible configurations C for each host, and a budget $b \in \mathbb{R}$.
Output: A list of paths.

1: // Initialization: Q is the queue of paths to process (initially empty); *solutions* is initially empty; G is a graph and contains only V_o
2: $\quad p \leftarrow emptyPath; p.addVertex(V_o); Q.add(p)$
3: **while** $Q \neq \emptyset$ **do**
4: $\quad\quad dataToExplore \leftarrow \emptyset$
5: $\quad\quad p \leftarrow Q.pop(); v \leftarrow p.getLast()$
6: $\quad\quad$ **if** $V.indegree <$ MAX_INDEGREE **then**
7: $\quad\quad\quad predecessors \leftarrow G.getPredecessors(V)$
8: $\quad\quad\quad newVertices \leftarrow getCombinations(V, predecessors, C)$
9: $\quad\quad\quad$ **for all** $V_n \in newVertices$ **do**
10: $\quad\quad\quad\quad G.addVertex(V_n)$
11: $\quad\quad\quad\quad G.addDirectedEdge(V, V_n, cost(V, V_n), \delta(V, V_n))$ //Add an edge from V to V_n
12: $\quad\quad\quad\quad verticesToLink \leftarrow getOneChangeVertices(G, V_n)$
13: $\quad\quad\quad\quad$ **for all** $V' \in verticesToLink$ **do**
14: $\quad\quad\quad\quad\quad G.addBidirectionalEdge(V, V_n, cost(V_n, V'), \delta(V_n, V'))$ //Add a bidirectional edge between V_n and V'
15: $\quad\quad\quad\quad$ **end for**
16: $\quad\quad\quad$ **end for**
17: $\quad\quad$ **end if**
18: $\quad\quad$ **for all** $V_n \in V.getDirectSuccessors()$ **do**
19: $\quad\quad\quad$ **if** $\neg V.isPrecedessor(V_n) \wedge \neg p.contains(V_n) \wedge (p.totalCost + cost(V, V_n) \leq b)$ **then**
20: $\quad\quad\quad\quad importance \leftarrow \delta(V, V_n)/cost(V, V_n) + estimate(V_n, p, C)$
21: $\quad\quad\quad\quad dataToExplore.add([V_n, p, importance])$
22: $\quad\quad\quad$ **end if**
23: $\quad\quad$ **end for**
24: $\quad\quad sort(dataToExplore)$
25: $\quad\quad numToEval \leftarrow \min(k, dataToExplore.size)$
26: $\quad\quad$ **if** $numToEval = 0$ **then**
27: $\quad\quad\quad solutions \leftarrow p$
28: $\quad\quad$ **end if**
29: $\quad\quad$ **for all** $i \in [1, numToEval]$ **do**
30: $\quad\quad\quad newPath \leftarrow dataToExplore[i].p$
31: $\quad\quad\quad newPath.addAsLeaf(V_n)$
32: $\quad\quad\quad Q.push(newPath)$
33: $\quad\quad$ **end for**
34: **end while**
35: **return** *solutions*

Active fingerprinting approaches are typically more sophisticated than *passive fingerprinting*. In the simplest scenario, the attacker does not resort to stealth techniques and gathers information about the OS by trying to connect to the host. For instance, while establishing a connection via the standard Telnet or SSH protocol, the OS version is often sent to the client as part of a welcome message. Moreover, some FTP server implementations allow to retrieve this information through the SYST command. In general, active fingerprinting techniques trigger the target into sending one or more responses, which are then analyzed by the attacker to infer the type and version of the OS installed on the remote host. Carefully crafted ICMP, TCP and UDP packets are sent to the target in order to observe how it responds to both valid and invalid packets. For instance, in the case of TCP probes, features that can be used to distinguish between different operating systems include: (1) the relative order of the TCP Options and (2) the total length of TCP Options. Additionally, the IP header may reveal some information about the nature of the OS.

Table 1 OS-dependent IP
and TCP parameters

Operating system	IP initial TTL	TCP window size
Linux Kernel 2.4/2.6	64	5, 840
Windows XP	128	65, 535
Windows 7	128	8, 192

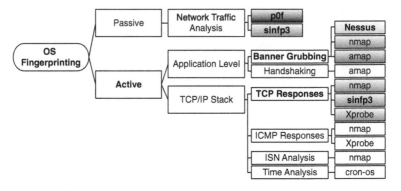

Fig. 6 Taxonomy of OS fingerprinting tools

Conversely, *passive fingerprinting* consists in using a packet sniffer to passively collect and analyze packets traveling between hosts. A simple passive method consists in inspecting the Time To Live (TTL) field in the IP header and the TCP Window Size of the SYN or SYN+ACK packet in a TCP session. The values of both the initial TTL and the TCP Window Size depend on the specific OS implementation, as shown in Table 1. One reason for this is that RFC specifications define intervals of values and recommended values, but do not mandate specific values. For instance, RFC 1700 recommends to initialize TTL to 64. Of course, relying only on the TTL value is not sufficient to determine the OS because, given the nature of this parameter, the TTL decreases as a packet traverses the network, and inferring the correct initial TTL may not always be possible.

Many different fingerprinting tools are available today. To better assess the impact of our approach, we have defined a taxonomy to classify existing fingerprinting tools based on the different approaches they implement. Figure 6 shows the proposed taxonomy. Given the variety of existing tools, it is practically impossible to develop a single technique that would defeat all of them. However, the proposed approach is effective against the most widely used fingerprinting tools. The tools that we explicitly target in our work are shown in boldface, whereas other tools that are at least partially impacted by our deception techniques are shown with a colored background. In the following sections, we provide a detailed description of some of the tools that we explicitly target in our work, namely, SinFP3, pOf, and Nessus. We refer the reader to [5] for a description of additional tools.

6.1 SinFP3

The development of SinFP was prompted by the need to reliably identify a remote host's operating system under worst-case network conditions [6]: (1) only one TCP port is open; (2) traffic to all other TCP and UDP ports is dropped by filtering devices; (3) stateful inspection is enabled on the open port; and (4) filtering devices with packet normalization are in place. In this scenario, only standard probe packets that use the TCP protocol can reach the target and elicit a response packet.

SinFP uses three probes: the first probe **P1** is a standard packet generated by the `connect()` system call, the second probe **P2** is the same as **P1** but with different TCP Options, and the third probe **P3** has no TCP Options and has the TCP `SYN` and `ACK` flags set. The first two probes elicit two TCP `SYN+ACK` responses from the target. The third probe has the objective of triggering the target into sending a TCP `RST+ACK` response. After the three probes have been sent and the three replies have been received, a signature is built from the analysis of the response packets. Then, a signature matching algorithm searches a database for a matching OS fingerprint.

The analysis of the responses considers both IP headers and TCP headers [6]. With respect to IP headers, the following fields are analyzed: TTL, ID, Don't Fragment Bit. With respect to TCP headers, the following fields are analyzed: Sequence Number, Acknowledgment Number, TCP Flags, TCP Window Size, and TCP Options. An example of SinFP report against a Windows 7 target is reported in Fig. 7. The packets exchanged during the scan are shown in Fig. 8.

Limitations When there are too few TCP Options in **P2**'s response, the signature's entropy becomes weak [6]. In fact, TCP Options are the most discriminant

```
...
score 100: Windows: Microsoft: Windows: Vista (RC1)
score 100: Windows: Microsoft: Windows: Server 2008
score 100: Windows: Microsoft: Windows: 7 (Ultimate)
score 100: Windows: Microsoft: Windows: Vista
```

Fig. 7 SinFP report against a Windows 7 host

Fig. 8 Packets exchanged during a SinFP scan

characteristics that compose a signature. That is because virtually no two systems implement exactly the same TCP Options, nor in the same order. Thus, when only the MSS option is in the TCP header, the risk of misidentification is high. SinFP also suffers from the same limitation of all knowledge-based fingerprinting tools: their capability to identify a system is limited by the existence of a corresponding fingerprint in the database.

6.2 pOf

pOf (v3) is a tool that utilizes an array of sophisticated, purely passive traffic fingerprinting mechanisms to identify the players behind any incidental TCP/IP communication [23]. Its fingerprint contains:

1. **ver:** IP protocol version.
2. **ittl:** Initial TTL used by the OS.
3. **olen:** Length of IPv4 options or IPv6 extension headers.
4. **mss:** Maximum Segment Size (MSS), if specified in TCP Options.
5. **wsize:** Window Size, expressed as a fixed value or a multiple of MSS, of MTU, or of some random integer.
6. **scale:** Window Scaling factor, if specified in TCP Options.
7. **olayout:** Comma-delimited layout and ordering of TCP Options, if any. Supported values: explicit end of options, no-op option, maximum segment size, window scaling, selective ACK permitted, timestamp.
8. **quirks:** Comma-delimited properties observed in IP or TCP headers.
9. **pclass:** Payload size.

Limitation The initial TTL value is often difficult to determine since the TTL value of a sniffed packet will vary depending on where it is captured. The sending host will set the TTL value to the OS's default initial TTL value, but this value will then be decremented by one for every router the packet traverses on its way to the destination. An observed IP packet with a TTL value of 57 can therefore be expected to be a packet with an initial TTL of 64 that has done 7 hops before it was captured. This tool also suffers from the TCP Options entropy issue described for SinFP.

6.3 Nessus

Nessus provides a comprehensive analysis of a target, including information about its OS and vulnerabilities. Tenable Research introduced a highly accurate form of operating system identification [14]. This method combines the outputs of various other plugins that execute separate techniques to guess or identify a remote operating system. It is worth noting that some of these techniques could also be adopted independently by an attacker. Each of these plugins reports a confidence level for their scan results. An example of Nessus output for OS identification is reported in Fig. 9.

```
The remote host is running Linux Kernel 2.4
Confidence Level : 70
Method : SinFP

The remote host is running Linux Kernel 2.6
Confidence Level : 60
Method : ICMP
```

Fig. 9 Nessus OS identification report

Limitation Nessus's approach to fingerprinting can be very effective when used during a "credentialed" scan. Otherwise, it will report partial information and in some cases it will not use all the plugins it is equipped with. Additionally, Nessus's approach to identify vulnerabilities is strictly dependent on service banners and welcome messages. Generally, Nessus merely checks if the service's version present in the service's banner belongs to a certain interval. For instance, if a vulnerability is know to be present in a service up until version 2.0, it is really simple to make Nessus generate false negatives by exposing a fake service banner claiming that the service version is higher than 2.0.

6.4   Solution Design

With respect to OS fingerprinting, our approach to deceive attackers relies on modifying outgoing traffic in a way that such traffic resembles traffic generated by a different protocol stack implementation. As we pointed out in Sect. 6, protocol specifications may leave some degrees of freedom to developers. The choices that a developer makes with respect to (1) default values (e.g., initial TTL, size of the TCP window), (2) length of TCP Options, or (3) order of the TCP Options may reveal the nature of the operating system or even the type of device (e.g., firewall, switch, router, printer or general purpose machine).

All the information required to impersonate a certain operating system or device can be extracted from the SinFP's signature database. All the outgoing packets that may reveal relevant information about the OS are modified to reflect the deceptive signature, as shown in Fig. 10.

The most critical step in this process is the manipulation of the TCP Options. In order to present the attacker with a deceptive signature, not only we need to modify some parameters, but we also need to reorder the options and correctly place no-operation[4] option codes to obtain the right options length. The TCP Options format can be inferred from the signature. Modifying the Options length requires to adjust the total length field in the IP header, the Offset value in the TCP header

---

[4] As specified in RFC 793, this option code may be used between options, for example, to align the beginning of a subsequent option on a word boundary.

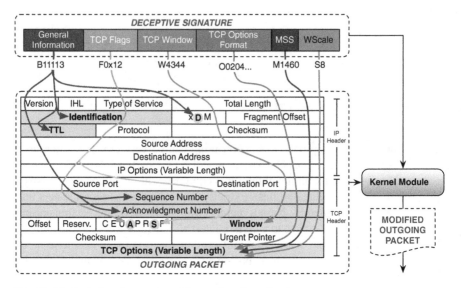

Fig. 10 Manipulation of outgoing traffic to reflect deceptive signatures

and subsequently adjust the sequence numbers. On the bright side, the majority of commonly used operating systems share the same length for TCP Options.

With respect to service fingerprinting, we need to consider the following two cases: (1) the service banner can be modified through configuration files, and (2) the service banner is hard-coded into the service executable. Being able to modify the packet carrying the identifying information before it leaves the host (or the network) enables to successfully address both scenarios. Moreover, even if services are under our control, we prefer to alter service banners in a completely transparent way. Our long term goal is to develop a network appliance that can be deployed at the network boundary and which is able to transparently manipulate services and operating system fingerprints.

It is worth noting that, when the original service banner is replaced with a deceptive banner of a different length, we need to: (1) adjust the size of the packet based on the difference in length between the two banners, (2) modify the total length field in the IP header, (3) modify the sequence numbers in order to be consistent with the actual amount of bytes sent, and (4) correctly handle the case of fragmented packets, which requires to reassemble a packet before modifying it.

However, this approach is not applicable to all categories of services. Services that actively use the banner information during the connection process (such as SSH) require us to use a non-transparent approach. For instance, the SSH protocol actively uses the banner information while generating hashes in the connection phase. The banner format is "SSH-protocolversion-softwareversion comments\r\n". Even though this approach can deceive tools like nmap and amap, modifying the banner will cause legitimate user to receive termination messages from the server[5] such as: (1) Bad packet length or (2) Hash Mismatch.

In summary, defeating passive tools requires to modify all outgoing packets, whereas defeating active tools only requires to alter those packets that are likely to be part of an attacker's probes.

6.5 Implementation

We implemented kernel modules that use the Netfilter POST_ROUTING hook to process and modify relevant information in the IP header, TCP header, and TCP payload. Netfilter is a packet handling engine introduced in Linux Kernel 2.4. It enables the implementation of customized handlers to redirect, reject or alter incoming and outgoing packets. Netfilter can be extended with hooks. A hook is a function handler that allows specific kernel modules to register callback functions within the kernel's network stack. A registered callback function is then called back for every packet that traverses the net filter stack. We used the POST_ROUTING hook to alter the packets just before they are finally sent out.

Specifically, we implemented an *operating system fingerprint module* to modify the responses to the SinFP's probes and a *service fingerprint module* to modify banner information for specific services.

6.5.1 Operating System Fingerprint Module

The hook function checks if the packet is a response to the first SinFP's probe (**P1**): an ACK+SYN packet with a length of 44 bytes (in the case of an underlying Linux Kernel 3.02). If this is the case, the packet is altered in order to mimic a particular operating system (more details are provided later), otherwise the module checks whether the packet is an ACK+SYN with a length of 60 bytes (in the case of an underlying Linux Kernel 3.02). This packet is used in most TCP connections and it might be a response to the second SinFP's probe (**P2**). If so, the packet is modified accordingly, based on the target OS fingerprint.

Additionally, we verified that this approach can deceive the **p0f** tool when we modify the TTL value of all IP packets and the window size value of all TCP packets.

---

[5]Errors occurs only during the connection phase, and altering the banner will not affect previously established connections.

During the packet manipulation stage, we track whether any of the following has been altered: the IP Header, the TCP header, the length of TCP Options, the TCP payload or its size. Based on this information, we:

- modify the *IP Total length* value, if the size of the TCP payload has changed;
- recompute the *TCP Offset* value in the TCP header and the *IP Total Length*, if the length of the TCP Options has changed;
- recompute the TCP checksum, if the TCP header and/or the TCP payload have been altered;
- recompute the IP checksum, if the IP header has been altered.

In order to modify the responses such that they appear to have been generated by a specific OS, we created a script that (1) extracts the required characteristics of the responses to the first and second probe from SinFP's signature database, and (2) generates the C code necessary to alter the responses. The script determines how the following policies should be implemented:

- **ID policy:** the ID could be a fixed value different from zero, zero or a random number.
- **Don't fragment bit policy:** the DF bit can be enabled or disabled.
- **Sequence Number policy:** the sequence number can be zero or not altered.
- **Ack Number policy:** the ack number can be zero or not altered.
- **TCP Flags policy:** the TCP flags value is copied from the signature.
- **TCP Window Size policy:** the Window size is copied from the signature.
- **TCP MSS Option policy:** the MSS value is copied from the signature.
- **TCP WScale Option policy:** the WScale is copied from the signature.
- **TCP Options policy:** the TCP Options layout is copied from the signature.

The generated code is then compiled in order to build the actual kernel module. The scheme of the resulting kernel module is presented in Listing 1.[6] We assume that all the *set* and *get* functions are able to access the packet and track if the IP or TCP header have been modified.

```
if (ip->protocol == TCP && ip->len == 44 && tcp->ack == 1 && tcp->syn == 1)
{
    //Probably 1st sinfp3's probe Response
    set_id();
    set_df_bit();
    set_ttl();
    set_tcp_window();
    set_tcp_flags();
    set_tcp_sequence();
    set_tcp_ack();

    if(new_option_len != option_len)
    {
        modify_packet_size(); //expands or shrinks packet and updates IP Length and Offset
    }

    set_tcp_options(MSS, WScale, Option_Layout);
}
else if (ip->protocol == TCP && ip->len == 60 && tcp->ack == 1 && tcp->syn == 1)
{
    //Probably 2nd sinfp3's probe Response
```

---

[6]We omit the code dealing with sequence numbers adjustment for reasons of space.

```
23   // Extract the timestamp from the packet and save it for re-injecting it in the right position later
24   timestamp = get_tcp_timestamp();
25
26   set_id();
27   set_df_bit();
28   set_ittl();
29   set_tcp_window();
30   set_tcp_flags();
31   set_tcp_sequence();
32   set_tcp_ack();
33
34   if (new_option_len != option_len)
35   {
36      modify_packet_size(); //expands or shrinks packet and updates IP Length and Offset
37   }
38   set_tcp_options(timestamp, MSS, WScale, Option_Layout);
39   }
40
41   if (tcpHeader_modified)
42   {
43      tcp->check = 0;
44      tcp->check = tcp_csum();
45   }
46
47   if (ipHeader_modified)
48   {
49      ip->check = 0;
50      ip->check = ip_csum();
51   }
```

Listing 1 OS deception kernel module

6.5.2 Service Fingerprint Module

In order to alter the service fingerprint, we modify the banner sent by the application either at the time of establishing a connection or in the header of each application-level protocol data unit. Packets matching the service source port one wants to protect are analyzed. If a packet contains data, the banner string is searched and subsequently replaced. When replacing the banner, the packet size can vary: the packet is then resized according to the specific case. Listing 2 shows the sample pseudo-code for the case of an Apache Server.[7]

```
1    #define FAKE_APACHE_BANNER "Apache/1.1.23"
2    ...
3    if (ntohs(tcph->source) == 80 && len > 0)
4    {
5       // Pointers to where to store the start and end address of the Apache Banner String for substitution
6       char *b = NULL, *l = NULL;
7
8       // Pointer to the TCP payload
9       char *p= (char *)((char *)tcph+(uint)(tcph->doff*4));
10
11      b = strstr(p, "\r\nServer: "); //String Search
12      if (b != NULL) l = strstr(((char *)b + 10), "\r\n");
13
14      if (b != NULL && l != NULL)
15      {
16         // b points to \r\nServer: x, so we add 10 to move to the beginning of x
17         uint8_t signature_len = l - (b + 10);
18
19         if (signature_len != (sizeof(FAKE_APACHE_BANNER)-1))
20         {
21            resize_packet();
22         }
23         copy(b + 10, FAKE_APACHE_BANNER, sizeof(FAKE_APACHE_BANNER)-1);
24      }
25      ...
26   }
```

Listing 2 Service deception kernel module

---

[7]For the sake of brevity, we omit the code for checksum recomputation.

7   Experimental Evaluation

In this section, we report the results of the experiments we conducted to validate the proposed approach. We evaluated the performance of algorithms *TopKDistance* (Sect. 7.1) and *TopKBudget* (Sect. 7.2) in terms of processing time and approximation ratio for different numbers of hosts and different numbers of admissible configurations. We also evaluated our approach for deceiving fingerprinting tools from the point of view of both legitimate users interacting with the system (Sect. 7.3) and attackers trying to determine the OS of a remote host or the type of services running on it (Sect. 7.4).

7.1   Evaluation of TopKDistance

First, we show that, as expected, the processing time increases when the number of admissible configurations for each host increases. Figure 11 shows processing time as a function of the number of hosts for $k = 7$ and a required minimum distance $d = 5$. The processing time is practically linear in the number of hosts in the case of three configurations per host but, as soon as the number of configurations increases, it becomes polynomial, as shown in the case of five configurations per host.

Figure 12 shows the processing time as a function of the graph size for different values of k. The graph size is measured as the number of nodes that have a distance from the internal view that is less than or equals to d.

Comparing the trends for $k = 3, 4, 5$ we can notice that the algorithm is polynomial for $k = 3$ and linear for $k = 4, 5$. This can be explained considering the

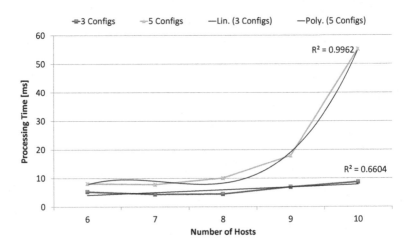

Fig. 11 Processing time vs. number of hosts

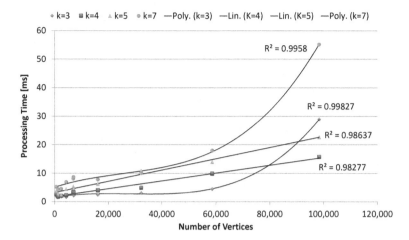

Fig. 12 Processing time vs. graph size

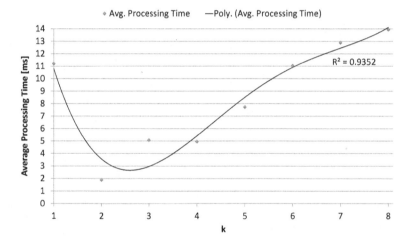

Fig. 13 Average processing time vs. k

fact that, for $k = 3$, it is necessary to explore the graph in more depth than in the case of $k = 4, 5$. Moreover, if we consider values of k bigger than 5, the trend is again polynomial due to the fact that the algorithm starts exploring the graph more broadly. Indeed, as we will show shortly, relatively small values of k provide a good trade-off between approximation ratio and processing time, therefore this result is extremely valuable. To better visualize the relationship between processing time and k, we plotted the average processing time against k (see Fig. 13). The trend can be approximated by a polynomial function and the minimum is between $k = 2$ and $k = 3$. For k greater than 4, the average time to process the graph increases almost linearly.

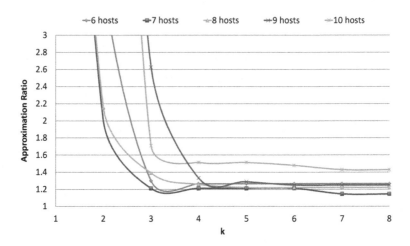

Fig. 14 Approximation ratio vs. k

Moreover, we evaluated the approximation ratio achieved by the algorithm. To compute the approximation ratio, we divided the cost of the algorithm's solution by the optimal cost. In order to compute the optimal solution, we exhaustively measured the shortest path (in term of cost) from the internal view to all the solutions with a distance greater than the minimum required d, and sorted those results by increasing cost. The optimal solution has the maximum distance and the minimum cost. When the algorithm could not find a solution (none of the discovered paths has a distance greater than the minimum required d), we considered an infinite approximation. Figure 14 shows how the ratio changes when k increases in the case of a fixed number of configurations per node (five in this case) and for increasing numbers of hosts. It is clear that the approximation ratio improves when k increases. Relatively low values of k (between 3 and 6) are sufficient to achieve a reasonably good approximation ratio in a time-efficient manner.

7.2 Evaluation of TopKBudget

As done for the *TopKDistance* algorithm, we show that, as expected, the processing time increases when the number of admissible configurations for each node increases. Figure 15 shows the processing time as a function of the number of hosts for $k = 2$ and a budget $b = 18$. The processing time is practically linear in the number of hosts in the case of three configurations per host. In this case, the minimum time (six hosts) is $\sim$150 ms and the maximum time (ten hosts) is $\sim$3,500 ms. When the number of configurations increases, the time rapidly increases due to the time spent in the generation of the graph.

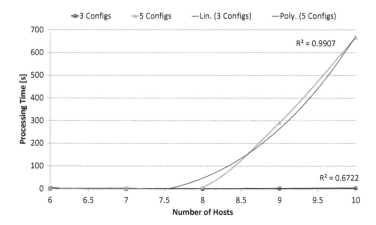

Fig. 15 Processing time vs. number of hosts

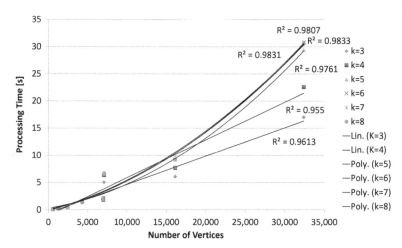

Fig. 16 Processing time vs. graph size

Figure 16, shows a scatter plot of average processing times against increasing graph sizes. This chart suggests that, in practice, processing time is linear in the size of the graph for small values of k. Similarly, Fig. 17 shows how processing time increases when k increases for a fixed budget $b = 18$. The trend is approximated by a polynomial function and tends to saturate for values of $k \geq 6$. This can be explained considering the fact that, for larger values of k, most of the time is spent in the graph generation phase and, starting from $k = 6$, the graph is generated almost completely. Even in this case, the important result is that small values of k achieve linear time. Moreover, for these values, the algorithm can achieve a good approximation ratio.

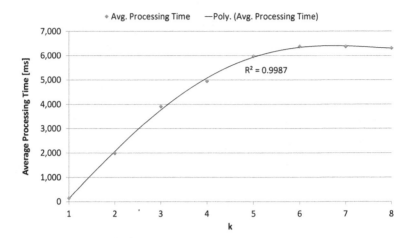

Fig. 17 Average processing time vs. k

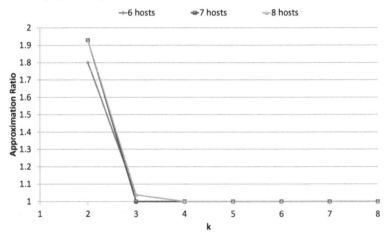

Fig. 18 Approximation ratio vs. k

To compute the approximation ratio we divided the optimal distance by the distance returned by the algorithm. In order to compute the optimal solution we exhaustively measured the shortest path (in term of distances) from the internal view to all the solutions in a given graph. Due to the fact that it would be unfeasible to generate an exhaustive graph, we generated a sub-graph up to a maximum number of nodes. We then ordered the paths by decreasing values of the distance and noted the cost needed to reach the solution. We then started the algorithm with a budget equal to this cost. Figure 18 shows how the ratio changes when k increases for a fixed number of configurations per node (five in this case) and for increasing numbers of hosts. The approximation ratio is good even for $k = 1$, but to a more accurate solution can be obtained for $k = 2, 3$. Larger values of k are not ideal in terms of time efficiency.

7.3 Legitimate User Perspective

In the next set of experiments, we evaluated our approach for deceiving fingerprinting tools from the point of view of legitimate users interacting with the system being defended. Our goal is to make manipulation of outgoing traffic completely transparent to users from both a functional and a performance perspective. To this end, we run performance tests with the Apache Benchmark, testing the server's ability to process 20,000 requests, with a maximum of 200 simultaneous active users. The results are shown in Fig. 19.

The tests we performed involved different system configuration scenarios: (i) the behavior of the system is not altered (Original); (ii) the kernel module to alter the OS fingerprint and deceive only *active fingerprinting tools* is enabled (Sinfp3); (iii) the kernel module to alter the service fingerprint is enabled (Apache); (iv) both modules from scenarios (ii) and (iii) are enabled (Sinfp3+Apache); (v) the kernel module to alter the OS fingerprint and deceive both *active* and *passive fingerprinting tools* is enabled (Sinfp3+p0f); and (vi) both modules from scenarios (iii) and (v) are enabled.

The performance degradation for scenario (ii) is negligible, as only two packets need to be altered for each connection. On the other hand, when the OS fingerprint kernel module alters all the outgoing packets (scenario (v) above), there is a slight delay in the response time due to the larger number of packets that need to be altered. When the Service Fingerprint Kernel Module is enabled (scenario (iii) above), the response time increases due to the string comparison operations performed to identify and replace the banner information. It is clear from Fig. 19 that the Service Fingerprint Kernel Module has the largest impact on the performance of the system. However, even in the worst-case scenario, the performance degradation is limited.

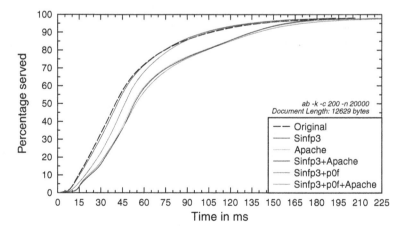

Fig. 19 Apache Benchmark

Fig. 20 FTP transfer rate
(500 MB file)

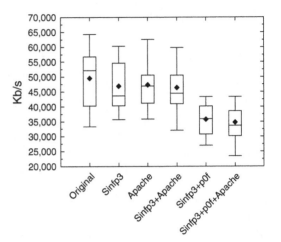

Considering the same scenarios, we have also tested the overhead introduced by the kernel modules on large data transfers by uploading a 500 MB file to an FTP server. Most packets will not be altered, but the conditions of the if statements in the kernel modules need to be evaluated, thus adding some overhead, which will eventually affect the net transfer rate. As we can see from Fig. 20, the more conditions need to be evaluated the greater the effect on performance is going to be.

7.4 Attacker Perspective

In the last set of experiments, we evaluated our approach from the point of view of an attacker trying to determine the operating system of a remote host or the type of services running on it.

In order to test how our approach can deceive attackers using Nessus, we audited the system with and without the deceptive Kernel Module enabled. Table 2 shows the results of the respective Nessus scans. The original system is a fully patched Ubuntu 12.04 server and has no known vulnerabilities. When no deception is used, the system is correctly identified, and all the information derived by Nessus is accurate. Next, we deceived both OS and service fingerprinting by exposing a Windows 7/Vista OS fingerprint and an Apache 2.2.1 service fingerprint. When the deception mechanism is enabled, the OS is misidentified accordingly, and the deceiving service fingerprinting leads to false positives in the identification of vulnerabilities.

Table 3 reports the results of scans performed with different fingerprinting tools. Clearly, our approach is able to effectively deceive several tools. For instance, we are able to alter the perception of the target system even when the attacker uses either nmap or Xprobe++, which adopt a different probing scheme.

Table 2 Results of Nessus scans

	Without deception	With deception
Device type	General purpose 85%	General purpose 65%
OS	Ubuntu 12.04 85%	Windows Vista 65%
Info	13	15[a]
Low	0	0
Medium	0	2 (100%: false positives)
High	0	2 (100%: false positives)
Critical	0	0

[a]60% True positives; 13.33%: false positives; 26.67%: induced info

In conclusion, by intelligently crafting responses to SinFP probes, it is possible to force attackers into misclassifying a remote host as any of a broad variety of networked assets. For instance, a conventional Linux-based server can be fingerprinted as a network switch, an ADSL gateway, or even a printer. Of course, these fingerprints will cause attackers to derive an inconsistent map of the target network. We have successfully created SinFP deceptions for different network monitoring appliances, firewalls and printers. A partial SinFP output for the case of an *HP Officejet 7200 Printer* is reported in Fig. 21, whereas Fig. 22 illustrates the steps involved in forcing the attacker to believe that the target device is a printer.

7.5 Drawbacks

Altering some parameters of the TCP header can affect the connection performance, when legitimate users actively use the protocol based on the modified parameters. In such cases, the proposed mechanism is not completely transparent, and drawbacks include those listed in the following.

Maximum Segment Size (MSS) This parameter defines the largest unit of data that can be received by the destination of the TCP segment. Modifying this value makes the host announce a different limit for its capabilities. Consider two hosts h_A and h_B, where h_B is the host being altered. Assume h_A sends a SYN packet with an MSS of 1,460 and h_B responds with a SYN/ACK that has an MSS of 1,480. Then, h_A will not send any segment larger than 1,480 bytes to h_B, even if h_B may actually be able to handle larger segments. Note that h_A is not required to send segments of exactly 1,480 but it is required not to exceed this limit. For the same reason, it is not possible to advertise a larger value than what hosts are actually able to handle.

Window and Windows Scale Factor These two parameters affect the TCP flow control, which regulates the amount of data a source can send before receiving an acknowledgment from the destination. A sliding window is used to make transmissions more efficient and control the flow so that the destination is not overwhelmed with data. The TCP Window scale factor is used to scale the window

Table 3 OS fingerprinting deception

Deception	Tool			
	Sinfp3	p0f	nmap	Xprobe++
None	Linux 3.0.x-3.2.x (94%) Linux 2.4.x-2.6.x (73%)	Linux 3.x	Linux 2.6.x-3.x	Linux 2.4.19-28 (94%)
Windows Server 2008	Server 2008/Vista7 (100%) FreeBSD 7.0-9.0 (73%)	Windows 7/8	Unknown	Linux 2.4.26 (78%)
Firewall Fortigate	Firewall Fortigate 100%	Unknown	Unknown	Linux 2.4.23 (94%)
NetBSD 5.0.2	NetBSD 5.0.2 (98%)	Unknown	Unknown	Linux 2.4.21 (92%)
Windows Server 2008 (partial)	Server 2008/Vista7 (98%) FreeBSD 7.0-9.0 (73%)	Windows 7/8	Unknown	Linux 2.4.26 (81%)
Windows Server 2008 (partial)	Server 2008/Vista7 (88%)	Unknown	Unknown	Linux 2.4.14 (81%)

```
...
score: 100: Printer: HP Officejet 7200
score: 100: Printer: HP Officejet Pro L7600
score: 73: Appliance: APC AP9319
...
```

Fig. 21 SinFP output for a remote host discguHP Officejet 7200 Printer

Fig. 22 SinFP printer
deception

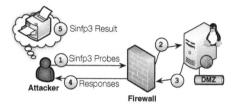

size by a power of 2. The window size may vary during the data transfer while the scale factor is determined once at the time of establishing a connection. Modifying the window size can alter the throughput: if the window is smaller than the available bandwidth multiplied by the latency then the sender will send a full window of data and then sit and wait for the receiver to acknowledge the data. This results in lower performance.

Selective ACK Selective acknowledgment allows the sender to have a better idea of which segments are actually lost and which have arrived out of order. If we disable the SACK option for a host that supports it, we may limit performance, depending on the capabilities of the counterpart.

8   Conclusions

In this chapter, we presented a principled approach for manipulating outgoing traffic so as to induce an external view of the system that will deceive potential intruders into making incorrect inferences about important system characteristics, including operating systems and active services. We demonstrated practical applicability of the proposed approach by presenting deception-based techniques for specifically defeating an attacker's effort to fingerprint operating systems and services on the target system. Although experimental results show that our approach can efficiently and effectively deceive an attacker, some limitations still exist and more work remains to be done. In addition to some of the limitations listed in Sect. 7.5, we need to consider that the proposed manipulation of outgoing traffic is limited to just some categories of traffic, and there might be other categories of traffic or other characteristics that attackers can use to infer the nature of the OS and services on a target host. We plan to extend our approach to address this scenario and make our solution more resilient.

Acknowledgements This work was partially supported by the Army Research Office under grants W911NF-13-1-0421, W911NF-09-1-0525, and W911NF-13-1-0317, and by the Office of Naval Research under grants N00014-12-1-0461 and N00014-13-1-0703.

References

1. F. H. Abbasi, R. J. Harris, G. Moretti, A. Haider, and N. Anwar. Classification of malicious network streams using honeynets. In *Proceedings of the IEEE Conference on Global Communications (GLOBECOM 2012)*, pages 891–897, Anaheim, CA, USA, December 2012. IEEE.
2. M. Albanese, S. Jajodia, A. Pugliese, and V. S. Subrahmanian. Scalable analysis of attack scenarios. In *Proceedings of the 16th European Symposium on Research in Computer Security (ESORICS 2011)*, pages 416–433, Leuven, Belgium, September 2011. Springer.
3. M. Albanese, A. De Benedictis, S. Jajodia, and K. Sun. A moving target defense mechanism for manets based on identity virtualization. In *Proceedings of the 1st IEEE Conference on Communications and Network Security (IEEE CNS 2013)*, pages 278–286, Washington, DC, USA, October 2013. IEEE.
4. M. Albanese, E. Battista, S. Jajodia, and V. Casola. Manipulating the attacker's view of a system's attack surface. In *Proceedings of the 2nd IEEE Conference on Communications and Network Security (IEEE CNS 2014)*, pages 472–480, San Francisco, CA, USA, October 2014.
5. M. Albanese, E. Battista, and S. Jajodia. A deception based approach for defeating OS and service fingerprinting. In *Proceedings of the 3rd IEEE Conference on Communications and Network Security (IEEE CNS 2015)*, pages 253–261, Florence, Italy, September 2015.
6. P. Auffret. SinFP, unification of active and passive operating system fingerprinting. *Journal in Computer Virology*, 6(3):197–205, August 2010.
7. D. Barroso Berrueta. A practical approach for defeating Nmap OS-Fingerprinting. http://nmap.org/misc/defeat-nmap-osdetect.html, January 2003.
8. V. Casola, A. De Benedictis, and M. Albanese. A moving target defense approach for protecting resource-constrained distributed devices. In *Proceedings of the 14th International Conference on Information Reuse and Integration (IEEE IRI 2013)*, pages 22–29, San Francisco, CA, USA, August 2013.
9. V. Casola, A. De Benedictis, and M. Albanese. *Integration of Reusable Systems*, chapter A Multi-Layer Moving Target Defense Approach for Protecting Resource-Constrained Distributed Devices. Advances in Intelligent and Soft Computing. Springer, 2013.
10. C.-M. Chen, S.-T. Cheng, and R.-Y. Zeng. A proactive approach to intrusion detection and malware collection. *Security and Communication Networks*, 6(7):844–853, July 2013.
11. Q. Duan, E. Al-Shaer, and H. Jafarian. Efficient random route mutation considering flow and network constraints. In *Proceedings of the 1st IEEE Conference on Communications and Network Security (IEEE CNS 2013)*, pages 260–268, Washington, DC, USA, October 2013. IEEE.
12. M. Dunlop, S. Groat, R. Marchany, and J. Tront. Implementing an IPv6 moving target defense on a live network. In *Proceedings of the National Moving Target Research Symposium*, Annapolis, MD, USA, June 2012.
13. Executive Office of the President, National Science and Technology Council. Trustworthy cyberspace: Strategic plan for the federal cybersecurity research and development program. http://www.whitehouse.gov/, December 2011.
14. R. Gula. Enhanced operating system identification with Nessus. http://www.tenable.com/blog/enhanced-operating-system-identification-with-nessus, February 2009.
15. J. H. Jafarian, E. Al-Shaer, and Q. Duan. OpenFlow random host mutation: Transparent moving target defense using software defined networking. In *Proceedings of the 1st Workshop on Hot Topics in Software Defined Networks (HotSDN 2012)*, pages 127–132, Helsinki, Finland, August 2012. ACM.

16. S. Jajodia, A. K. Ghosh, V. Swarup, C. Wang, and X. S. Wang, editors. *Moving Target Defense: Creating Asymmetric Uncertainty for Cyber Threats*, volume 54 of *Advances in Information Security*. Springer, 1st edition, 2011.

17. G. F. Lyon. *Nmap Network Scanning: The Official Nmap Project Guide to Network Discovery and Security Scanning*. Insecure, 2009.

18. P. K. Manadhata and J. M. Wing. An attack surface metric. *IEEE Transactions on Software Engineering*, 37(3):371–386, May 2011.

19. A. Rana. What is AMap and how does it fingerprint applications? http://www.sans.org/security-resources/idfaq/amap.php, March 2014.

20. G. Shu and D. Lee. Network protocol system fingerprinting - a formal approach. In *Proceedings of the 25th IEEE International Conference on Computer Communications (INFOCOM 2006)*. IEEE, April 2006.

21. C. Trowbridge. An overview of remote operating system fingerprinting. SANS Institute InfoSec Reading Room, July 2003.

22. D. Watson, M. Smart, G. R. Malan, and F. Jahanian. Protocol scrubbing: Network security through transparent flow modification. *IEEE/ACM Transactions on Networking*, 12(2):261–273, April 2004.

23. M. Zalewski. p0f v3 (version 3.06b). http://lcamtuf.coredump.cx/p0f3/, January 2012.

Embedded Honeypotting

Frederico Araujo and Kevin W. Hamlen

Abstract *Language-based software cyber deception* leverages the science of compiler and programming language theory to equip software products with deceptive capabilities that misdirect and disinform attackers. A flagship example of software cyber deception is *embedded honeypots*, which arm live, commodity server software with deceptive attack-response and disinformation capabilities. This chapter presents a language-based approach to embedded honeypot design and implementation. Implications related to software architecture, compiler design, program analysis, and programming language semantics are discussed.

1 Introduction to Software Cyber Deception

Throughout the history of warfare, obfuscation and deception have been widely recognized as important tools for leveling the ubiquitous asymmetry between offensive and defensive combatants. In the modern era of cyber warfare, this asymmetry has perhaps never been more extreme. Despite a meteoric rise in worldwide spending on conventional cyber defensive technologies and personnel, the success rate and financial damage resulting from cyber attacks against software systems has escalated even faster, rising dramatically over the past few decades. The challenges can be traced in part to the inherently uneven terrain of the cyberspace battlefield—which typically favors attackers, who can wreak havoc by finding a single weakness, whereas defenders face the difficult task of protecting all possible vulnerabilities.

The attack surface exposed by the convergence of computing and networking poses particularly severe asymmetry challenges for defenders. Our computing

This research was supported in part by AFOSR award FA9550-14-1-0173, NSF awards #1054629 and #1027520, ONR award N00014-14-1-0030, and NSA award H98230-15-1-0271. Any opinions, recommendations, or conclusions expressed are those of the authors and not necessarily of the AFOSR, NSF, ONR, or NSA.

F. Araujo (✉) • K.W. Hamlen
The University of Texas at Dallas, 800 W. Campbell Rd., EC31, Richardson,
TX 75080-3021, USA
e-mail: frederico.araujo@utdallas.edu; hamlen@utdallas.edu

© Springer International Publishing Switzerland 2016
S. Jajodia et al. (eds.), *Cyber Deception*, DOI 10.1007/978-3-319-32699-3_9

201

systems are constantly under attack, yet the task of the adversary is greatly facilitated by information disclosed by the very defenses that respond to those attacks. This is because software and protocols have traditionally been conceived to provide informative feedback for error detection and correction, not to conceal the causes of faults. Many traditional security remediations therefore advertise themselves and their interventions in response to attempted intrusions, allowing attackers to easily map victim networks and diagnose system vulnerabilities. This enhances the chances of successful exploits and increases the attacker's confidence in stolen secrets or expected sabotage resulting from attacks.

Deceptive capabilities of software security defenses are an increasingly important, yet underutilized means to level such asymmetry. These capabilities mislead adversaries and degrade the effectiveness of their tactics, making successful exploits more difficult, time consuming, and cost prohibitive. Moreover, deceptive defense mechanisms entice attackers to make fictitious progress towards their malicious goals, gleaning important threat information all the while, without aborting the interaction as soon as an intrusion attempt is detected. This equips defenders with the ability to lure attackers into disclosing their actual intent, monitor their actions, and perform counterreconnaissance for attack attribution and threat intelligence gathering.

The Deception Stack Deceptive techniques can be introduced at different layers of the software stack. Figure 1 illustrates this *deception stack* [31], which itemizes various deceptive capabilities available at the network, endpoint, application, and data layers. For example, computer networks known as *honeynets* [36, 37] intentionally purvey vulnerabilities that invite, detect, and monitor attackers; endpoint protection platforms [31] deceive malicious software by emulating diverse execution environments and creating fake processes at the application level to manipulate malware behavior; and falsified data can be strategically planted in decoy file systems to disinform and misdirect attackers away from high-value targets [38, 41].

The deception stack suggests that the difficulty of deceiving an advanced adversary increases as deceptions move up the stack. This is accurate if we compare, for instance, the work associated with emulating a network protocol with the challenge of crafting fake data that appear legitimate to the attacker: Protocols have clear and precise specifications and are therefore relatively easy to emulate, whereas there are many complex human factors influencing whether a specific datum is plausible and believable to a particular adversary.

Fig. 1 Gartner's *deception stack* [31], with examples of deceptive technologies inhabiting each layer

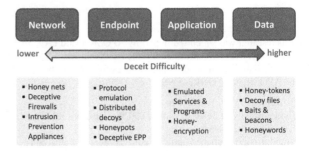

In general, deceptive software defenses must employ one or more forms of deception, and leverage all layers of the deception stack to some degree in order to be effective against a persistent and skilled adversary. This chapter focuses on application-level deceptive techniques, which offer critical mediation capabilities between the network, endpoint and data deception layers. For example, an application-level, deception-enabled web server can ask the network-level firewall to allow certain payloads to reach the application layer, where it can then offer deceptive responses that misdirect the adversary into attacking decoy machines within the endpoint layer. These decoys can appeal to the data deception layer to purvey disinformation in the form of false secrets or even malware counter-attacks against adversaries. Such scenarios demonstrate the ongoing need for tools and techniques allowing organizations to engineer applications with proactive and deceptive capabilities that degrade attackers' methods and disrupt their reconnaissance efforts.

Towards this end, the remainder of this chapter introduces a language-based methodology for arming live, legacy server software with deceptive attack-response and disinformation capabilities. We refer to such capabilities as *embedded honeypots*. Embedded honeypots differ from traditional honeypots in that they reside within the actual, mission-critical software systems that attackers are seeking to penetrate; they are not independent decoy systems. Thus, embedded honeypots offer advanced deceptive remediations against informed adversaries who can identify and avoid traditional honeypots. In order to be adoptable, embedded honeypots imbue production server software with deceptive capabilities without degrading its performance or intended functionality. These new capabilities mislead advanced adversaries into wasting time and resources on phantom vulnerabilities and decoy file systems, and pave the way for an emerging science of *deception-facilitating software engineering*.

Honeypots [35] are information systems resources conceived to attract, detect, and gather attack information. Figure 2 presents an abbreviated timeline of the extensive history of honeypot research. A more comprehensive survey on recent advances and future trends in honeypot research can be found in [7]. Modern honeypots are usually classified according to the interaction level provided to potential attackers. *Low-interaction* honeypots [32] present a façade of emulated services without full server functionality, with the intent of detecting unauthorized activity via easily deployed pseudo-services. In contrast, *high-interaction* honeypots provide a relatively complete system with which attackers can interact, and are designed to capture detailed information on attacks. Despite their popularity, both low- and high-interaction honeypots are often detectable by informed adversaries (e.g., due to the limited services they purvey, or because they exhibit traffic patterns and data substantially different than genuine services). Embedded honeypots are highest-interaction in the sense that they purvey genuine services, and have access to genuine data, unlike traditional low- and high-interaction honeypots.

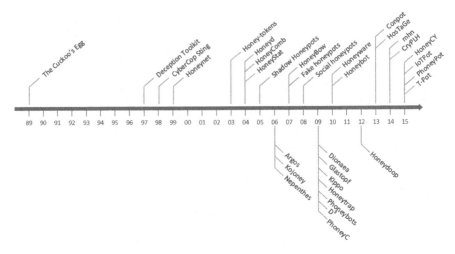

Fig. 2 Timeline of selected academic and industrial research on honeypots

2 Honey-Patching: A New Software Cyber Deception Technology

When a software security vulnerability is discovered, the conventional defender reaction is to quickly *patch* the software to fix the problem. This standard reaction can backfire, however, if the patch has the side-effect of disclosing and highlighting other exploitable weaknesses in the defender's network. Unfortunately, such back-fires are common; patches often behave in such a way that adversaries can reliably infer which systems have been patched, and therefore which are unpatched and vulnerable. The existence of at least some unpatched systems is almost inevitable, since patch adoption is rarely immediate—for example, testing is often required to ensure patch compatibility. Thus, most software security patches fix newly discovered vulnerabilities at the price of advertising to attackers which systems remain vulnerable. This has led to an adversarial culture for which *vulnerability probing* is a staple of the cyber killchain.

Cyber criminals easily probe today's Internet for vulnerable software, allowing them to focus their attacks on susceptible targets, in the following way. First, the attacker submits a malicious input (a *probe*) crafted to trigger a particular, known software bug in bulk to many servers across the network. Patched servers respond to the probe with a well-formed output, such as an error message; but unpatched servers behave erratically, such as by responding with a garbage string or crashing and restarting. Upon observing the latter response, the attacker next submits a more constructive malicious input to the unpatched servers, such as one that exploits the bug to hijack the victim software's control-flow, causing it to perform malicious actions on behalf of the attacker rather than merely crashing.

Honey-patching [5] is a game-changing alternative approach for anticipating and foiling these directed cyber attacks. The goal is to patch newly discovered software security vulnerabilities in such a way that future attempted exploits of the patched vulnerabilities appear successful to attackers even when they are not. This masks patching lapses, impeding attackers from easily discerning which systems are genuinely vulnerable and which are actually patched systems masquerading as unpatched systems. Detected attacks are transparently redirected to isolated, unpatched decoy environments that possess the full interactive power of the targeted victim server, but that disinform adversaries with honey-data and aggressively monitor adversarial behavior.

Deceptive honey-patching capabilities thereby constitute an advanced, language-based, active defense technique that can impede, confound, and misdirect attacks, and significantly raise attacker risk and uncertainty. In addition to helping protect networks where honey-patches are deployed, the practice also contributes to the public cyber welfare: Once a honey-patch for a particular software vulnerability has been adopted by some, attacks against all networks become riskier for attackers. This is because attackers can no longer reliably identify all the vulnerable systems and determine where to focus their attacks, or even assert whether they are gathering genuine data. Any ostensibly vulnerable network could be a honey-patch in disguise, and any exfiltrated secret could potentially be disinformation.

Threat Model Honey-patches add a layer of deception to confound exploits of known (patchable) vulnerabilities, which constitute the majority of exploited vulnerabilities in the wild. Previously unknown (i.e., zero-day) exploits remain potential threats, since such vulnerabilities are typically neither patched nor honey-patched. However, even zero-days can be potentially mitigated through cooperation of honey-patches with other layers of the deception stack. For example, a honey-patch that collects identifying information about a particular adversary seeking to exploit a known vulnerability can convey that collected information to a network-level intrusion detection system, which can then potentially identify the same adversary seeking to exploit a previously unknown vulnerability.

Although honey-patches primarily mitigate exploits of known vulnerabilities, they can effectively mitigate exploits whose attack payloads might be completely unique and therefore unknown to defenders. Such payloads might elude network-level monitors, and are therefore best detected at the software level at the point of exploit. Attackers might also use one payload for reconnaissance but reserve another for the final attack. Misleading the attacker into launching the final attack is therefore useful for discovering the final attack payload, which can divulge attacker strategies and goals not discernible from the reconnaissance payload alone.

Honey-patching is typically used in conjunction with standard access control protections, such as process isolation and least privilege. Attacker requests are therefore typically processed by a server possessing strictly user-level privileges, and must therefore leverage web server bugs and kernel-supplied services to perform malicious actions, such as corrupting the file system or accessing other users' memory to access confidential data. The defender's ability to thwart these and future attacks stems from his ability to deflect attackers to fully isolated decoys and perform counterreconnaissance (e.g., attack attribution and information gathering).

2.1 Honey-Patch Design Principles

Although the honey-patching concept is fairly straightforward, many significant security and performance challenges must be surmounted to realize it in practice. For example, a honey-patch that naïvely forks the entire server process to create a decoy clone process in response to attempted intrusions, inadvertently copies any secrets in the victim process's address space, such as encryption keys of concurrent sessions, over to the child decoy. Such an approach would be disastrous in practice, since the attack is allowed to succeed in the decoy, thereby giving the attacker potential access to secrets it may contain.

Moreover, practical adoption requires that honey-patches (1) introduce almost no overhead for legitimate users, (2) perform well enough for attackers that attack failures are not placarded, and (3) offer high compatibility with software that boasts aggressive multi-processing, multi-threading, and active connection migration across IPs. Solutions must therefore be sufficiently modular and generic that administrators require only a superficial, high-level understanding of each patch's structure and semantics to reformulate it as an effective honey-patch.

Specifically, effective honey-patching requires that remote forking of attacker sessions to decoys must happen live, with no perceptible disruption in the target application. This means that established connections—in particular, the attacker's connection—must not be broken. In addition, decoy deployment must be fast, to avoid offering overt, reliable timing channels that advertise the honey-patch. Finally, all sensitive data must be redacted before the decoy resumes execution to avoid giving the attacker potential access to user secrets.

Together, these requirements motivate three main design decisions: First, the required time performance precludes system-level cloning (e.g., VM cloning [9]) for session forking; instead, a lighter-weight, finer-grained alternative based on process migration through *checkpoint-restart* [25] is recommended. To scale to many concurrent attacks, *OS-level virtualization* should be leveraged to deploy forked processes to decoy containers, which can be created, deployed, and destroyed orders of magnitude faster than other virtualization techniques, such as full virtualization or para-virtualization [40].

OS-level virtualization allows multiple guest nodes (*containers*) to share the kernel of their controlling host. Linux containers (LXC) [23] implement OS-level virtualization, with resource management via process control groups and full resource isolation via Linux namespaces. This ensures that each container's processes, file system, network, and users remain mutually isolated. Fine-grained control of resource utilization prevents any container from starving its host. Furthermore, LXC supports containers backed by *overlayfs* snapshots, which is key for efficient container management and fast decoy deployment.

Second, transparent redirection of attacker sessions can be accomplished via a fine-grained, lightweight process migration technique based on checkpoint-restart, which facilitates the live migration of attacker sessions to decoys. This approach benefits from the synergy between mainstream Linux kernel APIs and user-space tools, allowing for a small freezing time of the target application and a special relocation mechanism for established connections that allows for transparent session migration.

Process migration through checkpoint-restart is the act of transferring a running process between two nodes by dumping its state on the source and resuming its execution on the destination. This problem is especially relevant for high-performance computing [15, 39]. As a result, several tools have been developed to support performance-critical process checkpoint-restart (e.g., BLCR [14], DMTCP [2], and CRIU [11]). Process checkpoint-restart plays a pivotal role in making the honey-patch concept viable. It provides a fast and seamless mechanism to enable transparent forking of attacker sessions, and scales well even in small environments due to its process-level granularity, which reduces the overall resources required to migrate the attacker process.

Third, to guarantee that successful exploits do not afford attackers access to sensitive data stored in application memory, honey-patches should implement a dynamic *secret redaction* mechanism that redacts the attacker process image during forking. This censors sensitive data from process memory before the forked (unpatched) session resumes. Forked decoys host a deceptive file system that omits all secrets, and that can be laced with disinformation to further deceive, delay, and misdirect attackers.

2.2 Architecture

The REDHERRING architecture [5], depicted in Fig. 3, embodies these design decisions by using process-level cloning and OS-level virtualization to achieve lightweight, resource-efficient, and fine-grained redirection of attacker sessions to sandboxed decoy environments in which secrets have been redacted with honey-data. Within this framework, developers use honey-patches to provide the same level of security as conventional patches, yet have the additional ability to deceive attackers.

Central to the system is a *reverse proxy* that acts as a transparent proxy between users and internal servers deployed as LXC containers. The *target* container hosts the honey-patched web server instance, and the *n* decoys form the pool of ephemeral containers managed by the *LXC Controller*. The decoys serve as

Fig. 3 REDHERRING system architecture overview

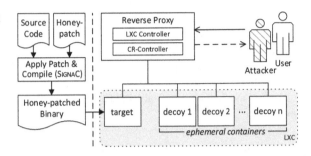

temporary environments for attacker sessions. Each container runs a *CR-Service* (Checkpoint/Restore) daemon, which exposes an interface controlled by the *CR-Controller* for remote checkpoint and restore.

Honey-Patching API To achieve low-coupling between target application and honey-patching logic, the honey-patch mechanism can be realized as a tiny library (e.g., implemented as a dynamically loadable C library). The library exposes three core API functions:

- `hp_init` $(pgid, pid, tid, sk)$: initialize honey-patch with the process group *pgid*, process *pid*, thread *tid*, and socket descriptor *sk* of the session.
- `hp_fork()`: initiate the attacker session remote forking process, implementing the honey-patching core logic.
- `hp_skip`(c): skip over block c if in a decoy.

Function `hp_init` initializes the honey-patch with the necessary information to handle subsequent session termination and resurrection. It is invoked once per connection, at the start of the session life cycle. For example, in the Apache web server, this immediately follows acceptance of an HTTP request and handing the newly created session off to a child process or worker thread; in Lighttpd and Nginx web servers, it follows the accept event for new connections.

> Lighttpd [22] and Nginx [27] are web servers whose designs are significantly different from Apache [3]. The most notable difference lies in the processing model of these servers, which employs non-blocking systems calls (e.g., select, poll, epoll) to perform asynchronous I/O operations for concurrent processing of multiple HTTP requests. In contrast, Apache dispatches each request to a child process or thread [30]. The ability to add deception capabilities to these three very different server architectures evidences the versatility of honey-patching as a software defense paradigm.

Listing 1 details the basic steps of `hp_fork`. Line 3 determines the application context, which can be either `target` (denoting the target container) or `decoy`. In a decoy, the function does nothing, allowing multiple exploits within a single attacker

Listing 1 hp_fork function

```
1  void hp_fork()
2  {
3      read_context();       // read context (target/decoy)
4      if (decoy) return;    // if in decoy, do nothing
5      register_handler();   // register signal handler
6      request_fork();       // fork session to decoy
7      wait();               // wait until fork process has finished
8      save_context();       // save context and resume
9  }
```

session to continue within the same decoy. In the target, a fork is initiated, consisting of four steps: (1) Line 5 registers the signal handler for session termination and resurrection. (2) Line 6 sends a *fork request* containing the attacker session's *pgid*, *pid*, and *tid* to the proxy's CR-Controller. (3) Line 7 synchronizes checkpoint and restore of the attacker session in the target and decoy, respectively, and guarantees that sensitive data is redacted from memory before the clone is allowed to resume. (4) Once forking is complete and the attacker session has been resurrected, the honey-patch context is saved and the attacker session resumes in the decoy.

The fork request (step 2) achieves high efficiency by first issuing a system fork to create a shallow, local clone of the web server process. This allows event-driven web servers to continue while attacker sessions are forked onto decoys, without interrupting the main event-loop. It also lifts the burden of synchronizing concurrent checkpoint operations, since CRIU injects a Binary, Large OBject (BLOB) into the target process memory space to extract state data during checkpoint.

The context-sensitivity of this framework allows the honey-patch code to exhibit context-specific behavior: In decoy contexts, hp_skip elides the execution of the code block passed as an argument to the macro, elegantly simulating the unpatched application code. In a target context, it is usually never reached due to the fork. However, if forking silently fails (e.g., due to resource exhaustion), it abandons the deception and conservatively executes the original patch's corrective action for safety.

LXC Pool The decoys into which attacker sessions are forked are managed as a pool of Linux containers controlled by the LXC Controller. The controller exposes two operations to the proxy: *acquire* (to acquire a container from the pool), and *release* (to release back a container to the pool). Each container follows the life cycle depicted in Fig. 4. Upon receiving a fork request, the proxy acquires the first available container from the pool. The acquired container holds an attacker session until (1) the session is deliberately closed by the attacker, (2) the connection's *keep-alive* timeout expires, (3) the ephemeral container crashes, or (4) a session timeout is reached. The last two conditions are common outcomes of successful exploits. In any of these cases, the container is released back to the pool and undergoes a recycling process before becoming available again.

Recycling a container encompasses three sequential operations: *destroy*, *clone* (which creates a new container from a template in which legitimate files are replaced with honeyfiles), and *start*. These steps happen swiftly for two main reasons.

Fig. 4 Linux containers pool
and decoys life cycle

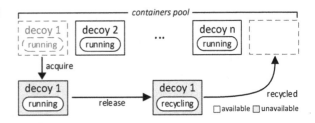

First, the lightweight virtualization implemented by LXC allows containers to be destroyed and started similarly to how OS processes are terminated and created. Second, ephemeral containers are deployed as overlayfs-based clones, making the cloning step almost instantaneous. The overlay file system is backed by a regular directory (the *template*) to clone new *overlayfs* containers (decoys), mounting the template's root file system as a read-only lower mount and a new private delta directory as a read-write upper mount. The template used to clone decoys is a copy of the target container in which all sensitive files are replaced with honey-files.

CR-Service The Reverse Proxy uses the CR-Controller module to communicate with CR-Service daemons running in the background of each container. The CR-Service uses an extended version of CRIU (Checkpoint/Restore In Userspace) [11] to checkpoint attacker sessions on the target and restore them on decoys. Each CR-Service implements a *façade* that exposes CR operations to the proxy's CR-Controller through a simple RPC protocol based on Protocol Buffers [16]. To enable fast, OS-local RPC communication between proxy and containers, IPC sockets (a.k.a., Unix domain sockets) are used.

However, because IPC sockets rely on the file system as an address namespace and the proxy runs on the host, establishing cross-container connections becomes difficult: host and containers file-systems are opaque to each other (due to namespace isolation). To overcome this issue, a directory located in the host is configured as bind mount to all containers to be used as bridge for IPC sockets, thus enabling the establishment of IPC connections between the CR-Controller running in the host and the CR-Service instances running inside containers.

Reverse Proxy The proxy plays a dual role in the honey-patching system, acting as (1) a *transport layer transparent proxy*, and (2) an *orchestrator* for attacker session forking.

As a transparent proxy, its main purpose is to hide the backend web servers and route client requests. To serve each client's request, the proxy server accepts a downstream socket connection from the client and binds an upstream socket connection to the backend server, allowing application-layer sessions to be processed transparently between the client and the backend server. To keep its size small, the proxy neither manipulates message payloads, nor implements any rules for detecting attacks. There is also no session caching. This makes it extremely innocuous and lightweight. The proxy is implemented as a transport-layer reverse proxy to reduce routing overhead and support the variety of protocols operating above TCP, including SSL/TLS.

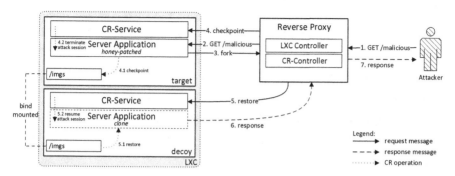

Fig. 5 Attacker session forking. Numbers indicate the sequential steps taken to fork an attacker session

As an orchestrator, the proxy listens for fork requests and coordinates the attacker session forking as shown in Fig. 5. Under legitimate load, the proxy simply routes user requests to the target and routes server responses to users. However, attack inputs elicit the following alternate workflow:

Step 1: The attacker probes the server with a crafted request (denoted by request GET /malicious in Fig. 5).

Step 2: The reverse proxy transparently routes the request to the backend target web server.

Step 3: The request triggers the honey-patch (i.e., when the honey-patch detects an attempted exploit of the patched vulnerability) and issues a fork request to the reverse proxy.

Step 4: The proxy's CR-Controller processes the request, acquires a decoy from the LXC Pool, and issues a checkpoint RPC request to the target's CR-Service. The CR-Service

   4.1: checkpoints the running web server instance to the /imgs directory; and

   4.2: signals the attacker session with a termination code, gracefully terminating it.

Step 5: Upon checkpoint completion, the CR-Controller commands the decoy's CR-Service to restore the dumped web server images on the decoy. The CR-Service then

   5.1: restores a clone of the web server from the dump images located in the /imgs directory; and

   5.2: signals the attacker session with a resume code, and cleans the dump data from /imgs.

Step 6: The attacker session resumes on the decoy, and a response is sent back to the reverse proxy.

Step 7: The reverse proxy routes the response to the attacker.

Throughout this workflow, the attacker's session forking is completely transparent to the attacker. To avoid any substantial overhead for transferring files between target and decoys, each decoy's /imgs folder is bind-mounted to the target's /imgs directory. After the session has been forked to the decoy, it behaves like an unpatched server, making it appear that no redirection has taken place and the original probed server is vulnerable.

Established TCP Connection Relocation Target and decoys are fully isolated containers running on separate namespaces. As a result, each container is assigned a unique IP in the internal network, which affects how active connections are moved from the target to a decoy. To realize this use case, an extension to CRIU is implemented as part of the honey-patching framework to support relocation of TCP connections during process restoration. In what follows, we will discuss important details of the implementation.

The reverse proxy always routes legitimate user connections to the target; hence, there is no need to restore the state of connections for these users when restoring the web server on a decoy. Legitimate connections can simply be restored to *drainer sockets*, since we have no interest in maintaining legitimate user interaction with the decoys. This ensures that the associated user sessions are restored to completion without interrupting the overall application restoration.

Conversely, the attacker connection must be restored to its dumped state when switching the attacker session to a decoy. This is important to avoid connection disruption and to allow transparent session migration (from the perspective of the attacker). To accomplish this, the proxy dynamically establishes a new backend TCP connection between proxy and decoy containers in order to hold the attacker session communication. Moreover, a mechanism based on *TCP repair options* [10] is employed to transfer the state of the original attacker's session socket (bound to the target IP address) into the newly created socket (bound to the decoy IP address).

Figure 6 describes the connection relocation mechanism, implemented as a step of the attacker's session restore process. At process checkpoint, the state information of the original socket sk is dumped together with the process image (not shown in the figure). This includes connection bounds, previously negotiated socket options, sequence numbers, receiving and sending queues, and connection state. During process restore, the connection is relocated to the assigned decoy by (1) connecting a new socket tsk to the proxy $port given in the restore request, (2) setting tsk to *repair mode* and silently closing the socket (i.e., no FIN or RST packages are sent to the remote end), and (3) transferring the connection state from sk to tsk in repair mode. Once the new socket tsk is handed over to the restored attacker session, the relocation process has completed and communication resumes, often with an HTTP response being sent back to the attacker.

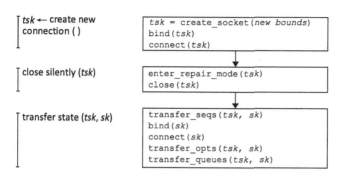

Fig. 6 Procedure for TCP connection relocation

3 Process Image Secret Redaction

To more effectively enforce data confidentiality and privacy concerns in security-sensitive, deceptive software environments, honey-patching leverages new compiler techniques that equip software with dynamic secret redaction capabilities. The resulting software responds to emerging cyber attacks by quickly and comprehensively substituting all secrets in its address space with honey-tokens that disinform attackers. Safe, efficient redaction of secrets from program address spaces has many applications, including the safe release of program memory dumps to software developers for debugging purposes, mitigation of cyber-attacks via runtime self-censoring in response to intrusions, and attacker deception through honeypotting.

Realizing such runtime process secret redaction in practice elicits at least two significant challenges. First, the redaction step must yield a runnable program process. Non-secrets must therefore not be conservatively redacted, lest data critical for continuing the program's execution be deleted. Secret redaction for running processes is hence especially sensitive to *label creep* and *over-tainting* failures.

Label creep and over-tainting are classic challenges in the data confidentiality enforcement literature. The former refers to the tendency of a datum's classification level to become ever more restrictive over its lifetime, until even those who own the data may lack the privileges to access it. The latter refers to the tendency of information security systems to over-classify even non-secrets as confidential—for example, due to non-secrets coming into brief contact with confidential data during information processing.

Second, many real-world programs targeted by cyber-attacks were not originally designed with information flow tracking support, and are often expressed in low-level, type-unsafe languages, such as C/C++. A suitable solution must be amenable to retrofitting such low-level, legacy software with annotations sufficient to distinguish non-secrets from secrets, and with efficient flow-tracking logic that does not impair performance.

This section summarizes a suitable dynamic secret redaction solution [4] that has been integrated into the LLVM compiler's [21] DataFlow Sanitizer (DFSan) infrastructure [13], which adds byte-granularity taint-tracking support to C/C++ programs at compile-time. At the source level, DFSan's taint-tracking capabilities are purveyed as runtime data-classification, data-declassification, and taint-checking operations, which programmers add to their programs to identify secrets and curtail their flow at runtime. Unfortunately, straightforward use of this interface for redaction of large, complex legacy codes can lead to severe over-tainting, or requires an unreasonably detailed retooling of the code with copious classification operations. This is unsafe, since missing even one of these classification points during retooling risks disclosing secrets to adversaries.

To overcome these deficiencies, our research has augmented LLVM with a declarative, type annotation-based secret-labeling mechanism for easier secret identification, and a new label propagation semantics, called *Pointer Conditional-Combine Semantics* (PC^{2}S). The semantics efficiently distinguishes secret data within C-style graph data structures from the non-secret structure that houses the data. This partitioning of the bytes greatly reduces over-tainting and the programmer's annotation burden, and proves critical for precisely and efficiently redacting secret process data whilst preserving process operation after redaction.

3.1   Sourcing and Tracking Secrets

Taint-tracking conceptually entails labeling each byte of process memory with a security label that denotes its classification level. At compile-time, a taint-tracking compiler instruments the resulting object code with extra code that propagates these labels alongside the data they label.

Extending such taint-tracking to low-level, legacy code not designed with taint-tracking in mind is often difficult. For example, the standard approach of specifying taint introductions as annotated program inputs often proves too coarse for inputs comprising low-level, unstructured data streams, such as network sockets. Listing 2 exemplifies the problem using a code excerpt from the Apache web server [3]. The excerpt partitions a byte stream (stored in buffer s1) into a non-secret user name and a secret password, delimited by a colon character. Naïvely labeling input s1 as secret to secure the password causes the compiler to over-taint the user name (and the colon delimiter, and the rest of the stream), leading to excessive over-tainting— everything associated with the stream becomes secret, with the result that nothing can be safely divulged.

A correct solution must more precisely identify data field `uptr->password` (but not `uptr->user`) as secret after the unstructured data has been parsed. This is achieved in DFSan by manually inserting a runtime classification operation after line 7. However, on a larger scale this brute-force labeling strategy imposes a dangerously heavy annotation burden on developers, who must manually locate all such classification points. In C/C++ programs littered with pointer arithmetic, the correct classification points can often be obscure. Inadvertently omitting even one classification risks information leaks.

Listing 2 Apache's URI parser function (excerpt)

```
1  /* first colon delimits username:password */
2  s1 = memchr(hostinfo, ':', s − hostinfo);
3  if (s1) {
4    uptr->user = apr_pstrmemdup(p, hostinfo,
5    s1−hostinfo);
6    ++s1;
7    uptr->password = apr_pstrmemdup(p, s1, s − s1);
8  }
```

Listing 3 Apache's session record (excerpt)

```
1  typedef struct {
2    NONSECRET apr_pool_t *pool;
3    NONSECRET apr_uuid_t *uuid;
4    SECRET_STR const char *remote_user;
5    apr_table_t *entries;
6    ...
7  } SECRET session_rec;
```

To ease this burden, a better solution is to introduce a mechanism whereby developers can identify secret-storing structures and fields *declaratively* rather than operationally. For example, to correctly label the password in Listing 2 as secret, users may add type qualifier SECRET_STR to the password field's declaration in its abstract datatype definition. A modified LLVM compiler responds to this static annotation by instrumenting the program with instructions that dynamically taint all values assigned to the password field. Since datatypes typically have a single point of definition (in contrast to the many code points that access them), this greatly reduces the annotation burden imposed upon code maintainers.

In cases where the appropriate taint is not statically known (e.g., if each password requires a different, user-specific taint label), parameterized type-qualifier SECRET⟨f⟩ identifies a function f that computes the appropriate taint label at runtime.

Unlike traditional taint introduction semantics, which label program input values and sources with taints, recognizing structure fields as taint sources requires a new form of taint semantics that conceptually interprets dynamically identified *memory addresses* as taint sources. For example, a program that assigns address &(uptr->password) to pointer variable p, and then assigns a freshly allocated memory address to *p, must automatically identify the freshly allocated memory as a new taint source, and thereafter taint any values stored at *$p[i]$ (for all indexes i).

To achieve this, DFSan's *pointer-combine semantics (PCS)* feature is extended to optionally combine (i.e., join) the taints of pointers and pointees during pointer dereferences. Specifically, when *PCS on-load* is enabled, read-operation *p yields a value tainted with the join of pointer p's taint and the taint of the value to which p points; and when *PCS on-store* is enabled, write-operation *p := e taints the value stored into *p with the join of p's and e's taints. Using PCS leads to a natural encoding of SECRET annotations as pointer taints. Continuing the previous example, PCS propagates uptr->password's taint to p, and subsequent dereferencing assignments propagate the two pointers' taints to secrets stored at their destinations.

PCS works well when secrets are always separated from the structures that house them by a level of pointer indirection, as in the example above (where `uptr->password` is a pointer to the secret rather than the secret itself). However, label creep difficulties arise when structures mix secret values with non-secret pointers. To illustrate, consider a singly linked list ℓ of secret integers, where each integer has a different taint. In order for PCS on-store to correctly classify values stored to ℓ->`secret_int`, pointer ℓ must have taint γ_1, where γ_1 is the desired taint of the first integer. But this causes stores to ℓ->`next` to incorrectly propagate taint γ_1 to the node's next-pointer, which propagates γ_1 to subsequent nodes when dereferenced. In the worst case, all nodes become labeled with all taints. Such issues have spotlighted effective pointer tainting as a significant challenge in the taint-tracking literature [12, 19, 33, 34].

To address this shortcoming, PC^2S semantics generalize PCS semantics by augmenting them with pointer-combine *exemptions* conditional upon the static type of the pointee. In particular, a PC^2S taint-propagation policy may dictate that taint labels are not combined when the pointee has pointer type. Hence, ℓ->`secret_int` receives ℓ's taint because the assigned expression has integer type, whereas ℓ's taint is *not* propagated to ℓ->`next` because the latter's assigned expression has pointer type. Empirical evaluation shows that just a few strategically selected exemption rules expressed using this refined semantics suffices to vastly reduce label creep while correctly tracking all secrets in large legacy source codes.

In order to strike an acceptable balance between security and usability, the solution only automates tainting of C/C++ style structures whose non-pointer fields share a common taint. Non-pointer fields of mixed taintedness within a single struct are not supported automatically because C programs routinely use pointer arithmetic to reference multiple fields in a struct via a common pointer (imparting the pointer's taint to all the struct's non-pointer fields)—for example, when copying structures, or when marshalling and demarshalling them to/from streams. This approach therefore targets the common case in which the taint policy is expressible at the granularity of structures, with exemptions for fields that point to other (differently tainted) structure instances. This corresponds to the usual scenario where a non-secret graph structure (e.g., a tree) stores secret data in its nodes.

With these new language extensions, users label structure datatypes as SECRET (implicitly introducing a taint to all fields within the structure), and additionally annotate pointer fields as NONSECRET to exempt their taints from pointer-combines during dereferences. Pointers to dynamic-length, null-terminated secrets get annotation SECRET_STR. For example, Listing 3 illustrates the annotation of `session_req`, used by Apache to store remote users' session data. Finer-granularity policies remain enforceable, but require manual instrumentation via DFSan's API, to precisely distinguish which of the code's pointer dereference operations propagate pointer taints. This solution thus complements existing approaches.

programs	$\mathcal{P} ::= \bar{c}$	locations	$\ell ::=$ memory addresses
commands	$c ::= v := e \mid \texttt{store}(\tau, e_1, e_2)$	environment	$\Delta : v \rightharpoonup u$
	$\mid \texttt{ret}(\tau, e) \mid \texttt{br}(e, e_1, e_0)$	prog counter	pc
	$\mid \texttt{call}(\tau, e, \overline{args})$	stores	$\sigma : (\ell \rightharpoonup u) \cup (v \rightharpoonup \ell)$
expressions	$e ::= v \mid \langle u, \gamma \rangle \mid \Diamond_b(\tau, e_1, e_2)$	functions	f
	$\mid \texttt{load}(\tau, e)$	function table	$\phi : f \rightharpoonup \ell$
binary ops	$\Diamond_b ::=$ typical binary operators	taint contexts	$\lambda : (\ell \cup v) \rightharpoonup \gamma$
variables	v	propagation	$\rho : \bar{\gamma} \rightarrow \gamma$
values	$u ::=$ values of underlying IR	prop contexts	$\mathcal{A} : f \rightarrow \rho$
types	$\tau ::= ptr\ \tau \mid \tau\ \bar{\tau} \mid$ primitive types	call stack	$\Xi ::= nil$
taint labels	$\gamma \in (\Gamma, \sqsubseteq)$ (label lattice)		$\mid \langle f, pc, \Delta, \bar{\gamma} \rangle :: \Xi$

Fig. 7 Intermediate representation syntax

3.2 Formal Semantics

For explanatory precision, the new taint-tracking semantics is formally defined in terms of the simple, typed intermediate language (IL) in Fig. 7, inspired by prior work [33]. The simplified IL abstracts irrelevant details of LLVM's IR language, capturing only those features needed to formalize the analysis.

Language Syntax Programs $\mathcal{P}$ are lists of commands, denoted $\bar{c}$. Commands consist of variable assignments, pointer-dereferencing assignments (stores), conditional branches, function invocations, and function returns. Expressions evaluate to value-taint pairs $\langle u, \gamma \rangle$, where u ranges over typical value representations, and γ is the taint label associated with u. Labels denote sets of taints; they therefore comprise a lattice ordered by subset ($\sqsubseteq$), with the empty set $\bot$ at the bottom (denoting public data), and the universe $\top$ of all taints at the top (denoting maximally secret data). Join operation $\sqcup$ denotes least upper bound (union) of taint sets.

Variable names range over identifiers and function names, and the type system supports pointer types, function types, and typical primitive types. Since DFSan's taint-tracking is dynamic, we here omit a formal static semantics and assume that programs are well-typed. Execution contexts are comprised of a store σ relating locations to values and variables to locations, an environment Δ mapping variables to values, and a tainting context λ mapping locations and variables to taint labels. Additionally, to express the semantics of label propagation for external function calls (e.g., runtime library API calls), function table ϕ maps external function names to their entry points, a propagation context $\mathcal{A}$ that dictates whether and how each external function propagates its argument labels to its return value label, and the call stack Ξ. Taint propagation policies returned by $\mathcal{A}$ are expressed as customizable mappings ρ from argument labels $\bar{\gamma}$ to return labels γ.

$$\frac{}{\sigma, \Delta, \lambda \vdash u \Downarrow \langle u, \bot \rangle} \text{ VAL} \qquad \frac{}{\sigma, \Delta, \lambda \vdash v \Downarrow \langle \Delta(v), \lambda(v) \rangle} \text{ VAR}$$

$$\frac{\sigma, \Delta, \lambda \vdash e_1 \Downarrow \langle u_1, \gamma_1 \rangle \quad \sigma, \Delta, \lambda \vdash e_2 \Downarrow \langle u_2, \gamma_2 \rangle}{\sigma, \Delta, \lambda \vdash \Diamond_b(\tau, e_1, e_2) \Downarrow \langle u_1 \Diamond_b u_2, \gamma_1 \sqcup \gamma_2 \rangle} \text{ BINOP}$$

$$\frac{\sigma, \Delta, \lambda \vdash e \Downarrow \langle u, \gamma \rangle}{\sigma, \Delta, \lambda \vdash \text{load}(\tau, e) \Downarrow \langle \sigma(u), \rho_{load}(\tau, \gamma, \lambda(u)) \rangle} \text{ LOAD}$$

$$\frac{\sigma, \Delta, \lambda \vdash e \Downarrow \langle u, \gamma \rangle \quad \Delta' = \Delta[v \mapsto u] \quad \lambda' = \lambda[v \mapsto \gamma]}{\langle \sigma, \Delta, \lambda, \Xi, pc, v := e \rangle \rightarrow_1 \langle \sigma, \Delta', \lambda', \Xi, pc+1, \mathcal{P}[pc+1] \rangle} \text{ ASSIGN}$$

$$\frac{\sigma, \Delta, \lambda \vdash e_1 \Downarrow \langle u_1, \gamma_1 \rangle}{\sigma, \Delta, \lambda \vdash e_2 \Downarrow \langle u_2, \gamma_2 \rangle \quad \sigma' = \sigma[u_1 \mapsto u_2] \quad \lambda' = \lambda[u_1 \mapsto \rho_{store}(\tau, \gamma_1, \gamma_2)]}{\langle \sigma, \Delta, \lambda, \Xi, pc, \text{store}(\tau, e_1, e_2) \rangle \rightarrow_1 \langle \sigma', \Delta, \lambda', \Xi, pc+1, \mathcal{P}[pc+1] \rangle} \text{ STORE}$$

$$\frac{\sigma, \Delta, \lambda \vdash e \Downarrow \langle u, \gamma \rangle \quad \sigma, \Delta, \lambda \vdash e_{(u?1:0)} \Downarrow \langle u', \gamma' \rangle}{\langle \sigma, \Delta, \lambda, \Xi, pc, \text{br}(e, e_1, e_0) \rangle \rightarrow_1 \langle \sigma, \Delta, \lambda, \Xi, u', \mathcal{P}[u'] \rangle} \text{ COND}$$

$$\frac{\sigma, \Delta, \lambda \vdash e_1 \Downarrow \langle u_1, \gamma_1 \rangle \quad \cdots \quad \sigma, \Delta, \lambda \vdash e_n \Downarrow \langle u_n, \gamma_n \rangle \quad \Delta' = \Delta[\overline{params_f} \mapsto \overline{u_1 \cdots u_n}]}{\lambda' = \lambda[\overline{params_f} \mapsto \overline{\gamma_1 \cdots \gamma_n}] \quad fr = \langle f, pc+1, \Delta, \overline{\gamma_1 \cdots \gamma_n} \rangle}{\langle \sigma, \Delta, \lambda, \Xi, pc, \text{call}(\tau, f, \overline{e_1 \cdots e_n}) \rangle \rightarrow_1 \langle \sigma, \Delta', \lambda', fr :: \Xi, \phi(f), \mathcal{P}[\phi(f)] \rangle} \text{ CALL}$$

$$\frac{\sigma, \Delta, \lambda \vdash e \Downarrow \langle u, \gamma \rangle \quad fr = \langle f, pc', \Delta', \overline{\gamma} \rangle \quad \lambda' = \lambda[v_{ret} \mapsto \mathcal{A}f\overline{\gamma}]}{\langle \sigma, \Delta, \lambda, fr :: \Xi, pc, \text{ret}(\tau, e) \rangle \rightarrow_1 \langle \sigma, \Delta'[v_{ret} \mapsto u], \lambda', \Xi, pc', \mathcal{P}[pc'] \rangle} \text{ RET}$$

Fig. 8 Operational semantics of a generalized label propagation semantics

Operational Semantics Figure 8 presents an operational semantics defining how taint labels propagate in an instrumented program. Expression judgments are large-step ($\Downarrow$), while command judgments are small-step ($\rightarrow_1$). At the IL level, expressions are pure and programs are non-reflective. Abstract machine configurations consist of tuples $\langle \sigma, \Delta, \lambda, \Xi, pc, \iota \rangle$, where pc is the program pointer and ι is the current instruction. Notation $\Delta[v \mapsto u]$ denotes function Δ with v remapped to u, and notation $\mathcal{P}[pc]$ refers to the program instruction at address pc. For brevity, we omit $\mathcal{P}$ from machine configurations, since it is fixed.

Rule VAL expresses the typical convention that hardcoded program constants are initially untainted ($\bot$). Binary operations are eager, and label their outputs with the join ($\sqcup$) of their operand labels. The semantics of $\text{load}(\tau, e)$ read the value stored in location e, where the label associated with the loaded value is obtained by propagation function ρ_{load}. Dually, $\text{store}(\tau, e_1, e_2)$ stores e_2 into location e_1, updating λ according to ρ_{store}. In C programs, these model pointer dereferences and dereferencing assignments, respectively. Parameterizing these rules in terms of abstract propagation functions ρ_{load} and ρ_{store} allows us to instantiate them with customized propagation policies at compile-time, as detailed in Sect. 3.2.

External function calls $\text{call}(\tau, f, \overline{e_1 \cdots e_n})$ evaluate arguments $\overline{e_1 \cdots e_n}$, create a new stack frame fr, and jump to the callee's entry point. Returns then consult propagation context $\mathcal{A}$ to appropriately label the value returned by the function based on the labels of its arguments. Context $\mathcal{A}$ can be customized by the user to specify how labels propagate through external libraries compiled without taint-tracking support.

$$\text{NCS} \quad \rho_{\{load, store\}}(\tau, \gamma_1, \gamma_2) := \gamma_2$$

$$\text{PCS} \quad \rho_{\{load, store\}}(\tau, \gamma_1, \gamma_2) := \gamma_1 \sqcup \gamma_2$$

$$\text{PC}^2\text{S} \quad \rho_{\{load, store\}}(\tau, \gamma_1, \gamma_2) := (\tau \text{ is } ptr) \ ? \ \gamma_2 : (\gamma_1 \sqcup \gamma_2)$$

Fig. 9 Polymorphic functions for no-combine, pointer-combine, and PC^2S propagation policies

Label Propagation Semantics The operational semantics are parameterized by propagation functions ρ that can be instantiated to a specific propagation policy at compile-time. This provides a base framework through which we can study different propagation policies and their differing characteristics. Figure 9 presents three polymorphic[1] functions that can be used to instantiate propagation policies. On-load propagation policies instantiate ρ_{load}, while on-store policies instantiate ρ_{store}. The instantiations in Fig. 9 define no-combine semantics (DFSan's on-store default), PCS (DFSan's on-load default), and our PC^2S extensions:

No-Combine The no-combine semantics (NCS) model a traditional, pointer-transparent propagation policy. Pointer labels are ignored during loads and stores, causing loaded and stored data retain their labels irrespective of the labels of the pointers being dereferenced.

Pointer-Combine Semantics In contrast, PCS joins pointer labels with loaded and stored data labels during loads and stores. Using this policy, a value is tainted on-load (resp., on-store) if its source memory location (resp., source operand) is tainted or the pointer value dereferenced during the operation is tainted. If both are tainted with different labels, the labels are joined to obtain a new label that denotes the union of the originals.

Pointer Conditional-Combine Semantics PC^2S generalizes PCS by conditioning the label-join on the static type of the data operand. If the loaded/stored data has pointer type, it applies the NCS rule; otherwise, it applies the PCS rule. The resulting label propagation for stores is depicted in Fig. 10.

This can be leveraged to obtain the best of both worlds. PC^2S pointer taints retain most of the advantages of PCS—they can identify and track aliases to birthplaces of secrets, such as data structures where secrets are stored immediately after parsing, and they automatically propagate their labels to data stored there. But PC^2S resists PCS's over-tainting and label creep problems by avoiding propagation of pointer labels through levels of pointer indirection, which usually encode relationships with other data whose labels must remain distinct and separately managed.

Condition $(\tau \text{ is } ptr)$ in Fig. 9 can be further generalized to any decidable proposition on static types τ. This feature is used to distinguish pointers that cross data ownership boundaries (e.g., pointers to other instances of the parent structure) from pointers that target value data (e.g., strings). The former receive NCS treatment

---

[1] The functions are polymorphic in the sense that some of their arguments are types τ.

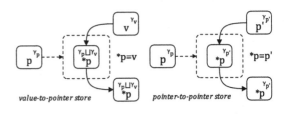

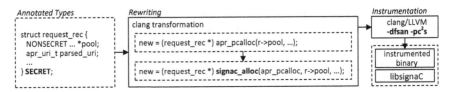

Fig. 10 PC²S propagation policy on store commands

by default to resist over-tainting, while the latter receive PCS treatment by default to capture secrets and keep the annotation burden low. In addition, PC²S is at least as efficient as PCS because propagation policy ρ is partially evaluated at compile-time. Thus, the choice of NCS or PCS semantics for each pointer operation is decided purely statically, conditional upon the static types of the operands. The appropriate specialized propagation implementation is then in-lined into the resulting object code during compilation.

Example. To illustrate how each semantics propagate taint, consider the following IL pseudo-code, which revisits the linked-list example informally presented in Sect. 3.1.

```
1  store(id, requestid, get(s, idsize));
2  store(key, p[requestid]->key, get(s, keysize));
3  store(ctx_t*, p[requestid]->next, queuehead);
```

Input stream s includes a non-secret request identifier and a secret key of primitive type (e.g., unsigned long). If one labels stream s secret, then the public *request_id* becomes over-tainted in all three semantics, which is undesirable because a redaction of *request_id* may crash the program (when *request_id* is later used as an array index). A better solution is to label pointer p secret and employ PCS, which correctly labels the key at the moment it is stored. However, PCS additionally taints the *next*-pointer, leading to over-tainting of all the nodes in the containing linked-list, some of which may contain keys owned by other users. PC²S avoids this over-tainting by exempting the next pointer from the combine-semantics. This preserves the data structure while correctly labeling the secret data it contains.

```
1  store(id, request_id, get(s, id_size));
2  store(key, p[request_id]->key, get (s, key_size) );
3  store(ctx_t*, p[request_id]->next, queue_head);
```

Fig. 11 Architectural overview of SIGNAC illustrating its three-step, static instrumentation process: (1) annotation of security-relevant types, (2) source-code rewriting, and (3) compilation with the sanitizer's instrumentation pass

3.3 An Integrated Secret-Redacting, Honey-Patching Architecture

Figure 11 presents the architecture of SIGNAC[2] (Secret Information Graph iNstrumentation for Annotated C) [4], which leverages compiler-instrumented secret-redaction to achieve secret-sanitized process migration for secure honey-patching. At a high level, it consists of three components: (1) a source-to-source preprocessor, which (a) automatically propagates user-supplied, source-level type annotations to containing datatypes, and (b) in-lines taint introduction logic into dynamic memory allocation operations; (2) a modified LLVM compiler that instruments programs with PC^2S taint propagation logic during compilation; and (3) a runtime library that the instrumented code invokes during program execution to introduce taints and perform redaction.

Type Attributes Server code maintainers first annotate data structures containing secrets with the type qualifier SECRET. This instructs the taint-tracker to treat all instantiations (e.g., dynamic allocations) of these structures as taint sources. Additionally, qualifier NONSECRET may be applied to pointer fields within these structures to exempt them from PCS. The instrumentation pass generates NCS logic instead for operations involving such members. Finally, qualifier SECRET_STR may be applied to pointer fields whose destinations are dynamic-length byte sequences bounded by a null terminator (strings).

Type Attribute Rewriting In the preprocessing step, the target application undergoes a source-to-source transformation pass that rewrites all dynamic allocations of annotated data types with taint-introducing wrappers. Implementing this transformation at the source level allows us to utilize the full type information that is available at the compiler's front-end, including purely syntactic attributes such as SECRET annotations. The implementation leverages Clang's tooling API [8] to traverse and apply the desired transformations directly into the program's AST.

Static Instrumentation The instrumentation pass next introduces LLVM IR code during compilation that propagates taint labels during program execution. The implementation extends DFSan with the PC^2S label propagation policy specified

---

[2]Named after *pointillism* co-founder Paul Signac.

Listing 4 Taint-introducing memory allocations

```
1   #define signac_alloc(alloc, args...) ({ \
2       void *__p = alloc ( args ); \
3       signac_taint(&__p, sizeof(void*)); \
4       __p; })
```

in Sect. 3.2. Taint labels are represented as 16-bit integers, with new labels allocated sequentially from a pool. DFSan maps (without reserving) the lower 32 TB of the process address space for *shadow memory*, which stores the taint labels of the values stored at the corresponding application memory addresses. At the front-end, compilation flags parametrize the label propagation policies for the store and load operations (viz., NCS, PCS, or PC^{2}S).

Runtime Library The source-to-source rewriter and instrumentation phases in-line logic that calls a tiny dedicated library at runtime to introduce taints, handle special taint-propagation cases (e.g., string support), and check taints at sinks (e.g., during redaction). The library exposes three API functions:

- `signac_init` (*pl*) : initialize a tainting context with a fresh label instantiation *pl* for the current principal.
- `signac_taint` (*addr*, *size*) : taint each address in interval [*addr*, *addr+size*) with *pl*.
- `signac_alloc` (*alloc*, ...) : wrap allocator *alloc* and taint the address of its returned pointer with *pl*.

Function `signac_init` instantiates a fresh taint label and stores it in a thread-global context, which function *f* of annotation SECRET⟨*f*⟩ may consult to identify the owning principal at taint-introduction points. In typical web server architectures, this function is strategically hooked at the start of a new connection's processing cycle. Function `signac_taint` sets the labels of each address in interval [*addr*, *addr+size*) with the label *pl* retrieved from the session's context.

Listing 4 details `signac_alloc`, which wraps allocations of SECRET-annotated data structures. This variadic macro takes a memory allocation function *alloc* and its arguments, invokes it (line 2), and taints the address of the pointer returned by the allocator (line 3).

Example. To instrument a particular server application, such as Apache, SIGNAC requires two small, one-time developer interventions: First, add a call to `signac_init` at the start of a user session to initialize a new tainting context for the newly identified principal. Second, annotate the security-relevant data structures whose instances are to be tracked. For instance, in Apache, `signac_init` is called upon the acceptance of a new server connection, and annotated types include `request_rec`, `connection_rec`, `session_rec`, and `modssl_ctx_t`. These structures are where Apache stores URI parameters and request content information, private connection data such as remote IPs, key-value entries in user sessions, and encrypted connection information. The redaction scheme instruments the server with PC^{2}S. At redaction time, it scans the resulting shadow memory for labels denoting secrets owned by user sessions other than the attacker's, and redacts such secrets. The shadow memory and taint-tracking libraries are then unloaded, leaving a decoy process that masquerades as undefended and vulnerable.

Apache has been the most popular web server since April 1996 [3]. Its market share includes 54.5 % of all active websites (the second, Nginx, has 11.97 %) and 55.45 % of the top-million websites (against Nginx with 15.91 %) [26]. It is a robust, commercial-grade, feature-rich open-source software product comprised of 2.27M SLOC mostly in C [29], and has been tested on millions of web servers around the world. These characteristics make it a highly challenging, interesting, and practical flagship case study to test compiler-assisted taint-tracking and embedded honeypotting.

Taint Spread Evaluation For comparing the taint spread properties of PC^2S and PCS, Apache's core modules for serving static and dynamic content, encrypting connections, and storing session data were annotated, omitting its optional modules. Altogether, approximately 45 type annotations need to be added to the web server source code, plus 30 SLOC to initialize the taint-tracker. Considering the size and complexity of Apache ($\sim$2.2M SLOC), PC^2S annotation burden is exceptionally light relative to prior taint-tracking approaches.

To test PC^2S's resistance to taint explosions, a stream of (non keep-alive) requests was submitted to each instrumented web server, recording a cumulative tally of distinct labels instantiated during taint-tracking. Figure 12a plots the results, comparing traditional PCS to PC^2S extensions. On Apache, traditional PCS is impractical, exceeding the maximum label limit in just 68 requests. In contrast, PC^2S instantiates vastly fewer labels (note that the y-axes are *logarithmic scale*). After extrapolation, an average 16,384 requests are required to exceed the label limit under PC^2S—well above the standard 10K-request TTL limit for worker threads.

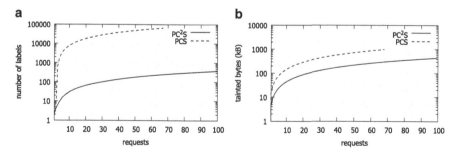

Fig. 12 Taint spread evaluation of PC^2S- and PCS-instrumented instances of the Apache web server. (**a**) Label creeping behavior. (**b**) Cumulative tally of bytes tainted

Listing 5 Abbreviated patch for CVE-2014-6271

```
1  + if ((flags & SEVAL_FUNCDEF) && command->type != cm_function_def)
2  + {
3  +     internal_warning ("%s:_ignoring_function_definition_attempt", ...);
4  +     should_jump_to_top_level = 0;
5  +     last_result = last_command_exit_value = EX_BADUSAGE;
6  +     break;
7  + }
```

Taint spread control is equally critical for preserving program functionality after redaction. To demonstrate, the experiment is repeated with a simulated intrusion after $n \in [1, 100]$ legitimate requests. Figure 12b plots the cumulative tally of how many bytes received a taint during the history of the run on Apache. In all cases, redaction crashed PCS-instrumented processes cloned after just 2–3 legitimate requests (due to erasure of over-tainted bytes). In contrast, PC^{2}S-instrumented processes never crashed during the experiment; their decoy clones continued running after redaction, impersonating vulnerable servers. This demonstrates the approach's facility to realize effective taint-tracking in legacy codes for which prior approaches fail.

4 Case Study: A Honey-Patch for Shellshock

The Shellshock GNU Bash remote command execution vulnerability (CVE-2014-6271) [28] was one of the most severe vulnerabilities in recent history, affecting millions of then-deployed web servers and other Internet-connected devices. This high impact combined with its ease of exploitation makes it a prime candidate for a practical application and evaluation of honey-patching.

Honey-Patching Shellshock Listing 5 shows an abbreviated, vendor-released patch in diff style for Shellshock. The patch introduces a conditional that validates environment variables passed to Bash, declining function definition attempts. Prior to this patch, attackers could take advantage of HTTP headers as well as other mechanisms to enable unauthorized access to the underlying system shell of remote targets. This patch exemplifies a common vulnerability mitigation: dangerous inputs or program states are detected via a boolean test, with positive detection eliciting a corrective action. The corrective action is typically readily distinguishable by attackers—in this case, a warning message is generated and the function definition is ignored.

Listing 6 presents an alternative, honey-patched implementation of the same patch. In response to a malicious input, the honey-patched application forks itself onto a confined, ephemeral, decoy environment, and behaves henceforth as an unpatched, vulnerable version of the software. Specifically, line 3 forks the user session to a decoy container, and macro `hp_skip` in line 4 elides the rejection in the decoy container so that the attack appears to have succeeded. Meanwhile, the attacker session in the original container is safely terminated (having been forked to the decoy), and legitimate, concurrent connections continue unaffected.

Listing 6 Honey-patch for CVE-2014-6271

```
1   if ((flags & SEVAL_FUNCDEF) && command->type != cm_function_def)
2   {
3 +   hp_fork();
4 +   hp_skip(
5       internal_warning ("%s:_ignoring_function_definition_attempt", ...);
6       should_jump_to_top_level = 0;
7       last_result = last_command_exit_value = EX_BADUSAGE;
8       break;
9 +   );
10  }
```

As a result, adversaries attempting to exploit Shellshock in a victim server that has been honey-patched receive server responses that seem to indicate that the exploit has succeeded. However, the shell commands they inject are actually executing in a decoy environment stocked with disinformation for attackers to explore. Observe that the differences between the patch and the honey-patch are quite minor, except for the fixed cloning infrastructure that the honey-patch code references, and that can be maintained separately from the server code. This allows information technology and security administrators to easily reformulate patches into honey-patches after a new vulnerability disclosure is received, facilitating a quick, aggressive response to the threat. In general, only a superficial understanding of many patches is required to convert them to honey-patches of this form.[3]

Decoy Monitoring Decoys host software monitors that allow defenders to collect rich and detailed fine-grained attack information. To minimize the performance impact on decoys, two powerful and highly efficient monitoring tools are implemented in REDHERRING [6]: *inotifywait* (to track modifications made to the file system), and *tcpdump* (to monitor ingress and egress of network packets). To avoid possible tampering with the collected data, all logs are stored outside the decoy environments. In addition, both monitoring tools are tuned to avoid generating spurious outputs (e.g., by excluding certain directories and limiting the monitored network traffic). As illustrative examples, attack data collected at decoys can be used to inform perimeter defenses to block exploit attempts against unpatched machines in the network, and to provide detailed threat intelligence to aid analysts in incidence response scenarios.

Performance Benchmarks To evaluate honey-patching, one must determine the performance overhead imposed upon sessions forked to decoys (i.e., the impact on malicious users), and estimate the impact of honey-patching on the overall system performance (i.e., its impact on legitimate users). To obtain baseline measurements that are independent of networking overhead, the experiments in this section are executed locally on a single host using default Apache settings. Performance is measured in terms of HTTP request round-trip time.

---

[3]Araujo et al. [5] present a more systematic study of honey-patchable patches for all official security patches released for the Apache web server from 2005 to 2013. Overall, the analysis shows that roughly 65 % of the patches analyzed are easily transformable into honey-patches.

Experimental Setup The target server was honey-patched against Shellshock and hosted a CGI shell script deployed atop Apache for processing user authentication in a web application created for this evaluation. All experiments were performed on a quad-core virtual machine (VM) with 8 GB RAM running 64-bit Ubuntu 14.04 (Trusty Tahr). Each LXC container running inside the VM was created using the official LXC Ubuntu template. We limited resource utilization on decoys so that a successful attack does not starve the host VM. The host machine is an Intel Xeon E5645 desktop running 64-bit Windows 7.

Session Forking Overhead To evaluate the performance impact on attackers, benchmark results are reported for three honey-patched Apache deployments: (1) a baseline instance without memory redaction, (2) brute-force memory sweep redaction, and (3) our PC²S redactor. Apache's server benchmarking tool (*ab*) is used to launch 500 malicious HTTP requests against each setup, each configured with a pool of 25 decoys. Figure 13 shows request round-trip times for each deployment. PC²S redaction is about 1.6× faster than brute-force memory sweep redaction [5]; the former's request times average 0.196 s, while the latter's average 0.308 s. This significant reduction in cloning delay considerably improves the technique's deceptiveness, making it more transparent to attackers. To mask residual timing differences, the reverse proxy in a honey-patching architecture artificially delays all the non-forking responses to legitimate requests so that their round-trip times match those of the malicious requests that trigger the honey-patch.

Overall System Overhead To complete the evaluation, REDHERRING was also tested on a wide variety of workload profiles consisting of both legitimate users and attacker sessions on a single node. In this experiment, users and attackers trigger legitimate and malicious HTTP requests, respectively. The request payload size is 2.4 KB, based on the median of KB per request measured by Google web metrics [17]. To simulate different usage profiles, the system is tested with 25–150 concurrent users, with 0–20 attackers. Figure 14a plots the results. Observe that for the various profiles analyzed, the HTTP request round-trip times remain approximately constant (ranging between 1.7 and 2.5 ms) when increasing the number of concurrent malicious requests. This confirms that adding honey-patching

Fig. 13 Request round-trip times for attacker session forking on honey-patched Apache

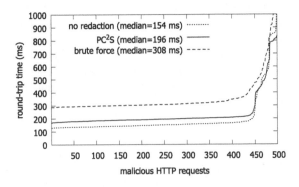

a

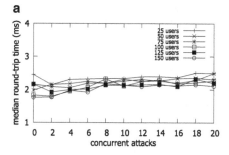

b

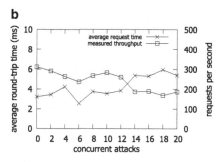

Fig. 14 Performance benchmarks. (a) Effect of concurrent attacks on legitimate HTTP request round-trip time on a single-node VM. (b) Stress test illustrating request throughput for a three-node, load-balanced REDHERRING setup (workload ≈ 5K requests)

capabilities has negligible performance impact on legitimate requests and users relative to traditional patches, even during concurrent attacks. This also shows that REDHERRING can cope with large workloads. This experiment assesses its baseline performance considering only one instance of the target server running on a single node virtual machine. In a real setting, several similar instances can be deployed using a web farm scheme to scale up to thousands of users, as presented next.

Stress Testing In this experiment, *ab* (Apache HTTP server benchmarking tool) is used to create a massive workload of legitimate users (more than 5000 requests in 10 threads) for different attack profiles (0–20 concurrent attacks) against a three-node load-balanced setup of REDHERRING. Each VM is configured with a 2 GB RAM and one quad-core processor. The load balancer and the benchmark tool run on a separate VM on the same host machine. Apache runs with default settings (i.e., no fine tuning has been performed). As Fig. 14b illustrates, the system can handle the strenuous workload imposed by the test suite. The average request time for legitimate users ranged from 2.5 to 5.9 ms, with measured throughput ranging from 169 to 312 requests per second. In typical production settings this delay is amortized by the network latency (usually on the order of several tens of milliseconds). This result is important because it demonstrates that honey-patching can be realized for large-scale, performance critical software applications with minimal overheads for legitimate users.

5   Is Honey-Patching *Security Through Obscurity*?

"Security through obscurity" (cf., [24]) has become a byword for security practices that rely upon an adversary's ignorance of the system design rather than any fundamental principle of security. History has demonstrated that such practices offer very weak security at best, and are dangerously misleading at worst, potentially offering an illusion of security that may encourage poor decision-making [1].

Security defenses based on deception potentially run the risk of falling into the "security through obscurity" trap. If the defense's deceptiveness hinges on attacker ignorance of the system design—details that defenders should conservatively assume will eventually become known by any suitably persistent threat actor—then any security offered by the defense might be illusory and therefore untrustworthy. It is therefore important to carefully examine the underlying basis upon which embedded honeypotting can be viewed as a security-enhancing technology.

Like all deception strategies, the effectiveness of honeypotting relies upon withholding certain secrets from adversaries (e.g., which software vulnerabilities have been honey-patched). But secret-keeping does not in itself disqualify honeypotting as obscurity-reliant. For example, modern cryptography is frequently championed as a hallmark of anti-obscurity defense despite its foundational assumption that adversaries lack knowledge of private keys, because disclosing the complete implementation details of crypto algorithms does not aid attackers in breaking cyphertexts derived from undisclosed keys.

Juels [18] defines *indistinguishability* and *secrecy* as two properties required for successful deployment of honey systems. These properties are formalized as follows:

Consider a simple system in which $S = \{s_1, \ldots, s_n\}$ denotes a set of n objects of which one, $s^* = s_j$, for $j \in \{1, \ldots, n\}$ is the true object, while the other $n - 1$ are honey objects. The two properties then are:

Indistinguishability: To deceive an attacker, honey objects must be hard to distinguish from real objects. They should, in other words, be drawn from a probability distribution over possible objects similar to that of real objects.

Secrecy: In a system with honey objects, j is a secret. Honey objects can, of course, only deceive an attacker that doesn't know j, so j cannot reside alongside S. Kerckhoffs' principle therefore comes into play: the security of the system must reside in the secret, i.e., the distinction between honey objects and real ones, not in the mere fact of using honey objects.

Embedded honeypotting as a paradigm satisfies both these properties by design:

Indistinguishability derives from the inability of an attacker to determine whether an apparently successful attack is the result of exploiting an unpatched vulnerability or a honey-patch masquerading as an unpatched vulnerability. While absolute, universal indistinguishability is probably impossible to achieve, many forms of distinguishability can nevertheless be made arbitrarily difficult to discern. For example, honey-servers can exhibit response delay distributions that mimic those of unpatched servers to arbitrary degrees of precision (e.g., by artificially delaying legitimate, non-forking requests to match the distribution of malicious, forking requests, as described in Sect. 4).

Secrecy implies that the set of honey-patched vulnerabilities should be secret. However, full attacker knowledge of the design and implementation details of honey-patching does not disclose *which* vulnerabilities a defender has identified and honey-patched. Adapting Kerckhoffs' principle [20] for deception, a honey-patch is not detectable even if everything about the system, except the honey-patch, is public knowledge.

This argues that embedded honeypotting as a paradigm (and language-based software cyber deception in general) does not derive its security value from obscurity. Rather, its deceptions are based on well-defined secrets—specifically, the set of honey-patched vulnerabilities in target applications. Maintaining this confidentiality distinction between the publicness of honeypot design and implementation details, versus the secrecy of exactly which vulnerabilities instantiate those details, is important for crafting robust, effective deceptions.

6 Conclusion

This chapter introduced and formulated the concept of *embedded honeypots* as a language-level approach for arming production software with deceptive capabilities that mislead adversaries into wasting time and resources on phantom vulnerabilities and embedded decoy file systems. Embedded honeypots employ *honey-patches* to conceal from attackers the information of which software security vulnerabilities are patched, thereby degrading attackers' methods and disrupting their reconnaissance efforts.

To realize efficient, precise honey-patching of production web servers, a new statically-instrumented, dynamic taint analysis built upon the LLVM compiler infrastructure is highlighted. The implementation significantly improves the feasibility of dynamic taint-tracking for low-level legacy code that stores secrets in graph data structures. To ease the programmer's annotation burden and avoid taint explosions suffered by prior approaches, it introduces a novel pointer-combine semantics that resists taint over-propagation through graph edges. Deceptive servers self-redact their address spaces in response to intrusions, affording defenders a new tool for attacker monitoring and disinformation.

Embedded honeypots differ from traditional honeypots in that they reside within the actual, mission-critical software systems that attackers are seeking to penetrate, and not as independent decoy systems. Thus, embedded honeypots offer advanced deceptive remediations against informed adversaries who can identify and avoid traditional honeypots. In order to be adoptable, embedded honeypots imbue production server software with deceptive capabilities without impairing its performance or intended functionality. These new capabilities make cyber attacks significantly more costly and risky for their perpetrators, and give defenders more time and opportunity to detect and thwart incoming attacks.

References

1. ANDERSON, R. Why information security is hard – an economic perspective. In *Proceedings of the 17th Annual Computer Security Applications Conference (ACSAC)* (2001), pp. 358–365.
2. ANSEL, J., ARYA, K., AND COOPERMAN, G. DMTCP: Transparent checkpointing for cluster computations and the desktop. In *Proceedings of the 23rd IEEE International Parallel and Distributed Processing Symposium (IPDPS)* (2009), pp. 1–12.
3. APACHE. Apache HTTP server project. http://httpd.apache.org, 2014.
4. ARAUJO, F., AND HAMLEN, K. W. Compiler-instrumented, dynamic secret-redaction of legacy processes for attacker deception. In *Proceedings of the 24th USENIX Security Symposium* (2015).
5. ARAUJO, F., HAMLEN, K. W., BIEDERMANN, S., AND KATZENBEISSER, S. From patches to honey-patches: Lightweight attacker misdirection, deception, and disinformation. In *Proceedings of the 21st ACM Conference on Computer and Communications Security (CCS)* (2014), pp. 942–953.
6. ARAUJO, F., SHAPOURI, M., PANDEY, S., AND HAMLEN, K. Experiences with honey-patching in active cyber security education. In *Proceedings of the 8th Workshop on Cyber Security Experimentation and Test (CSET)* (2015).
7. BRINGER, M. L., CHELMECKI, C. A., AND FUJINOKI, H. A survey: Recent advances and future trends in honeypot research. *International Journal of Computer Network and Information Security 4*, 10 (2012).
8. CLANG. clang.llvm.org. http://clang.llvm.org.
9. CLARK, C., FRASER, K., HAND, S., HANSEN, J. G., JUL, E., LIMPACH, C., PRATT, I., AND WARFIELD, A. Live migration of virtual machines. In *Proceedings of the 2nd Symposium on Networked Systems Design & Implementation (NSDI)* (2005), vol. 2, pp. 273–286.
10. CORBET, J. TCP Connection Repair. http://lwn.net/Articles/495304, 2012.
11. CRIU. Checkpoint/Restore In Userspace. http://criu.org, 2014.
12. DALTON, M., KANNAN, H., AND KOZYRAKIS, C. Tainting is not pointless. *ACM/SIGOPS Operating Systems Review (OSR) 44*, 2 (2010), 88–92.
13. DFSAN. Clang DataFlowSanitizer. http://clang.llvm.org/docs/DataFlowSanitizer.html.
14. DUELL, J. The design and implementation of Berkeley Lab's Linux checkpoint/restart. Tech. Rep. LBNL-54941, U. California at Berkeley, 2002.
15. GEROFI, B., FUJITA, H., AND ISHIKAWA, Y. An efficient process live migration mechanism for load balanced distributed virtual environments. In *Proceedings of the IEEE International Conference on Cluster Computing (CLUSTER)* (2010), pp. 197–206.
16. GOOGLE. Protocol Buffers. https://code.google.com/p/protobuf, 2014.
17. GOOGLE. Web metrics. https://developers.google.com/speed/articles/web-metrics, 2014.
18. JUELS, A. A bodyguard of lies: the use of honey objects in information security. In *Proceedings of the 19th ACM Symposium on Access Control Models and Technologies* (2014), ACM, pp. 1–4.
19. KANG, M. G., MCCAMANT, S., POOSANKAM, P., AND SONG, D. DTA++: Dynamic taint analysis with targeted control-flow propagation. In *Proceedings of the 18th Annual Network & Distributed System Security Symposium (NDSS)* (2011).
20. KERCKHOFFS, A. La cryptographie militaire. *Journal Sciences Militaires IX* (1883), 5–38.
21. LATTNER, C., AND ADVE, V. S. LLVM: A compilation framework for lifelong program analysis & transformation. In *Proceedings of the 2nd IEEE/ACM International Symposium on Code Generation and Optimization: Feedback-directed and Runtime Optimization (CGO)* (2004), pp. 75–88.
22. LIGHTTPD. Lighttpd server project. http://www.lighttpd.net, 2014.
23. LXC. Linux containers. http://linuxcontainers.org, 2014.
24. MERKOW, M. S., AND BREITHAUPT, J. *Information Security: Principles and Practices*. Pearson Education, 2014.

25. MILOJIČIĆ, D. S., DOUGLIS, F., PAINDAVEINE, Y., WHEELER, R., AND ZHOU, S. Process migration. *ACM Computing Surveys 32*, 3 (2000), 241–299.

26. NETCRAFT. Are there really lots of vulnerable Apache web servers? http://news.netcraft.com/archives/2014/02/07, 2014.

27. NGINX. Nginx server project. http://nginx.org, 2014.

28. NIST. The Shellshock Bash Vulnerability. https://web.nvd.nist.gov/view/vuln/detail?vulnId=CVE-2014-6271, Sep. 2014.

29. OHLOH. Apache HTTP server statistics. http://www.ohloh.net/p/apache, 2014.

30. PAI, V. S., DRUSCHEL, P., AND ZWAENEPOEL, W. Flash: An efficient and portable web server. In *Proceedings of the USENIX Annual Technical Conference (ATEC)* (1999), pp. 15–15.

31. PINGREE, L. Emerging Technology Analysis: Deception Techniques and Technologies Create Security Technology Business Opportunities. *Gartner, Inc.* (July 2015). ID:G00278434.

32. PROVOS, N. A virtual honeypot framework. In *Proceedings of the 13th USENIX Security Symposium* (2004), vol. 173.

33. SCHWARTZ, E. J., AVGERINOS, T., AND BRUMLEY, D. All you ever wanted to know about dynamic taint analysis and forward symbolic execution (but might have been afraid to ask). In *Proceedings of the 31st IEEE Symposium on Security & Privacy (S&P)* (2010), pp. 317–331.

34. SLOWINSKA, A., AND BOS, H. Pointless tainting?: Evaluating the practicality of pointer tainting. In *Proceedings of the 4th ACM SIGOPS/EuroSys European Conference on Computer Systems (EuroSys)* (2009), pp. 61–74.

35. SPITZNER, L. *Honeypots: Tracking Hackers*. Addison-Wesley Longman, 2002.

36. SPITZNER, L. The honeynet project: Trapping the hackers. *IEEE Security & Privacy, 2* (2003), 15–23.

37. THONNARD, O., AND DACIER, M. A framework for attack patterns' discovery in honeynet data. *Digital Investigation 5, Supplement* (2008), S128 – S139. The Proceedings of the 8th Annual DFRWS Conference.

38. VORIS, J., JERMYN, J., BOGGS, N., AND STOLFO, S. Fox in the trap: thwarting masqueraders via automated decoy document deployment. In *Proceedings of the 8th ACM European Workshop on System Security* (2015), p. 3.

39. WANG, C., MUELLER, F., ENGELMANN, C., AND SCOTT, S. L. Proactive process-level live migration in HPC environments. In *Proceedings of the ACM/IEEE Conference on Supercomputing* (2008).

40. WHITAKER, A., COX, R. S., SHAW, M., AND GRIBBLE, S. D. Constructing services with interposable virtual hardware. In *Proceedings of the 1st Symposium on Networked Systems Design and Implementation (NSDI)* (2004), pp. 169–182.

41. YUILL, J., ZAPPE, M., DENNING, D., AND FEER, F. Honeyfiles: Deceptive files for intrusion detection. In *Proceedings of the 5th IEEE International Workshop on Information Assurance* (2004), pp. 116–122.

Agile Virtual Infrastructure for Cyber Deception Against Stealthy DDoS Attacks

Ehab Al-Shaer and Syed Fida Gillani

Abstract DDoS attacks have been a persistent threat to network availability for many years. Most of the existing mitigation techniques attempt to protect against DDoS by filtering out attack traffic. However, as critical network resources are usually static, adversaries are able to bypass filtering by sending stealthy low traffic from large number of bots that mimic benign traffic behavior. Sophisticated stealthy attacks on critical links can cause a devastating effect such as partitioning domains and networks. Our proposed approach, called MoveNet, defend against DDoS attacks by proactively and reactively changing the footprint of critical resources in an unpredictable fashion to deceive attacker's knowledge about critical network resources. MoveNet employs virtual networks (VNs) to offer constant, dynamic and threat-aware reallocation of critical network resources (VN migration). Our approach has two components: (1) a correct-by-construction VN migration planning that significantly increases the uncertainty about critical links of multiple VNs while preserving the VN properties, and (2) an efficient VN migration mechanism that identifies the appropriate configuration sequence to enable node migration while maintaining the network integrity (e.g., avoiding session disconnection). We formulate and implement this framework using Satisfiability Modulo Theory (SMT) logic. We also demonstrate the effectiveness of our implemented framework on both PlanetLab and Mininet-based experimentations.

1 Introduction

Network assurance is essential for maintaining service availability and quality of service against increasingly sophisticated cyber threats, such as DDoS [1]. The indispensable nature of these services for today's world make them mission critical

This research was supported in part by National Science Foundation under Grants No. CNS-1320662 and CNS-1319490. Any opinions, findings, conclusions or recommendations stated in this material are those of the authors and do not necessarily reflect the views of the funding sources.

E. Al-Shaer (✉) • S.F. Gillani
University of North Carolina, Charlotte, NC, USA
e-mail: ealshaer@uncc.edu; sgillan4@uncc.edu

© Springer International Publishing Switzerland 2016
S. Jajodia et al. (eds.), *Cyber Deception*, DOI 10.1007/978-3-319-32699-3_10

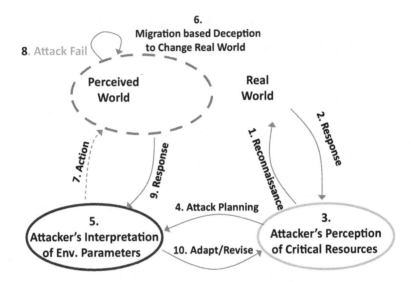

Fig. 1 Deception logic

and any disruption to them could be catastrophic. A DDoS attack can be either direct or indirect. In direct DDoS, the attack traffic is sent directly to the victim destinations to flood the last mile link. In indirect DDoS, the attack traffic is sent to the geographical neighbors of the victim destinations to flood critical links in the network shared between the neighbors and the victims. We concern ourselves with indirect DDoS attacks. Existing DDoS mitigation techniques [2, 3] attempt to defend by filtering out attack traffic. However, as the attacks are going more stealthy (like sending low traffic per bot to mimic normal user) [1], distinguishing benign from attack is not feasible in most cases. Nonetheless, a common network based DDoS attacks is to identify critical links which usually represent a small set of links carrying the most traffic [1, 4]). The attacker performs reconnaissance to identify such critical links and then plan the DDoS attack. Meanwhile, Internet ossification ensures the integrity of the reconnaissance knowledge by keeping the critical status-quo of the real world same as it is. However, If the critical resources are rendered non-critical by changing the real-world then attacker would go after the perceived critical resources, as illustrated in Fig. 1. This will cause the DDoS attack to miss the critical resources and will not be able to inflict devastation on the network.

Virtual networks (VNs) [5–10] provide the flexibility within a communication infrastructure. When VNs are deployed, physical (substrate) resources can be dynamically allocated to a service (VN placement) and if a physical resource is rendered unavailable, due to fault or attack, it can be replaced with a different resource to ensure service availability (VN migration). We call the capability to perform VN migration in an orchestrated fashion to deceive or evade attackers as *VN agility*. Even though virtualization has now been offered as a commercial service through companies posing central authority [11, 12], the existing VN placement techniques [6–10] do not provide for VN agility because (1) they are simply static (no migration support), and (2) the existing placement control mechanisms are not attack-aware. Apart from providing one time resource assignment for the VN, no technique considers migration as a cyber deception based defense to evade reconnaissance or DDoS attacks. Therefore, this static status quo of VN placement still enables adversaries to discover and plan devastating DDoS attacks. In this work, we propose an agile VN framework, called *MoveNet*, that proactively and actively defend against sophisticated DDoS attacks without requiring the distinguishing of attack traffic from benign traffic. We achieve this by actively reassigning the VN to new physical resources to deceive attackers without disrupting the service or violating the VN properties.

Persistent cyber threats demand continuous VN movement which requires a provably correct-by-construction VN placement technique to ensure intact service functionality at all times. We formalize such VN placement technique as a constraint satisfaction problem using Satisfiability Modulo Theory (SMT) [13] based formal methods. We adopt the same VN placement requirements used in the literature [6–10] and define them formally as constraints in the model.

Proactive defense requires identifying critical resources that can be replaced with non-critical resources to deceive attackers. Our VN framework is generic for all link-based DDoS and reconnaissance attacks. But to show the effectiveness of our approach, we use, as a case study, Crossfire attack [1] which is the most devastating and stealthy attack to-date. We call this proactive defense MoveNet Agility. Proactive defense, nonetheless effective, may be costly for some applications. Therefore, *MoveNet* also offers reactive deception called MoveNet Sensing that actively investigates the reconnaissance attempts in the network and uses this information to replace the target critical resources with non-critical resources to deceive the adversary.

The purpose of VN agility is to frequently replace potentially discovered critical substrate with *valid* new threat safe resources such that (1) the selection process of the new resources is unpredictable by the adversary, and (2) the migration to these new resources will be faster than the reconnaissance time of the adversary. This, therefore, will constantly deceive the adversaries by invalidating their assumptions about the VN critical resources placement, discovered in the reconnaissance phase. All these constraints are formally defined as a part of the agile VN framework model using SMT [13]. Advanced SMT solvers such as Z3 [14] and Yices [15] can solve tens of thousands of constraints and millions of variables [13]. Therefore, our approach is scalable to large networks with multiple VNs.

We also develop a migration mechanism that executes VN migration strategy generated by the framework on a virtualized physical network. It handles all the logistics of the movement including the exact sequence/schedule of the steps to be executed in order to complete the move and the timing of such steps. We use Planet-Lab [16] as our virtualized infrastructure and extend the functionality of an existing PlanetLab based controller, PL-VNM [17], to implement our migration mechanism. We use this PlanetLab implementation to rigorously evaluate the effectiveness of this framework. We also use simulation and Mininet-based experiments to evaluate other features (like scalability) which were not possible on Planetlab.

Our proposed solution only requires provision in the network to have redundant paths that can be used to switch critical role to non-critical. Furthermore, the proposed framework model can be extended by adding more functional, security or threat related constraints.

In the rest chapter, we explain related work in Sect. 2 followed by our approach in Sect. 3. Threat modeling is explained in Sect. 4 followed by agile migration in Sect. 6. The migration mechanism and evaluation are presented in Sects. 7 and 8, respectively.

2 Related Work

The existing DDoS remedies, all requiring the ability to distinguish attack traffic from benign, can be divided into two major categories, infrastructure based and virtual network based approaches. Some infrastructure based approaches proposed adding authentication bits in packet headers to assist routers drop attack traffic [18–20]. Others, suggested routers to signal each other through a signaling protocol to filter DDoS traffic [21–23]. The major limitation of these approaches is to expect major changes in existing hardware and network protocols, which is hard to come by. The VN based approaches [2, 24–28] suggest deploying a secret virtual layer on top of the substrate to route traffic only through designated routers. For example Akamai's SiteShield service [3], SOS [2, 28] and Mayday [24] use a shared secret (e.g., IP of the target) to route traffic through specialized routers. The major shortcoming with such approach is that if the secret (virtual layer) is revealed then the adversary can bypass the virtual network to directly flood the target. Some countermeasure approaches [2, 24] were also proposed preventing malicious overlay components to disclose shared secret, but they require expensive mechanisms such as anonymous routing. However, none of these approaches can counter the Crossfire attack [1] because this attack does not directly attack the target.

For VN agility, the work done in [7] focuses on one time VN placement but allows a limited moving of ensemble of VNs deployed on the substrate under resource contention. The work in [29] considers VN link reconfiguration in response to changes in traffic demands. That work does not allow moving the placement of the virtual components. Recent work [30] has developed and implemented a scheme for moving a single virtual router in a virtualized separated router environment. This work has recently been used in [31] to design entire network migration mechanisms. A recent effort [32] has also considered this VN migration with software defined networks (SDN). All these recent work deals with low level migration configuration functions with no investigation of strategy and higher-level mechanism questions.

An alternative solution would be to dilate (spread) the critical footprint by dividing the traffic equally across all available routes. Major shortcoming of such approach will be a significant increase in the power consumption [33] and network management cost of the network. Also, it would require changing routing protocols as they automatically converges to a power-law (shortest path) based path logic after a failure.

3 Agile VN Framework

The objective of agile VN framework is to develop a migration strategy that determines *what* resources should be reassigned and, *where* and *when* migration should take place to defy hostile environment. In the following of the paper, we explain all the modules of agile VN framework, as in Fig. 2.

3.1 Modeling VN Placement

The VN placement is all about finding the best combination of substrate (physical) resources satisfying all the requirements of one or more services, where each service will have its own VN. These requirements are enforced as constraints of the VN placement model, explained in the following. Let the substrate network be a directed graph, $G = <C, L>$, where C is the set of all components (any network device) and L is the set of all links. Similarly, let υ be a VN placed for a service. By definition, VN placement is about connecting a source to a destination through a network path, that satisfies all constraints. This path can further be restricted to pass through one of more designated components (e.g., components capable of hosting virtual routers). But in environments such as SDN, all components are considered the same. Our model is applicable to both scenarios. A source can be an ingress router, some aggregating node (like proxy node) or some corporate office [11] in the network. Similarly, a destination can be an egress router, some aggregating node (like proxy) or the actual destination (e.g., web server or data center). And, C_{src} is the set of all source components in the network and C_{dst} is the set of all destination components. Both components represent a small fraction of the network size. Then, the VN placement is about finding optimal paths between these source-destination pairs. Where u_i represents such a source-destination pair, $u_i = <c_j, c_k>$ and $u_i \in U = \{C_{src} \times C_{dst}\}$. The shortest distance between any node c_i to a destination

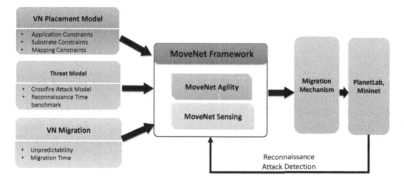

Fig. 2 Abstract MoveNet framework diagram

node c_j is represented as d_{ij}, where, $c_j \in C_{dst}$. And, b_{ik} is a Boolean variable, which is true if component c_k is carrying the traffic for a source-destination pair u_i. I_{c_j} is the set of neighboring components of c_j and κ_j is the traffic capacity of a node c_j. Deploying a VN for a service requires the estimated traffic that needs to be handled by the network, e.g., between a corporate office to a data center and λ_i is the estimated traffic load of a pair u_i. All thresholds in our model are represented as $T^{(\cdot)}$.

The following placement model assigns substrate resources to multiple VNs. Let A be the set of all VNs, i.e., $A = \{v_1, v_2, \ldots, v_m\}$. Then, the binary assignment variable b_{ij} will be updated as, $b_{ij}^{v_l}$, which means, node c_j is assigned to pair u_i of VN v_l. For the simplification of notations and understanding, we explain the VN placement modeling for one VN in the following section.

From the existing literature [6–10] we are using following requirements as constraints of our model.

3.1.1 Reachability Constraints

According to this constraint, data generated from a source must reach the destination. Assuming that each node will forward the traffic, this constraint is about connecting components so that from source to destination there is a connected path. This constraint is formally defined as: $b_{ij} \rightarrow \bigvee_{\exists c_k \in I_{c_j}} (b_{ij} \wedge b_{ik})$ which says, if a node is selected i.e., b_{ij} is true, then it must be connected to one of its neighboring components. In other words, a node receiving traffic must forward this to one of its neighboring component. However, forwarding a packet to one of the neighboring components does not guarantee packet delivery to the destination. So, in this constraint we need to add a sense of direction from source to destination. This sense can be added by using the distance from the current node to the destination node. The updated reachability constraint is defined in the following as: $b_{ij} \rightarrow \bigvee_{\exists c_k \in I_{c_j}} ((b_{ij} \wedge b_{ik}) \wedge (d_{kl} < d_{jl}))$ where, d_{jl} is the distance from node c_j to destination node c_l of the pair $u_i = < ., c_l >$. Now, what if the next node is itself the destination node, then traffic must not be forwarded to any other node. This completes the reachability constraint and the final version of this constraint is as follows:

$$b_{ij} \rightarrow (c_j \in C_{dst}) \vee \bigvee_{\exists c_k \in I_{c_j}} ((b_{ij} \wedge b_{ik}) \wedge (d_{kl} < d_{jl})) \tag{1}$$

Although, we assume mostly stable routing because according to [1] 72 % of links tend to be stable. However, our model can be incrementally updated to recalculate placement in case of topology changes.

In a commercial products like Virtela [11] and Aryaka [12], they expect to estimate this traffic as well.

3.1.2 Load Satisfaction Constraint

According to this constraint, every node along the path must be able to carry traffic load of the pair assigned to it. In practical, this decision must be made by comparing traffic load with leftover capacity of the node. If the aggregate traffic supported by node c_j is σ_j, which is calculated in the Sect. 3.1.5, then formal definition of this constraint is as followed:

$$\forall u_i \in U, c_j \in C, b_{ij} \rightarrow (\kappa_j - \sigma_j) \geq \lambda_i \tag{2}$$

3.1.3 Middle-ware Device Constraint

An application might require some specialized service from the substrate (e.g., the IPSec, etc.). Mostly, only a handful of components can provide such requested feature. This constraint ensures that the VN traffic must go through such specialized node. Let $C_{sp} \subset C$ be a set of such specialized components. Then, for a pair u_i we enforce this constraint by explicitly setting the node assignment variable b_{ij} to true, where, $c_j \in C_{sp}$. The formal definition is as followed:

$$\exists c_j \in C_{sp}, b_{ij} \tag{3}$$

The VN placement model, as a whole, will automatically connect this component with other components to satisfy all constraints.

3.1.4 Quality of Service Constraint

For simplicity, we assume relatively homogeneous network where hop count can be proportional to network latency. This assumption has been widely used in existing literature [34].

$$\forall u_i \in U, (\sum_{c_j \in C} b_{ij}) \leq T^{qs} \tag{4}$$

where, T^{qs} is the affordable delay threshold for any pair in the system. It can be the same for the entire system or different for each VN. In addition, our framework allows for more complex constraints to define bounded delays.

3.1.5 Node Stress Constraint

According to this constraint, each node must have some leftover capacity to accommodate the traffic fluctuations due to load dynamics or uncertainty. Therefore, the leftover capacity can ensure to handle such incorrect traffic load estimations. This is formally defined as:

$$\forall c_j \in C, \{(\kappa_j - (\sum_{u_i \in U} b_{ij} * \lambda_i)) \geq T^{lo}\} \tag{5}$$

where, $\sum_{u_i \in U} b_{ij} * \lambda_i$ is the aggregate traffic σ_j at node c_j and T^{lo}, is the leftover capacity threshold set for each component. This can be provided as a percentage of the capacity of a node.

3.1.6 Pair Mapping Constraint

According to this constraint, each pair must be assigned some substrate resources. This is formally defined as:

$$\forall u_i \in U, \sum_{\forall c_j \in C} b_{ij} > 1 \tag{6}$$

3.1.7 Loop Avoidance Constraint

According to this constraint, a substrate component cannot be assigned to a pair twice because this would create a loop within the path. Our model enforces this by checking how many links are enabled for each component? In the model, a link is represented logically as $b_{ij} \wedge b_{ik}$, i.e., both components c_j and c_k are assigned to u_i and they are connected. If the number of links are more than 2, this means there is a loop. E.g., $c_j \wedge c_k$ and $c_k \wedge c_i$, if $c_i \wedge c_j$, this makes a loop that is not allowed. It is formally defined as follow:

$$\forall u_i \in U, c_j \in C, (\sum_{\forall c_k \in I_{c_j}} (b_{ij} \wedge b_{ik} \rightarrow 1)) \leq 2 \tag{7}$$

In this equation the condition is not modeled as exactly equal to 2 but less than or equal to 2. It is because, the source and destination components will have only one link.

4 Modeling Threat

The planning phase of all link-level DDoS attacks is to perform network reconnaissance to identify critical components. For proactive defense, we have to model this reconnaissance attack and incorporate it as a module of the agile VN framework. This module proactively identifies potential targets so that these can be replaced with threat safe components to evade attacks. Our framework can handle general classes of link-level DDOS and reconnaissance attacks. It is general enough to consider new attack by only updating threat model to reflect new logic of identifying critical components without changing any other parts of the framework in Fig. 2. However, in this paper we are using the most devastating and stealthy DDoS attack to-date, i.e., Crossfire [1] as a case study threat model.

In the Internet, data connectivity follows power law distribution [4], which means a small set of links carry most of the traffic. The Crossfire attack identifies such links that carry most of the traffic to the victim. Then, it selects a list of surrounding public servers (decoy servers) that share these links with the target. The attack uses botnets to send enough traffic to these selected decoy servers that it throttles the shared links, thus causing denial of service to the victim, indirectly. The static critical status quo of these links enables the adversary to launch such attack.

Under the hood of a Crossfire attack, bots send traceroute probes to all victim machines and decoyer servers. As opposed to manually searching decoy servers in the original attack, we use the geographical coordinates of the victim and select initial set of decoy servers who are within certain miles from the victim. We use geographical database *GeoLite* [35] to get all such coordinates and use geographical distance formula [36] to calculate distance. The final set of decoy servers is selected after observing the sharing of infrastructure between victim and decoy servers (Fig. 3). Let the set of bots be O, decoy servers be Y and victim machines be V. Each traceroute probe provides a data path, and P is the set of all paths from bots to decoy servers and victim machines. And, P^Y and P^V are the set of paths from bots to decoy servers and victim machines, respectively, i.e., $P = \{P^Y \bigcup P^V\}$. A path is simply a set of components, i.e., $p = \{c_1, c_2, \ldots, c_n\}$, where, $c_n \in C$ is from Sect. 3.1. All those paths that do not share anything between victim and decoy servers are removed from the set, i.e., $P' = \{\neg \exists p_i \in P^Y, p_j \in P^V, p_i \cap p_j = \emptyset\}$.

For a packet to be delivered successfully from origin to destination, every component along the path must forward the packet. Let $F(.)$ be a Boolean function, which is true if the component successfully forwards the packet and false otherwise. This path forwarding property can be expressed as a logical formula as, $F(p_i) = \bigwedge_{c_j \in p_i} F(c_j)$. However, from the adversary's perspective, she just has to compromise any single component along the path to violate the forwarding property of the

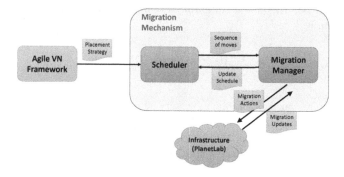

Fig. 3 Migration mechanism on PlanetLab

path. Let $B(.)$ be a Boolean function that is true if a component along the path is compromised and false otherwise. Now, the same path forwarding property can be flipped to formulate path breaking property, which says a path is broken if any component along the path is compromised, i.e., $B(p_i) = \bigvee_{c_j \in p_i} B(c_j)$. Due to limited resources of the adversary, compromising the entire network may not be a possibility. Therefore, the model must also be equipped with a choice to disrupt any path or not based on the budget of the adversary. This choice is added as a binary variable χ in the path breaking property, i.e., $B(p_i) = \bigvee_{c_j \in p_i} B(c_j) \rightarrow \chi_i$. The value of this choice variable, χ, will be one if the path can be broken or zero otherwise. This all depends upon the budget of the adversary. We can extend this path breaking property for the entire network to find a set of components, if broken, can cause the desired devastation in the network. Its formal definition is as follows:

$$B(P') = \bigwedge_{p_i \in P'} B(p_i) = \bigwedge_{p_i \in P'} \{ \bigvee_{c_j \in p_i} B(c_j) \rightarrow \chi_i \} \tag{8}$$

This formula provides a list of components, if attacked, can inflict the desired devastation. This is an NP complete problem and we formalize it as a SAT [13] problem. Normally, the adversary has limited budget (cost) in terms of attack traffic that can be generated to throttle components. Therefore, finding only those components that can be throttled with that budget is the prime focus of adversary. Let $|B(.)|$ be a function that provides number of true components in Eq. (8). If the adversary does not know the capacity of target components, then cost will represent the number of components that can be throttled assuming them to have same capacity. This constraint is encoded as follows:

$$|B(P')| \leq Cost \tag{9}$$

$$\sum_{\forall p_i \in P'} \chi_i \leq T^{dv} \tag{10}$$

where T^{dv} is the desired disruption of the adversary. If we assume that the adversary has the leftover capacity information of each component then Eq. (9) can be optimized. This last assumption is optional and we do not use this in our implementation.

$$\sum_{\forall c_j \in C} ((\kappa_j - \sigma_j) * B(c_j)) \leq Cost \tag{11}$$

During reconnaissance attack, the adversary cannot afford to send too much traffic to decoy servers or victim machines to avoid detection. Therefore, by simply adding more bots will not guarantee that the reconnaissance attack will complete in small time window. The experiments explained in Sect. 8.2.5 provide the reconnaissance time benchmark of this attack model.

4.1   Defender's View vs. Attacker's View

When an adversary runs a traceroute probe for a target machine, she views the same data plane that is deployed during the VN placement for the service running on that target machine. Similar observation is made in [1] that bots probing a target from the Internet would end up viewing the same critical footprint with respect to the target. Therefore, the threat model discussed in this section can be viewed as both defender's and attacker's perspective of potential targets. Because the decoy server information is public and available to both. Similarly, the adversary would rely on bots to decoy server relation and the defender would rely on source to decoy server relation to identify critical components. Bots and sources are logically the same just with different roles. In Sect. 8.2.6, we have demonstrated this fact using simulation to show that both see the same components as potential targets.

5   Modeling MoveNet Sensing

Although, proactive defense effectively render the reconnaissance knowledge stale for the attacker, however, continuous migration may not be an affordable solution in all circumstances. Therefore, MoveNet also offers reactive solution, called MoveNet Sensing, that identifies the critical network components by actively monitoring the reconnaissance attempts, if any, in the network. An overlay service provider can instruct its sources (ingress nodes, aggregating nodes) to analyze the traceroute based probing behavior and flag the presence of reconnaissance attack in the network. Furthermore, MoveNet controller can utilize these reconnaissance based traceroute probes to identify the critical targets likely to be attacked and then apply appropriate cyber deception approach. In the following, we explain these steps of MoveNet Sensing in detail.

5.1   Detecting Reconnaissance Attack

Crossfire attack is considered a stealthy DDoS attack because it uses thousands of bots and each bot is sending low-rate attack traffic mimicking benign users. Although, this attack traffic is able to bypass any behavioral based detection. However, as the adversary launches millions of traceroute probes for reconnaissance from thousands of bots to the victim destinations and decoy servers, this results into a unique probing behavior in the network. In reconnaissance phase, the adversary sends multiple traceroute probes from each bot to a destination (victim or decoy server). Similarly, each bot sends these probes to all destinations to find out the bots that can exploit the shared link between victim and decoy servers. This process develops two distinct behaviors, (1) thousands of probes will be seen originating from the same source in some period of time, and (2) each destination is probed by thousands of distinct end-users (bots) in some period of time.

One important thing to note here is that benign users also perform traceroute to some servers and similarly traceroute is also used as a diagnostic tool for network administrator. In the former case, a benign user will never probe a large number of destinations and would never launch a large number of probes. Similarly, in the letter case, administrators can add exceptions in the source nodes to not ignore the probes originated from trusted IPs.

Let R be the set of all traceroute probes, collected from all sources, where each probe $r_m \in R$ is originated from some source $o_j \in O$ and destined to a destination node $c_k \in C_{dst}$. As this probe will be received by a source component $c_l \in C_{src}$, we can always trace its path p_i for pair $u_i = <c_l, c_k>$. Moreover, R_{o_j} represents all probes originated from bot source o_j and R_{c_k} represents all probes destined to $c_k \in C_{dst}$. Similarly, $|R_{o_j}^t|$ and $|R_{c_k}^t|$ represents source probe frequency and destination probe frequency in t time window respectively. Currently, we use one hour as our detection time window but it can be varied depending upon the situation. In the following, we only use $|R_{o_j}|$ and $|R_{c_k}|$ to represent frequencies for simplicity.

From above, we can calculate the number of distinct bots probing a destination represent as O_{c_k} and $|O_{c_k}|$ represents its count. It can be calculated for a destination component c_k as:

$$O_{c_k} = \{\forall o_j \in O, R_{o_j} \cap R_{c_k} \neq \emptyset\} \tag{12}$$

We assume that there is always a benign period where probing exhibit benign behavior. During this time, network administrators can calculate the benign thresholds for above mentioned parameters. In our detection technique, we calculate these benign thresholds as:

- *Source Probing Frequency Threshold*: We calculate the minimum and maximum source probe frequencies and use their difference over multiple days to calculate the average normal probing difference represented as T^{diff} per hour.
- *Destination Probing Frequency Threshold*: We calculate the maximum times a destination is probed during an hour and calculate the average of this value over multiple days to be used as threshold, represented as T_{dest}.

If during any hour the current average exceeds twice the T^{diff} value, we can safely assume that probing behavior is not normal and it may be an indication of reconnaissance attack. Next threshold is used to identify which destinations are being probed by adversary. All bot end-users involved in this probing are considered as likely bots and their future probes are reported to the MoveNet controller.

We use these bot and traceroute information to identify critical network component and change their mapping to stale reconnaissance knowledge. Otherwise, we do not use this information to block or filter out any traffic therefore any false-positives will not result into any connection termination or failed communication.

5.2 Identifying Critical Targets

Each traceroute probe r_m results into a path p_i and multiple probes may result into the same path. For example, if bot users reside outside the ISP network then multiple bots probing same destination may end-up going through same ingress node (component) in the network and we only use path from ingress (source) component to the destination. Furthermore, traceroute probes may not use all the paths in the network therefore, we represent each path observed in a probe as evidences of reconnaissance attack. We represent set of all evidences as E where an evidence e_i represents path p_i but with a positive integer w_i representing how many times evidence e_i is observed. E_{c_j} represents all evidences that passes through the component c_j and $\beta_j = \sum_{\forall e_i \in E_{c_j}} w_i$ represents the aggregate probes went through the component c_j.

As mentioned in the Sect. 4, the adversary only consider those paths that are shared between victim and decoy servers to identify and overwhelm the minimum set of critical links (components) for the victim. We use a variant of Eq. (8) to identify such critical set of components as:

$$B(E) = \bigwedge_{e_i \in E} \{ \bigvee_{c_j \in e_i} B(c_j) \} \tag{13}$$

Where $B(E)$ provides a set of components (links) if flooded can cause complete break down for the victim traffic. However, adversary would be interested in only those links that cause significant damage for the victim, so we classify those links as critical using following equation.

$$Critical = \{ \forall c_j \in B(E), (B(c_j) * \beta_j) > T^{sig} \} \tag{14}$$

Where T^{sig} represents a percentage, e.g., 50 %, which means that a component (link) should be selected as critical if its unavailability can affect 50 % traffic going to the victim.

5.3 Cyber Deception Based Defense

The purpose of cyber deception is to deceive the adversary by ensuring that the reconnaissance knowledge gets stale when she gets to use this knowledge to build-up the attack. There are two possible ways to achieve this deception, (1) *path stability*, the crossfire attack requires paths to be stable before exploiting these in launching the DDoS attack. Because, some routers along the path employ load balancing that may cause the path to change regularly. Such paths are not good for ensuring a successful DDoS attack. Therefore, the adversary sends multiple probes from the same bot to same destination. If she does not observe same path in all probes then the bot to decoy-server combination is considered unsuitable. We can utilize this limitation for our benefit and ensures that the critical components do not stay along a path for longer period of time. (2) *reconnaissance knowledge*: The adversary builds the critical knowledge database during reconnaissance and then plan a stealthy attack by instructing each bot to send traffic to particular decoy servers. This returns into a gap between probing and launching actual DDoS attack. Once, reconnaissance attempt is detected then such gap can be considered as the preparation for DDoS attack which can trigger the change of critical physical footprint of the virtual network.

In the following section, we discuss the threat-aware migration to replace the critical components with threat-safe components for both `MoveNet Agility` and `MoveNet Sensing`.

6   Modeling Threat Aware Migration

The purpose of agility is to replace critical components with only new threat safe components. Following are the two constraints that ensure threat safe agility.

6.1   Migration Disturbance Constraint

Simply replacing critical components with randomly selected new distinct resources is not enough because the adversary can start building a dictionary of learned critical components of the network. So, if we only move to distinct components then next time the adversary will not consider previously learned critical components and this will reduce the sample space for the adversary to find new critical targets. Instead, we assign a random counter to a component once flagged as critical. This component cannot be reused in the next iterations of migration unless its counter is expired. This uniformly distributes the reuse probability of the components in the upcoming iterations, leaving adversary to keep guessing. Let R be the set of critical nodes and h_j is a Boolean variable. If it is true, this means component c_j was selected as a critical node. And, q_j is the index of the iteration in which the component c_j was selected as a critical component. And, β_j is the number of iterations (counter) for component c_j before it can be repeated. Now, the original assignment variable b_{ij} will be constrained to assign only allowed resources in the next placement. This constraint is formally defined in the following equation.

$$\forall u_i \in U, c_j \in C, b_{ij} \wedge ((q_j = 0) \vee ((t + 1) - q_j > \beta_j \rightarrow h_j)) \qquad (15)$$

Where, $t + 1$ represents the index of the next iteration.

After each iteration, the counter is decremented by one.

6.2 Migration Distance Constraint

Migration should complete before the adversary can finish reconnaissance and launch the attack. During migration [30] following steps take place, (1) setting up tunnels between current and new components, (2) copying control plane from old to new components, (3) populating data plane in the new component, and (4) establishing routes between the new components and the neighbors of the old components. In our implementation on PlanetLab, we use controller based architecture [17] like in Software Defined Networking (SDN). In such architecture, the relative positions of the current and new components do not matter. Instead, control plane resides at the controller, which populate the forwarding plane in the new components by sending control packets carrying forwarding rules. The controller already has established tunnels with each network component. So, the migration time will be calculated as: how much time will it take to send control packets from the controller to the new substrate components. Let α_j be the distance from the controller to a component c_j, τ be the time spent per hop and b_{ij}^t, where t represents the index of the iteration of migration. Now, we can define our migration distance constraint as:

$$[\sum_{\forall u_i \in U} \{ \sum_{\forall c_j, c_k \in C} ((b_{ij}^t \wedge b_{ik}^{t+1}) \wedge (j \neq k)) \rightarrow \tau * \alpha_k \}] < \omega \qquad (16)$$

where, b_{ij}^{t+1} represents the assignment of component c_j to pair u_i in the new placement. To quantify this equation, we use calculations from our PlanetLab experiments. In our PlanetLab experiments, we observed the average traceroute probe time to be almost 250 ms and the average path length observed is almost 10 hops. We use these measures to develop following heuristics. The traceroute time represents a round trip time, so, one way time will be 125 ms. This means it takes almost 13 ms per hop along the path. Assuming the controller can initiate populating forwarding tables at once for all nodes, then using a more conservative hop time i.e., $\tau = 20$ ms we can calculate the migration distance in Eq. (16). We assume that the size of forwarding tables is the same and bandwidth from controller to node will remain the same. Therefore, we do not need to put these as the part of the model as they are not going to change for different controller-node pairs.

7 Migration Mechanism

The migration mechanism executes VN migration strategy generated by the framework on a virtualized physical network. Given the destination for a particular VN to be moved, it handles all the logistics of the movement including the exact sequence/schedule of the steps to be executed in order to complete the move and the timing of such steps. It is also responsible for interacting with specific substrate and its virtualization technology to accomplish the move. You can find the details of this component in our Infocom paper [37]. Instead of developing the entire mechanism from scratch, we leverage an existing PlanetLab based controller, PL-VNM [17]. This controller offers the core migration functionality and implements single virtual node migration. An abstract level view of our extensions to PL-VNM are briefly described as follows.

7.1 Implementing VN Placement

We have implemented our VN placement module inside the controller.The controller can perform two key tasks, (1) given a substrate topology of PlanetLab, the controller finds the initial suitable placement, and (2) using the threat model it generates a migration strategy to initiate a threat aware migration.

7.2 Implementing Threat Model

We have implemented the Crossfire threat model as the part of the controller. For a complete threat simulation, we designate different PlanetLab nodes (allocated to us) as bots and decoy servers manually. As each node might be located in a different country, there is no way to automating the decoy server selection using geographical coordinates. The threat module instructs these bots to send traceroute probes to each decoy server and collects the path information to identify the critical footprint. Then, the bots are instructed to start sending attack traffic using already configured UDP based iperf.

7.3 Implementing Partial Migration

The controller was lacking the partial VN migration capability. In the existing implementation, all end users connect to specialized nodes that are called gateways. These gateways were responsible to switch from old substrate to new substrate. This method was used to migrate entire VN from one substrate to another. Our framework requires each node to act as a gateway, as any node can be flagged as critical. Once a node is identified as critical, all of its neighboring nodes must act as gateways to move traffic away from this critical node. We have extended the role of each node to act as a gateway in the controller.

Due to limited space, we skip the technical details of our controller extensions and implementation.

8   Implementation and Evaluation

We have used PlanetLab, simulation and Mininet-based experiments to perform a rigorous evaluation.

8.1   Experiment Setup Discussion

8.1.1   PlanetLab Based Experiment Setup

The experiment topology that we have developed within the PlanetLab is showed in Fig. 4. In our experiments, we classify the components into four categories: (1) fixed components, the components that do not change during migration, black ones in the figure; (2) Gateway components, that switch traffic between new and previous components, respectively. These are the yellow components c_6 and c_{17} in the figure; (3) Critical components, that are needed to be changed, these are c_9, c_8 and c_{10} in migration step 1, 2 and 3, respectively; (4) Origin and sink components in the network, e.g., c_1, c_{18} and y_1 are the source, the destination and the decoy server components, respectively.

The same color of bots, components and links in Fig. 4 represents that these were active during the same migration iteration. For example, during iteration 1, the component c_9 was selected as critical (only one carrying traffic to the target) and only bots o_1 through o_4 were activated in the attack because these are the only

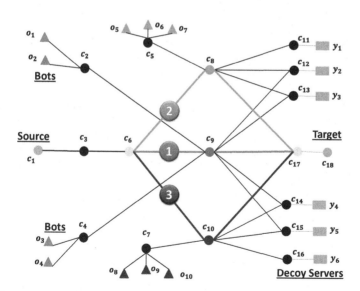

Fig. 4 Topology diagram of the experiments

bots sharing critical component c_9 with the destination. Similarly, the path from source c_1 to destination c_{18} was $\{c_3, c_6, c_9, c_{17}\}$ and as the links between c_6–c_9 and from c_9 to c_{17} were carrying the traffic during iteration 1, they all have the same color. Another pivotal concept observable in this example is that the changing of the critical footprint also changes the association between bots and decoy servers. This means that the adversary has to repeat reconnaissance to learn this new association again.

8.1.2 Simulation Based Experiment Setup

We have used BRITE [38] to generate topologies with random connectivity and power law based preferential connectivity. In each topology, a small fraction of nodes (5 %) with least degree are designated as sources and destinations. The shortest paths between these nodes simulate network data plane. First, VN placement module finds different placement, second, threat model finds critical footprint in each placement. We randomly select least degree nodes as bots and near by nodes to destination (based on hop distance) as decoy server for the threat model. We use Z3 theorem prover to formulate a constraints satisfaction model [14]. The Z3 is a state-of-the art theorem prover from Microsoft Research and it is used to check the satisfiability of logical formulas over one or more theories. We have used a machine with Core i7, 2.4 GHz processor and 8 GB RAM to run all the simulation based experiments.

8.1.3 Mininet-Based Experiment Setup

Mininet-based experiments are used to observe packet loss during the migration. We have used POX as a reference controller for Mininet. We create similar topology as in Fig. 4, but without any bots and decoy servers. We have extended the POX controller by adding a switching module for our implementation. We have used the iperf functionality built-in the Mininet to observe packet loss during switching.

8.2 Agile VN Framework Evaluation

8.2.1 Evaluating Depth of Defense

The depth of defense measures the over provisioning available in terms of alternate routes and placements available in the Internet. Alternate route provisioning is measured using PlanetLab experiments and alternate VN placements is calculated by simulation. In our simulation, we let the model calculate all possible placements for each network type and size, and results are showed in Fig. 8b. The results

Table 1 Depth of defense

East coast	West coast	Europe
4	5	4

demonstrate that in every network, we always have a decent number of alternate VN placements available, e.g., almost 100 for a network of size 700 nodes, and that increases with the increasing size of the network.

To calculate the depth of defense in terms of alternate routes, we selected three PlanetLab university sites, from East Coast of USA, West Coast of USA and Europe. Then, we select 100 PlanetLab nodes scattered through the world to send traceroute probes to these university sites. Each probing node sent exactly 6 probes (following same model as in [1]). After extracting the paths from these probes, we calculate alternate routes by simply identifying the minimum number of links that need to be broken to compromise all paths. The results showed in Table 1 demonstrate that there are always 4–5 alternate paths available to each destination. Now, as commercial overlay virtualization providers [11, 12] are in the picture, exhibiting central authority, this over provision will be more because in probing we do not necessarily view all links e.g., backup links.

8.2.2 Evaluating the Evasion Effectiveness

In evaluating evasion effectiveness experiment on topology in Fig. 4, we calculated link bandwidth between source (c_1) and destination (c_{18}) nodes using TCP based Iperf. It was observed sometimes as 2.0 Mbps and sometimes 1.6 Mbps due to bandwidth limit enforced by PlanetLab, results are available in [37]. At around 13th second, the Crossfire attack was launched from bots o_1 through o_4 that crippled the available bandwidth by limiting it to 300 Kbps within seconds. At time 27th, the agility module of the PL-VNM is activated to initiate evasion through migrating to a different path and that instantly restored the bandwidth back to 2.0 Mbps. We let this experiment run for sometimes to calculate the migration time in switching between nodes. The migration time stayed around 1–2 s which is way better than the reconnaissance time observed in Fig. 7b. During these experiments we kept the SSH channel open between the controller and the nodes.

8.2.3 Evaluating the Disruptiveness of Migration

We design an experiment on PlanetLab using the same topology from Fig. 4 to analyze the packet loss caused by the migration. Without any attack, we let the controller keep migrating between nodes in every 2 min and we calculated the average bandwidth available in each minute. The cumulative distributed function

---

Due to anonymous submission, we are not disclosing this information.

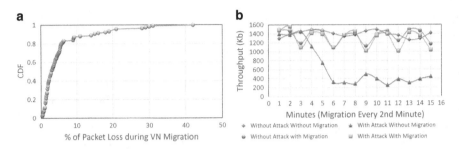

Fig. 5 These diagrams present the effectiveness and disruptiveness of agility results. (**a**) Packet loss during migration; (**b**) Bandwidth analysis in PlanetLab

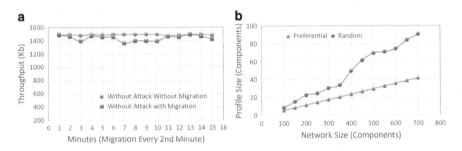

Fig. 6 These diagrams present the effectiveness and disruptiveness of agility results. (**a**) Bandwidth analysis in Mininet; (**b**) Critical footprint size

(CDF) results in Fig. 5a show only 5 % packet loss for almost 85 % of the times and higher for just 15 % of the times. Further investigation revealed that PlanetLab uses unicast reverse path forwarding. It matches the arriving interface of the packet with the departure interface, if it does not match, it simply drops packets. That is why all in-flight packets were lost every time the migration happens. This is the limitation with PlanetLab architecture and there is nothing that can be done. Therefore, we tested the same migration experiment in a more migration friendly virtualized infrastructure, i.e., Mininet. The combined results in PlanetLab and Mininet experiments, under all possible scenarios, are presented in Figs. 5b and 6a, respectively. And the Mininet results in Fig. 6a clearly show that there is almost no or minimal packet loss observed during migration. This proves that if the network is supportive of the migration then packet loss will not be an issue.

8.2.4 Evaluating the Overhead of Migration

Because on PlanetLab such large scale experiments were not possible, we have calculated migration overhead in terms of nodes to be migrated and extra traffic to be generated with simulation. Figures 6b and 7a demonstrate the overhead results in terms of the size of critical footprint and its percentage w.r.t. the network

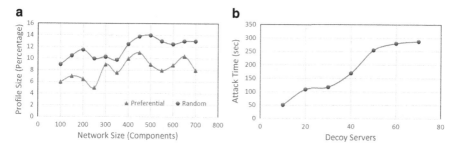

Fig. 7 These diagrams present the effectiveness and disruptiveness of agility results. (**a**) Critical footprint size (%); (**b**) Reconnaissance time

Table 2 Migration overhead

Network size	Preferential network		Random network	
	No. of rules	No. of MBs	No. of rules	No. of MBs
100	60,000	145.2	90,000	217.8
150	90,000	217.8	157,500	381.15
200	120,000	290.4	230,000	556.6
600	360,000	871.2	750,000	1815

size respectively, for different networks. We have used both types of networks (preferential and random) to evaluate the critical footprint size. For networks with preferential connectivity, the size of critical profile represents around 5–10 % of the network, whereas, for random network this percentage is 9–14 %. Intuitively, preferential connectivity based networks have a smaller critical footprint than random ones. The amount of overhead traffic is calculated by multiplying average routing table size, which is 24.4 MB or 10 K rules [30], with the number of components to be migrated. Results with different network sizes are showed in Table 2. For a network of 100 nodes, in both preferential and random networks, overhead traffic size is only 145 MB and 217 MB, respectively. For a large network of the size 600 nodes, it is 871 MB and 1.8 GB for preferential and random networks, respectively. In large networks, this amount of traffic is not significant because of the over provisioning for handling DDoS traffic amounting to hundreds of GB, e.g., in 2014, CloudFair's customer was hit with a massive 400 Gbps DDoS attack [39]. Furthermore, reconnaissance in large networks also takes more time and this helps to reduce the frequency of migration.

8.2.5   Benchmarks of Reconnaissance Time

The VN agility is bounded by reconnaissance time, which we calculate through a PlanetLab experiment. In this experiment, we host a web server on one of the USA based university node of PlanetLab and select 10–70 decoy server around

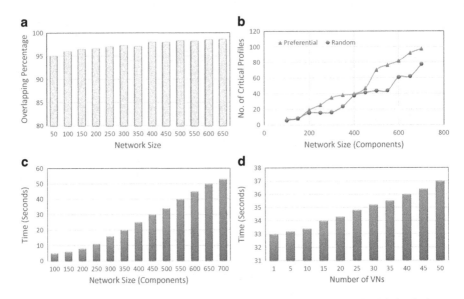

Fig. 8 This diagram shows Migration overhead, Framework scalability results. (**a**) Attaker's vs. defender's view; (**b**) Depth of defense; (**c**) Time w.r.t. network size; (**d**) Time w.r.t. Multiple VNs

that university. We use only one PlanetLab node as bot to perform reconnaissance because, in actual Crossfire reconnaissance, each node has to perform same amount of probing. Furthermore, increasing bots will not increase time because of limited probing traffic that can be sent to decoy servers and targets to avoid detection. The results in Fig. 7b show that even for a small scaled experiment with just 70 decoys, it took us 5 min to complete reconnaissance, that provide a big margin to maneuver.

8.2.6   Evaluating Attacker's View vs. Defender's View

We use simulation environment, explained in Sect. 8.1.2, to calculate disparity, if any, between the defender's view and the attacker's view of the critical footprint. They both view the same network data plane and decoy server information is publicly available. We start by finding critical footprint sets, firstly, by just using bots (for attacker) and secondly, by just using sources (for defender). Then, we calculate overlapping ratio of two sets. The results in Fig. 8a clearly show that almost 95 % of the time both end up selecting the same critical footprint that coincide the finding of the existing literature [1].

8.2.7 Evaluating Scalability

We tested the scalability of our approach in terms of time complexity by varying the size of the networks and number of VNs within a network. The results of varying network size and number of VNs are showed in Figs. 8c, d respectively. The time complexity increases linearly with network size and for network of the size 700 components, it is well below 1 min. Whereas, it does not change much by just increasing the number of VNs within same network e.g., it changes from 33 to 38 s for one VN to 50 VNs in a network of size of 500.

9   Conclusion

All DDoS attacks focus only on a small set of critical network components. If we can change the role of these critical components to non-critical ones, we can evade DDoS attacks. Virtual networks provide this flexibility by dynamically assigning and reassigning physical resources to a service. In this paper, we have proposed a correct-by-construction agile VN framework that proactively defended against sophisticated DDoS attacks like Crossfire by actively reassigning the VN to new threat safe physical resources and without breaking the service or violating the VN properties. We have implemented that framework on PlanetLab and our experiments showed the effectiveness of restoring the downgraded bandwidth (80 %) due to DDoS attack by migrating to a threat safe placement in just seconds. Furthermore, the existing provision of redundant paths in the Internet, 4–5 as found in our experiments, is enough to defend against large scale DDoS attack.

References

1. M. S. Kang, S. B. Lee, and V. D. Gilgor, "The crossfire attack." in *Proceedings of IEEE Symposium on Security and Privacy*, 2013.
2. A. D. Keromytis, V. Misra, and D. Rubenstein, "Sos: Secure overlay services." in *Proc. ACM SIGCOMM*, August 2002.
3. "Akamai," http://www.akamai.com.
4. M. Faloutsos, P. Faloutsos, and C. Faloutsos, "On power law relationships on the internet topology." in *In Proc. ACM SIGCOMM*, 1999.
5. T. Anderson, L. Peterson, S. Shenker, and J. Turner, "Overcoming the internet impasse through virtualization," *IEEE Computer*, 2005.
6. A. Gupta, J. Kleinberg, A. Kumar, R. Rastogi, and B. Yener, "Provisioning a virtual private network: a network design problem for multicommodity flow." in *Proc. ACM symposium on Theory of computing (STOC),*, 2001, pp. 389–398.
7. Y. Zhu and M. Ammar, "Algorithms for assigning substrate network resources to virtual network components." in *INFOCOM*, 2006.
8. A. Haque and P.-H. Ho, "Design of survivable optical virtual private networks (o-vpns)." in *Proc. 1st IEEE International Workshop on Provisioning and Transport for Hybrid Networks*, 2004.

9. W. Szeto, Y. Iraqi, and R. Boutaba, "A multi-commodity flow based approach to virtual network resource allocation." in *Proc. GLOBECOM: IEEE Global Telecommunications Conference,,* 2003.

10. M. Demirci, S. Lo, S. Seetharaman, and M. Ammar, "Multi-layer monitoring of overlay networks," in *Proceedings of the PAM,* 2009.

11. "Virtela," http://www.virtela.net/platforms/virtualized-overlay-networking/.

12. "Aryaka," http://www.aryaka.com/.

13. L. D. Moura and N. Bjorner, *Satisfiability Modulo Theories: Introduction and Applications.* CACM, 2011.

14. "Z3 theorm prover," http://research.microsoft.com/en-us/um/redmond/projects/z3/.

15. "Yices: An smt solver," http://yices.csl.sri.com/.

16. "Planetlab," http://www.planet-lab.org.

17. S. Lo, M. Ammar, E. Zegura, and M. Fayed, "Virtual Network Migration on Real Infrastructure: A PlanetLab Case Study," in *Proceedings of the 12th International IFIP TC 6 Conference on Networking,* 2014.

18. T. Anderson, T. Roscoe, and DavidWetherall, "Preventing internet denial-of-service with capabilities." in *Proceedings of Hotnets-II,* November 2003.

19. A. Yaar, A. Perrig, and D. Song, "An endhost capability mechanism to mitigate ddos flooding attacks." in *Proceedings of the IEEE Symposium on Security and Privacy,,* May 2004.

20. X. Yang, D. Wetherall, and T. Anderson, "An endhost capability mechanism to mitigate ddos flooding attacks." in *Proc. ACM SIGCOMM,,* August 2005.

21. J. Ioannidis and S. M. Bellovin, "Implementing pushback: Router-based defense against ddos attacks." in *In Proc. Network and Distributed System Security Symposium (NDSS),* February 2002.

22. R. Mahajan, S. M. Bellovin, S. Floyd, J. Ioannidis, V. Paxson, and S. Shenker, "Controlling high bandwidth aggregates in the network." *Computer Communication Review,* vol. 32(3), pp. 62–73, 2002.

23. A. C. Snoeren, C. Partridge, L. A. Sanchez, C. E. Jones, F. Tchakountio, B. Schwartz, S. T. Kent, and W. T. Strayer, "Single-packet ip traceback." *IEEE/ACM Transactions on Networking,* vol. 10(6), pp. 295–306, December 2002.

24. D. G. Andersen, "Mayday: Distributed filtering for internet services." in *Proc. 4th USENIX Symposium on Internet Technologies and Systems (USITS),* March 2003.

25. J. Kurian and K. Sarac, "Fonet: A federated overlay network for dos defense in the internet," University of Texas at Dallas, Technical Report, 2005.

26. K. Lakshminarayanan, D. Adkins, A. Perrig, and I. Stoica, "Taming ip packet flooding attacks." in *In Proceedings of the HotNets-II,* 2003.

27. A. Stavrou and A. D. Keromytis, "Countering dos attacks with stateless multipath overlays." in *CCS 05: Proceedings of the 12th ACM conference on Computer and communications security,* 2005, pp. 249–259.

28. A. Stavrou, D. L. Cook, W. G. Morein, A. D. Keromytis, V. Misra, and D. Rubenstein, "Websos: an overlay-based system for protecting web servers from denial of service attacks." *Computer Networks,* 2005.

29. J. Fan and M. H. Ammar, "Dynamic topology configuration in service overlay networks: A study of reconfiguration policies," in *Proc. IEEE INFOCOM,* 2006.

30. Y. Wang, E. Keller, B. Biskeborn, J. van der Merwe, and J. Rexford, "Virtual routers on the move: Live router migration as a network-management primitive," in *SIGCOMM,* Seattle, WA, Aug. 2008.

31. S. Lo, M. Ammar, and E. Zegura, "Design and analysis of schedules for virtual network migration," *Georgia Institute of Technology SCS Technical Report,* vol. GT-CS-12-05, July 2012.

32. E. Keller, D. Arora, D. P. Botero, and J. Rexford, "Live migration of an entire network (and its hosts)," *Princeton University Computer Science Technical Report,* vol. TR-926-12, June 2012.

33. S. Nedevschi, L. Popa, G. Iannaccone, S. Ratnasamy, and D. Wetherall, "Reducing network energy consumption via sleeping and rate-adaptation." in *NSDI,* vol. 8, 2008, pp. 323–336.

34. B. Peng, A. H. Kemp, and S. Boussakta, "Qos routing with bandwidth and hop-count consideration: A performance perspective," *Journal of Communications*, vol. 1, no. 2, pp. 1–11, 2006.
35. "Geolite free geo IP database." http://dev.maxmind.com/geoip/legacy/geolite/.
36. "Geographical distance." http://en.wikipedia.org/wiki/Geographical_distance.
37. F. Gillani, E. Al-Shaer, S. Lo, Q. Duan, M. Ammar, and E. Zegura, "Agile virtualized infrastructure to proactively defend against cyber attacks." in *Infocom*. IEEE, 2015.
38. "Brite topology generator," http://www.cs.bu.edu/brite/.
39. "Technical details behind a 400gbps ntp amplification ddos attack." http://blog.cloudflare.com/technical-details-behind-a-400gbps-ntp-amplification-ddos-attack.

Exploring Malicious Hacker Forums

Jana Shakarian, Andrew T. Gunn, and Paulo Shakarian

Abstract For consumers the increasingly widespread consumer-grade connected ("smart") devices; growing use of cloud-storage and globally still expanding use of Internet and mobile phones; mobile payment options will pose increasing risk of becoming a victim of cyber-attack. For companies and institutions of all kinds, matters regarding the protection of Intellectual Property (IP) and Personally Identifiable Information (PII) from cyber-breaches and -leaks will demand higher financial investment. With the discovery of Stuxnet, offensive and defensive cyber-capabilities have already become an acknowledged tool in military arsenals worldwide and are at the cusp of shifting the global landscape of military power. With the expanding yield of cyber-related activities, understanding the actors creating, manipulating, and distributing malicious code becomes a paramount task. In this chapter we report on the results of an exploration of black hat hacker forums on both the Internet and crypto-networks (in particular those accessed via the Tor-browser). We report on the structure, content, and standards of behavior within these forums. Throughout we highlight how these activity augment the activities of the black hat hackers who participate.

1 Introduction

In this chapter we provide an account of our exploration of malicious hacker forums. The observed English-language forums are accessible through the Tor-network only, while the web forums addressing Russian speakers are more often hosted on the surface-layer Internet. These platforms of communication between malicious hackers allow insights into their concerns, motivations and goals as well as the environment in which they spend considerable time. An intimate understanding of this community will greatly aide pro-active cyber-security [6]. While the structure of these forums in large resembles similar platforms, it is in the content and members that they differ. Many English-language forums set up for malicious hackers are hosted on Tor.

J. Shakarian (✉) • A.T. Gunn • P. Shakarian
Arizona State University, Tempe, AZ 85281, USA
e-mail: jshak@asu.edu; andrewgunn@asu.edu; shak@asu.edu

© Springer International Publishing Switzerland 2016
S. Jajodia et al. (eds.), *Cyber Deception*, DOI 10.1007/978-3-319-32699-3_11

In this chapter we focus on forums frequented by malicious hackers, because this is where techniques and code are created, shared [41] and distributed [10, 17]. Furthermore, these platforms constitute the arena where rules of conduct are enforced, the legitimacy of certain undertakings and targets are negotiated [6, 24]. As such the forums aide in the propagation of hacking, cracking and the ethics associated with these activities [12, 17]. Concerns, ambitions and modi operandi of malicious hackers are showcased in forums as their main medium of communication in regard to the particular practice of network and computer intrusion. A profound understanding of these factors will aide early detection of cyber-attacks and allows for identification of future cyber-threats—and the study in this chapter represents initial research in this direction.

The remainder of this chapter is organized as follows. In Sect. 2 we present background material. We then review our exploration methodology in Sect. 3. Structure and content of these online community areas are described in Sects. 4 and 5 respectively. Finally, we conclude in Sect. 6.

2   Background

Many of the individuals behind cyber-operations—originating outside of government run labs or military commands—rely on a significant community of hackers, preferably interacting through a variety of online forums (as means to both stay anonymous and to reach geographically dispersed collaborators). The distribution of *MegalodonHTTP* Remote Access Trojan (RAT) utilized the amateur black hat platform, HackForum. Five people accused of the malware's creation and/or distribution resided in three European countries requiring law enforcement to cooperate internationally in pursuit of the malicious hackers' arrest [27]. The international nature of the cyber-domain—the organization of cooperating malicious hackers as well as their international targets—transcends not only territorial executive powers, but adds to the importance of virtual communication platforms. Oftentimes—as in the case with *LulzSec* [39] and *MegalodonHTTP*—the hackers are very unlikely to ever meet each other in person. As we will describe in the section on contents /refcntSec, malicious hackers frequently suggest that providing any personally identifiable information is regarded as unsafe practice. This might hint at the benefits online communities and even more so anonymizing services provide for netizens who want their physical existence to remain hidden and separate from their online persona and activities.

2.1   Darknet and Clearnet Sites

"The Onion Router" (Tor) is free software dedicated to protect the privacy of its users by obscuring traffic analysis as a form of network surveillance [13]. After downloading and installing the small software package, the Tor browser may be started like any other application. The network traffic is guided through a number of volunteer-operated servers (also called "nodes") rather than making a direct connection. Each node of the network encrypts the information it blindly passes on registering only the node immediately prior and the one immediately following [13], disallowing any tracking. Each session surfing the Internet through Tor is utilizing different pathways through virtual tunnels. Tor further allows to publish websites without ever giving away the location of the hosting server. Addresses to websites hosted on this hidden service use the ".onion" extension and do not render outside the Tor-network. Effectively, this allows not only for anonymized browsing (the IP-address revealed will only be that of the exit node), but also for circumvention of censorship.[1] For journalists, activists and individuals living under repressive regimes Tor allows a private and safe manner to communicate. Here we will refer to "darknet" as the anonymous communication services crypto-networks like Tor provide. It stands in contrast to "deepweb" which commonly refers to websites hosted on the open portion of the Internet, but not indexed by search engines. Corporate websites supporting employees and library catalogs are good examples for deep web presences. Additionally, dynamic websites which display content according to user requirements are also difficult to index [44]. Websites hosted on crypto-networks like Tor, such as Freenet, I2P, Hyperboria Network, M-Web, Shadow Web and others are collectively referred to as "Darkweb". Although the "Darknet" describes a less populous, less expansive antecedent of the "Darkweb", it is the term most commonly found in use by the people who frequent it. In contrast, "Clearnet" is a term employed by Tor-users pointedly exposing the lack of anonymity provided by the surface-layer Internet [29].

2.2   Malicious Hacking

Hackers as subculture has been subject of many publications amongst them Steven Levy's seminal "Hackers - Heroes of the Computer Revolution" [30], which outlines ideological premises which many early computer geeks and programmers shared. The machines comprising the early computers were extensions of the self [47], which might compliment the creative ownership and the demand for free software that permeated Levy's account [30]. The term "hacker" in recent use (and especially in popular media) has become restricted to individuals who seek unauthorized access to computers and computer networks not their own with the purpose to manipulate, steal, log or alter data or structures [16, 41]. In this limitation the term becomes synonymous with "crackers"—a label more befitting to those who indeed appear to solely pursue destruction and havoc. These are the activities more

---

[1]See the Tor Project's official website (https://www.torproject.org/) and Tor's "About"-page (https://www.torproject.org/about/overview.html.en) for more details.

likely to be linked to criminal activities [22]. The term "crackers" itself derives from the practice of cracking passwords or levels in online games. Likewise, black hat hackers (or "black hats") employ their sometimes significant skills and knowledge towards illicit goals (e.g. financial fraud and identity theft). However, the hacker community is much more diverse. From meticulous tinkerers, phreakers [30], technology savvy libertarians [12], and ideology-driven script kiddies or "vandals" as Jaishankar [22] calls hacktivists [45]—"hacker" signifies everybody who uses his or her computer in innovative and creative ways [30]. Yet the hacker population encompasses divergent skill levels, motivations and purposes as well as various modi operandi. New to programming and computer technology, "n00bs" or "newbies" stand in the beginning of their could-be hacking career. Script Kiddies (also called "Skiddies" or "Skids") utilize programs (scripts, tools, kits, etc.) created by more highly skilled crafters [16]. Whereas highly knowledgeable and experienced hackers see themselves as part of an elite ("leet") or in hacker argot replacing letters with numerals "1337" [16].

2.3   Online Communities

However, the focus on web-forums so far was limited to online discussion forums on the Clearnet [2, 6, 19, 29, 31, 38] and more often on a diversity of social media outlets in respect to activism and social organization as well as online games [7, 20, 23, 25, 35]. Many aspects of Internet forums have been subject of academic research, for example in regard to their usability in social science or psychology research [17, 40] or as form of technologically enabled communication amongst individuals [5, 47]. In the field of Social and Cultural Anthropology an entirely new field, "Netnography" [11, 28, 40], is dedicated to conducting research in various online settings on most diverse Internet-communities.

Hackers of all skills and motivations experience scholarly scrutiny in manifold aspects [12, 24, 30]. The emergence of cyber-activism [14, 32] culminated in the heyday of Anonymous and associated hacktivist, earned not only headlines, but also scholarly attention [39, 42]. Malicious hackers are the subject particularly of criminological studies [10, 15–19, 21, 31, 46, 48]. While scholars of diverse social sciences have gained insights into online behavior, the "Darkweb" so far elicited a lot less attention than the "Clearnet". Beyond a technical introduction to the "Darkweb" [8], Hsinchun Chen's group examines terrorism as subject of communication on suspected jihadi-websites [9]. Work on black hat-forums hosted on the darknet focus on trust in the anonymized environment of crypto-networks or concentrate on the social relationships in the presence of mutual distrust.

3   Methodology and Scope

Tor-hosted platforms are often shorter lived than their Clearnet counterparts: sites migrate frequently or alternate through multiple addresses and their availability (or uptime) is unreliable. Through search engines and spider services on the Tor-network we were able to find more than sixty forums populated by malicious hackers. Other platforms were discovered through links posted on forums either on the Tor-network or on the Clearnet. About half of these forums use English to communicate (33), but French (8), Russian (4), Swedish (2), and various other languages (5) also served as communication vehicles. On the Clearnet, we found more than seventy forums for black hat hackers, the majority of which are English-speaking (52), eighteen are in Russian, and two in other languages. To gain access most require registration and agreement to forum-rules. Registration and log-in often includes completing CAPTCHA verification codes or images or solving puzzles and answering questions to prevent automated entry and DDoS-attacks, thus requiring manual action.

Our initial non-participant observation [40] in different black hat-forums hosted in various languages had the purpose of exploring community structures, the interactions and communication amongst the members of the forum as well as the contents. Non- or minimal participant observation on these forums extend up to more than seven months at the time of writing and is ongoing. The structural organization of forums into boards, child boards, threads and posts therein may resemble more familiar platforms dedicated to various other subcultures. The technical environment, social interaction, and predominant topics become especially interesting with black hat hackers however. The members—at least to some degree—are profoundly familiar with the very same technology that constitutes the boundaries and limitations of their environment. They have the skills to negotiate and manipulate their technical environment should they wish to do so. All observed forums thus require agreement to a set of rules as part of the registration process. Darknet-hosted-marketplaces are repeatedly hacked to test them for their security and anonymity. Yet the majority of the forum-members appear to follow the rules. Perhaps this is due to the requirement to accept the forum-regulations upon registration which is absent in most marketplaces. Also, forums constitute community through interaction with like-minded others more than trading on marketplaces. Nonetheless, frequently discussions arise on how to improve the forum experience. Many forums feature different access levels, granting more information and activity options to vetted members. For our research we conducted observations in forums that required registration, but were otherwise unrestricted. Other platforms grant access only with invitation code, after payment or after a formal application, questionnaire and online-interview (Fig. 1).

Fig. 1 Screenshot from an (paid) invite-only English-language Tor-hosted malicious hacker forum, taken December 2015. The BTC-amount indicated exchanges for USD100

4   Forum Structure and Community Social Organization

The structure of online-forums entails two aspects: the structure of the technical environment as well as the social structure (and dynamics) each member finds himself in. Online-forums constitute technical environments with potentially malleable restraints. These user-oriented platforms have the sole purpose of enabling communication while providing the opportunity for the emergence of a community of like-minded individuals—regardless of their geophysical location.

4.1   Technical Structure

Administrators set up Darkweb-forums with communication safety for their members in mind. During registration (though not necessarily with every login) every prospective member has to complete CAPTCHAS, answer simple questions, solve puzzles or complete simple arithmetic operation, presumably to prevent automated access.

Further safety measures are member-specific and member-supplied data such as a PIN-number, mnemonics and often the member's public PGP-key.[2] Another precaution the individual user is urged to take is the employment of *Tails* (The amnesiac incognito live system). This free operating system (based on Debian Gnu/ Linux) is run from a CD or USB stick and wipes the RAM once the computer shuts

---

[2]In English-language forums typically PGP-encryption is encouraged, online communities hosted in other languages (e.g. French and Italian) occasionally suggest the use of GPG (Gnu Privacy Guard).

down[3]—leaving no trace of the past session. Tails is portable so that it can be used even on the library computer or on a cell phone. The pre-installed browser is thought to intensify the level of privacy when browsing the Tor- or I2P-network.

It is further recommended, that the person at the keyboard disable JavaScript via a button in the browser. Not only is JavaScript known as being particularly vulnerable to software exploits [4], but it is also considered to be a weak spot which can easily be exploited and used to track and identify Tor-users. A technique dubbed (device-) "fingerprinting" is employed by third parties to track Internet-users by querying the browser for a variety of characteristics among which JavaScript is but one that informs on the browser used as well as screen properties [3, 37]. Many Darkweb-hosts therefore avoid JavaScript, which is a practice which appears to be widely regarded as enhancing security and also seems to improve the image of the respective forum. In another attempt to add security measures and gain members' trust, some forums' administrators will avoid time-stamping posts.

While forums need to feature techniques and mechanisms to enhance anonymity and privacy to gain support, it appears that strategies users employ to protect themselves are mainly limited to disguising the traces they leave on the Darkweb in such a way that they cannot be connected to the real person. In order to diffuse the online persona, an individual could establish different monikers, with alternating interests (as indicated by browsing habits), use fake (written) accents, and register a multitude of anonymous email accounts. Irregular Tor-usage and connecting through ever-changing routing locations and Tor-bridges help defy tracking attempts and pattern-detection. A Darkweb second life—in order for it to be considered safe—needs to become a puzzle with pieces being widely dispersed, thus rendering the puzzle unsolvable. However, such a focus on security makes participation in a an online-community impossible.

4.2 The Process of Forum Registration

The means and tools of communication in the observed forums are public threads or private messages ("pm"). Plug-ins for instant chat platforms like Jabber and ICQ are sometimes provided and are gateways to platforms beyond.[4] The first step in the registration process consists of the agreement to official forum rules.[5] Most often that means to abstain from posting child pornography, to refrain from expressing any form of discrimination and racism. Trolls—provocative virtual trouble seekers who seemingly enjoy replying to other members' posts in a mere slanderous manner,

---

[3]Tails only uses RAM storage space, which is automatically erased after the computer shuts down.

[4]While Jabber allows members to contact each other independently of the forum, ICQ provide (private) chatrooms and access to groups not associated with the forum.

[5]The broadcasting of rules also helps to describe the scope of the forum to the most general audience.

are also unwelcome. Moreover, the registration process focuses on the essentials of what the user might need (username, password) as well as on safety. During registration the prospective user may decide whether she/he wants to receive emails from other users and the administrators. The chosen username then becomes the handle by which the user will be recognized by other members of the forum and which he/ she can enhance graphically. Other forms of personalization are signature block-like short statements that are printed at the bottom of each of the user's posts. It can be suggested that online handle and signature block serve to add character dimension to the virtual persona.

4.3 Forums' Boards and Their Content

Good *OpSec*—it is often stressed on darknet-forums' *Introduction* board and tutorials—recommends to employ different usernames on each and every website or online entity. All of a user's online handles, passwords and posts—the online persona—should be as different from the real-world individual creating them as possible. Members find buttons to their account, private messages, and one to log off in the header of the page. Within the account, settings like email-address, password and other customization tools are found (depending on the forum).

Discussion forums addressing English-speakers on the darknet consist of boards and sub-boards (also called "child-boards") filled with threads concerned with different topics. Heading the page there will usually be an "introduction" board where new members are encouraged to present themselves first. A *General Discussion*-board hosts threads of importance to every member, e.g. "forum news", "news" (mainstream media), "privacy". The English-speaking forums observed feature boards concerned with carding (financial fraud), hacking, cyber-security, dumps or leaks (release of hacked data), and sometimes doxing (release of personally identifiable information). Members of the forums can either start discussions, i.e. new threads, in any of these boards or post comments on existing ones. The newest threads will most often appear at the top of the list of discussion features on any board. The same is true for the replies within individual threads.

While the structural environment of Russian hacker forums remain similar to their English counterparts (Fig. 2), there are several marked differences that are worth noting. Many Russian forums reside on the Clearnet. Whether Russian hackers are less worried about *OpSec*,[6] or if they simply prefer the more stable environment the Clearnet offers,[7] we do not yet know. A limited availability

---

[6]"Operation Security" refers to the protection of personal identifiable information recommended for everyone on the Darkweb.

[7]The availability of Tor-hosted websites ("uptime") is much less reliable than those hosted on the surface-layer Internet. Due to the tunneling through multiple nodes, the loading of Tor-hosted sites also takes longer than direct connections. To evade monitoring many administrators migrate between a number of web-addresses, though that practice is more common with Darkweb-marketplaces.

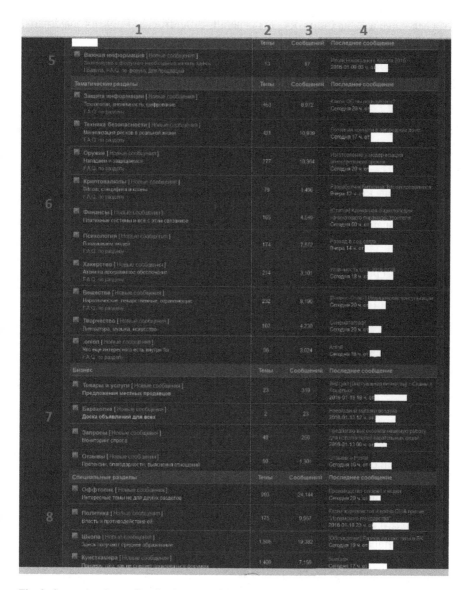

Fig. 2 Screenshot from a Russian-language Clearnet-forum, taken January 2016

of bandwidth might be a technological barrier to using Tor. Furthermore, in some countries the website of the Tor-project and IP-addresses known as Tor-exit nodes are blocked and inaccessible to the general public. Still, the majority of Russian Clearnet-forums closely resemble the Darkweb-forums in their structure, with features such as invite-only access, CAPTCHA, and heavy monitoring by administrators enforcing forum rules. Figure 2 shows that the structural organization of a Russian black-hat forum does not generally differ from its English counterparts. The page breakdown is as follows:

1. Темы ("Themes"). Shows the number of original posts for each thread topic.
2. Сообщений ("Messages"). Shows the total number of messages for each thread topic.
3. Последнее сообщение ("Last Messages"). Shows the title of the last thread, its date, and the user that posted it.
4. Важная информация ("Important Information"). This category contains important information about the forum, and new users are encouraged to read this first.
5. Тематические разделы ("Thematic Sections"). Contains a list of thread topics including: Protection of information, Technical Safety, Weapons, Cryptocurrency, Finances, Psychology, Hacking, Substances, Creation (literature, music, art), .onion (other as interesting perceived Tor-hosted websites).
6. Бизнес ("Business"). This category or section contains all of the topics related to the sale of products or services. Topics include: Wares and Services, Flea Market (a public board for all to use), Requests (for among others hacking services), Reviews (of forum-/marketplace-vendors)
7. Специальные разделы ("Special Sections"). Contains a list of topics not related to the aforementioned, including: "Off Topic"[8] "Politics", "School" (tests, exams, cheating)—hinting at the presumed age of many members, and finally "Curiosities".

4.4   The Social Structure of Black Hat-Forums

Aside from the graphical interface, social hierarchies can be observed within these forums. Like Clearnet-forums the social organization of all of the observed darknet-black hat forums feature a multi-layered social structure with status-respective tasks [2]. The vetting process itself differs between forums, but it generally entails frequent participation and showcasing of skills. The former is frequently underlined in its importance in the threat, that accounts of insufficiently active members will be suspended. This requirement may suggest that much effort is dedicated to community-building as non-sociability is punished. The latter, showcasing of skills-while currently not an ubiquitous demand- generally serves multiple purposes. For the forum it serves in identifying skillful members and aids in building up resources to be shared within the community. For the providing individual it aides in improving their social status and ranking within the community.

---

[8]"Off Topic"-sections are often spaces provided to both answer the apparent need of members to sometimes discuss topics unrelated to the forum's general theme as well as to maintain the integrity of the other hosted boards.

Every forum needs a person to create this virtual gathering space in the first place, keep it online and maintain the user database. Therefore, as the creator and host at the top of most every online forum one or more administrators ("admin(s)") are found. The site administrators reside at the top of the social ladder with the powers to admit new members, delete accounts, and change the structural development as well as the scope of the forum. Administrators can decide about the level of accessibility of the forum to the larger community by setting or removing the hurdles of invite-only or other means of restricted access. On the next lower-level moderators are often responsible for specific boards, specializing in specific topics. They censor the textual contributions of members by deciding and deleting off-topic posts, thus enforcing forum-rules. In forums where assets [38] such as tutorials and malware are provided, moderators also filter and condone community resources. Quality control of resources provided includes checking for the functionality and benefits of code. Distinction among members is accomplished through rankings, which are established through ratings from others, activity level and seniority, i.e. duration of membership. Aside from the regard paid to members of higher social status in the form of trust in the truthfulness and accuracy of information and resources they share on the forum, members may apply to become moderators (or short "mods") in response to "job" announcements by the administrator or they may be "hand-selected". It appears as though at least some forums- administrators value subject matter expertise in the selection of moderators. However, other moderators seem to be selected based on their popularity, which is indicated by higher member ratings, i.e. other members are voting on posts or accomplishments. As noted for hackers elsewhere [17, 24] online communities appear to constitute a meritocratic system in that skills and success are the bases of a hacker's reputation and social status within the community. At the bottom of a forums' social spectrum, "newbies" (or "noobs") are restricted to their own boards and are expected to learn before advancing to more specific topics (often aggressively reprimanded should they decide to post questions deemed as inadequate for threads of more highly skilled hackers). Social mobility lies for many forum-participants in the sharing of knowledge or in the showcasing of expertise. As mentioned earlier, the organization of hacker communities follows the principles of meritocracy [17, 24], forcing the semi-private publication of hacks.

4.5   The Double-Edged Sword of the Hacker Meritocracy

Since social status is gained through proofs of expertise [17, 24], but upload-ing evidence of malicious activity puts malicious hackers at risk of attracting the attention of law enforcement, malicious hackers vying social status face a dilemma. Mediums allowing the concealment of personally identifiable information are therefore important. Negotiating the precious balance between notoriety and possible legal repercussions is aggravated by the use of monikers as near single markers of recognition. PGP-keys also frequently serve—aside from their purpose as guarantors of privacy—as proof of authenticity in online communication, but as

Fig. 3 Screenshot from an
English-language Tor-hosted
malicious hacker forum,
taken July 2015

such are invariably tied to the moniker. Therefore, as mentioned before, malicious hackers who use identical handles across multiple platforms as a way to "carry" their reputation with them are at an elevated risk of being tracked by law enforcement. Should they be identified on one platform all their activities can easily be traced across the networks. On the other hand, advantages of being recognized include instant access to arenas closed to "nobodies". For vendors standing in high regard translates directly into profits.

Establishing a reputation is time-intensive and laborious, perhaps even more so in an environment that is as discouraging of trust as the Internet. The mechanisms of establishing good standing within the online-community of hackers relies on the number of posts and their relevance. Members showcasing high levels of activity on the forum and expertise improve their status fastest. Contributors rate each others posts according to content and perceived attitude. The rank earned is publicly displayed in Fig. 3 alongside the moniker. Seniority indicates a members' loyalty to the forum and might encourage the administrator to offer moderator positions. Occasionally different levels of access are granted to vetted members. Under these circumstances the wish of popular *darknetizens* to transfer earned reputation across different platforms becomes conceivable. We observed highly regarded black hat hackers and carders use the same moniker in multiple forums and social media sites.

Considering that PGP-keys frequently serve as proof of identity, this practice appears to be an audacious *OpSec*-fault. For example, the developer of the Phoenix exploit kit—a software toolkit designed to compromise various pieces of common software—went by the moniker of "AlexUdakov" across several forums [49], one of which was the seized crime-forum "Darkode" [1, 26]. Russian forum members will often refer to other forums and marketplaces where they have sold hacking products, or where they have an established profile. This aids in the discovery of new forums and marketplaces, but does not lend itself to protecting the anonymity of the user. In this way it would be easy to trace members, regardless of their handle across multiple forums, across marketplaces and forums through their advertised pieces of software or malware. Similarly, both Clearnet- and Tor-hosted Russian-language forums utilize some form of member ranking system, whereby users may elevate their status on a forum by making a certain number of posts, providing especially valuable knowledge or even samples of free malware.

4.6   The Russian Forum-cum-Marketplaces

One of the most fascinating aspects of Russian forums is the use of Internet slang "'Padonkaffsky Jargon'" or "'Olbanian'" that permeates every aspect of the Darkweb, Deepweb and Clearnet alike. Russian forums also differ from their English counterparts in that many forums also act as marketplaces. It is crucial to take the social and legal landscape of Russia into account to gain a better understanding of the inner workings of Russian forums, and the users that frequent them. The (relative) impunity for hacking-related offenses may explain the Clearnet-hosting of many forums addressing Russian-speakers. Although Russia does have laws that govern cybercrime, they generally fail to prosecute hackers unless offenses are committed inside the Russian Federation. A further complicating factor is the fact that the U.S. and Russia have no extradition treaty—a circumstance that is effectively creating a safe haven for malicious[9] hackers. Coupled with the over-educated and underemployed population, we find a breeding ground for malicious hackers seeking to earn a living with their technical skills. Evidence of this can be seen in 17-year-old hacker Sergey Taraspov, from St. Petersburg. Mr. Taraspov, along with a small team of hackers, allegedly wrote a piece of malware that targeted point-of-sale (POS) software, and sold it for USD2000 (about 2 month's salary for a programmer) on a Russian forum/marketplace. This malware was, in turn, used by around forty individuals to steal over 110 million American credit card numbers in the "Target"-data breach of 2013 [34]. This is one example of the more notorious uses of Russian malware to exploit American companies, but attacks like these happen on a daily basis. It is not the intent of this chapter to exhaustively discuss the legal landscape for hackers in Russia, but it is worth noting some of the idiosyncrasies.

We have observed that Russian forums vary greatly in exclusivity. Some are very easy to gain access into, or even allow guests to peruse the forum topics.

Fig. 4 Screenshot of "Welcome"-page of a closed Russian-language forum, taken December 2015

---

[9]Oleg Luzyanin, Alexander Andreyev and Renat Irmagombetov were arrested in Moscow for for hacking into Russian payment system Rapida in April 2015 (http://rapsinews.com/news/20150417/273569361.html).

Others allow the creation of a forum-account by simply signing up and providing an email address. The exclusive Russian forums, on both the Darknet and Clearnet, are much more suspicious of prospective users and use certain countermeasures against unwanted entry. Many demand a code that can only be obtained by forum admins or other vetted users. Some require that a prospective user submit a sample of their code or malware as part of the vetting process. Several forums will even fluctuate between open and closed registration, presumably to keep up with user registration and to thwart unwanted access. These exclusive forums are home to highly skilled hackers. As you can see in Fig. 4, and its translation into English, the administrators believe their forum to be suitable only for true professionals, and advise less sophisticated users to seek alternative places to gain knowledge. These forums, which are easy to gain access to, are teaming with script kiddies, aspiring black hat hackers. It is possible to discuss or even purchase many different types of malware on these less-exclusive forums, but the products might at times be outdated. Apparently much less effort is extended towards uncluttering these lower-tier marketplace-forums making noteworthy and valuable products harder to reach. The more exclusive the forum is, the greater the chance that it will provide newer, high-quality malware and exploit kits. Access to exclusive forums must be carefully obtained and can often take a lengthy process. Forum administrators will conduct online-interviews through private messages, either utilizing the exchange of PGP keys for identification purposes or through a messaging protocol like Jabber or ICQ. It is the job of the administrators to keep forum-members safe from the prying eyes of law enforcement, researchers, and inactive lurkers. Administrators take pride in spotting different types of prospective members. Russian hackers behave and communicate in a certain way, identifying themselves as hackers, and obfuscating their communication—as well as their code.

Forum depicted in Fig. 4 is closed to all "non-specialists". The direct translation is as follows: "Information". This is a closed forum for specialists on the themes of hacking and malware. This discrimination is motivated [to deter] inadequate and unqualified people [trying to gain access]. We are focused on individuals with experience. For beginners there other places like *VIR* and *WEB* (both links to forums known to be populated by script-kiddies). Carding and commerce is strictly forbidden. On this forum nothing is sold or purchased. To resolve contentious issues, please write to JId (Jabber id). As also observed in English forums, highly skilled malicious hackers sometimes appear to prefer a community of similarly sophisticated individuals and consequently discriminate against "newbies".

5   The Content of Observed Forums

While structure and organization of darknet-hosted forums might be very similar to more familiar web-forums, the topics and concerns of the users vary distinctly. In the English clandestine darknet people interested in cats, steampunk, and the latest conspiracy theories convene, but an abundance of arenas dedicated to child pornography (CP), drugs, and weapons can also be found. Other forums appear to be venues for sharing erotic images—whether involving real persons or cartoon characters. Trading places are also very popular among *darknetizens* and in many forums marketplaces are discussed. Lengthy threads seek information on the reliability of individual vendors and marketplaces in general. Links to other darknet-sites and information on potentially fraudulent Web sites are especially useful in the absence of pervasive search engines and can be found on many forums. Forums addressing malicious hackers feature discussions on programming, hacking, and cyber-security. Threads are dedicated to security concerns like privacy and online-safety—topics which plug back into and determine the structures and usage of the platforms.

5.1 The Common Boards

Oftentimes the *Introduction*-board is dedicated to explain the functions, extensions and buttons of the forum. Beyond technical advice, proper manners as well as measures of *OpSec* are discussed. The latter includes tips on staying anonymous and safe on the Darknet in general, which is a near ubiquitous topic on forums populated by malicious hackers. Threads concerned with the protection of members' anonymity are often pinned to the top of the list of threads within a (child-)board ("sticky threads"). Members are expected to adhere to manners including to abstain from "trolling" (derisive posts for the mere purpose of provoking angry responses) and from posting prohibited material (most often child pornography). On many forums "spamming" (posts of no informative value) and "grave-digging" (reviving of posts long abandoned) are considered ill-mannered. *General Discussion*-boards often feature news from the administrators, suggestions for the improvement of the forum, membership policies, information on the recruitment of moderators, and a relay of mainstream-news (especially in connection with news concerning the community). The sentencing of Ross Ulbricht, a.k.a. *Dread Pirate Roberts* (DPR), to life in prison for pioneering the concept of darknet-markets (DNMs) as the creator and administrator of the original Silkroad-Market (SR1) reverberated heavily on *General Discussion*-boards. However, this is also the place to deliberate about new movies, software as well as other products and popular items. In some cases new members are required to contribute a set number of posts before being granted access to other parts of the forum. This board is also the place where to find answers to frequently asked questions (FAQs).

5.2 The Flavored Boards

Hereafter the titles of boards and their content strongly varies between forums according to their general theme. Specific to forums addressing malicious hackers space is afforded for boards offering tutorials and guides for both novice and experienced malicious hackers, and carders. In many forums old malicious code is shared for practice purposes. This conforms to the hacker-ideal of self-education [24]. Members challenge each other by posting claims of their accomplishments, sometimes publishing either code or links to "dumps" (repositories, pastebins or other data-sharing sites) containing hacked data. General information and discussions on software, malware, exploits and vulnerabilities are ubiquitous. Discussions relating to hacking certain systems, advertisements for services, and the bartering of stolen data are also commonly encountered. Posters inquire about how to crack specific applications or they are seemingly working on a particular project and seek help (often eliciting responses). Though expressly prohibited by many forums, others do feature boards for doxxing—where personally identifiable and private information is shared for the purpose of harassing the exposed individuals (and their families and friends) in the real (physical) world. Comparatively harmless forms thereof typically include endless prank phone calls and the delivery of unordered food, but can be taken to extremes with events involving physical attacks and surprise visits by teams of law enforcement officers (also called "swatting") [36]. On almost all Russian forums we observed rampant requests for hacking as a service (HaaS) (Fig. 5). For a novel amount of digital currency, a professional hacker offers a variety of tasks—from hacking a friends email- or Facebook-account, to shutting down an entire website (DDoS). In yet another way, hacking is being democratized, the sale or provision of complete malicious code on forums being one. But other than with malware, where minimal modification to target-specific needs still demands some basic skills, hiring a hacker does not require a basic understanding of coding, hacking, or malware. Merely investing thirty minutes on general research and installing the Tor-browser, newbies effectively become cyberthreats. For many Russians (and citizens of the former Soviet Block) this is a way to earn an income. Whether they built the malware tool or just purchase it as an investment, they are able to turn around and use it to make money.

The post demonstrated in Fig. 5 is a hacker listing services for sale. The sections include:

1. Topic title—Взлом почты и социальных сетей. Дешево, Взлом ("Hacking of Email and Social Media Accounts. Cheap, Hacking").
2. Forum user information, including his or her forum rating, number of posts written, date of post, user number, post classification, and reputation information.
3. Post—Translation:
   Hi All! I want to provide hacking services for email and social media. Price for hacking VK and OK:

   Hacking VK 2500 rubles
   Hacking Classmates (similar to VK) 2000 rubles

   Price for hacking email:

   Hacking Yandex email 2000 rubles
   Hacking mail.ru, list.ru, bk.ru, inbox.ru 2000 rubles
   Hacking gmail.com 2500 rubles

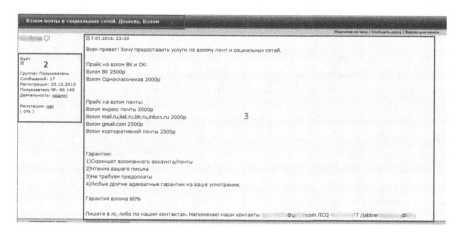

Fig. 5 Screenshot of post in Russian forum/marketplace advertising Hacking-as-a-Service (HaaS), taken January 2016

Hacking corporate email 2500 rubles

Guarantee:

1) Screenshot of the hacking of the account/email
2) Reading your letter
3) We don't require payment in advance
4) Any other adequate safeguards at your discretion

Guarantee of hacking 80%
Send a private message or to one of the following contacts: (ICQ/Jabber/email)

Additional boards dedicated to political topics suggest potential future cyber-operations as well as reinforce hacker culture. Topics in these areas include censorship, surveillance, and privacy on the Internet as well as freedom of information. Here members engage in heated debates over basic rights and core values or collectively condemn infringements on Internet anonymity and authorities' actions in support of censorship and surveillance. Privacy and anonymity are perceived to be imperiled by techniques and code like cookies and fingerprinting allowing ISPs and corporations to track user behavior for purposes like targeted advertising and customized search functions. Perhaps the popularity of these topics on hacker-forums is not surprising since members vacated to networks granting a higher degree of anonymity. Dark-netizens seek reprieve from monitoring mechanisms whether instituted by corporations or government, even if they may not and do not plan to engage in illegal or unlawful activities.

5.3 Sentiments and Concerns

Many posts (not just in black hat forums) are laced with anti-corporate and anti-government sentiments or carry a negative connotation in regard to law enforcement and their efforts to stem unlawful activities on the darknet. A member of an English-language darknet-forum, for example, assured that "corporate media is an elitist-controlled brainwashing apparatus where truth is irrelevant" (observed December 2015). In fact, members and administrators of black hat-forums are aware of the probable presence of law enforcement officers—right along with "nosy researchers and journalists" (observed September 2016). Suggestions to rid the forum of such unwanted guests include, but are not limited to closing the forum off to new members and instituting invite-only access. Forcing members to contribute to the forum community sometimes exceeds a preset number of required posts per time frame and calls for tests of heart, such as proof of criminal activity or the fulfillment of a specific illegal task demanded for by a moderator or administrator. However, the effectiveness of these measures in excluding law enforcement agents is a matter of discourse in the respective communities. Furthermore, not all members favor invite-only forums out of fear the topics and livelihood of the community might grow stale with a near static population. Though apparently in the minority, some appear to realize the futility of measures aiming at exclusivity and prefer to face the inevitable risks associated with their actions with good personal OpSec and a trust-no-one doctrine in order to evade criminal prosecution.

In English-language forums cultural references permeate platforms. Underdogs and antiheroes such as Edward Norton's nameless character in *Fightclub*, Robert DeNiro's character in *Taxi Driver*, Jeff Bridges' "The Dude" in *The Big Lebowski* (Fig. 6), Heath Ledger's *Joker* ("The Dark Knight") and Kevin Spacey's "Verbal/Keyser Soze" from *The Usual Suspects* are well liked profile pictures. Other recited pop-cultural icons are the "Cheshire Cat" of *Alice in Wonderland* and *Seinfeld*'s "Kramer". Images or short video replays of protagonists are displayed near online handles or citations appear in signature blocks. Monikers themselves sometime are seen to hijack meaning in reference to these counter-popular characters. Many seem to be adopted because they find themselves standing apart from what normally busies the world (e.g. *Seinfeld*'s "Kramer" and "The Dude" (*The Big Lebowski*)). They are underdogs, seemingly weak and powerless, who either emerge to be the lone survivors (e.g. "Verbal" in *The Usual Suspects*, "Mr. Pink" in *Reservoir Dogs*) (Fig. 6) or escalate into outbursts of violence (e.g. the protagonist in *Taxi Driver*). Statements expressing political views or the author's perspective on society are often found in signature blocks, which are automatically displayed with every posting. Occasionally links and invite-codes to alternate forums or marketplaces can also be found here.

5.4 Linguistic Characteristics

The self-identification of hacker—whether malicious or not—is strongly reflected and transmitted through the use of a specific jargon ("leet"-speak). In English, as mentioned before, letters are replaced with numbers and characteristic abbreviations are used. All users of Russian forums express themselves in a sophisticated and popular Internet slang known as "Padonkaffsky Jargon" or "Olbanian". For example, Padonkaffsky Jargon utilizes many aspects of the Russian language, culture, and subculture. It entails a sophisticated system of alternate orthography. Alternate representations of vowels and consonant are based on word-pronunciation and not on standard-modern Russian orthography. Voiced consonants are pronounced, and thus written in Internet slang, as their devoiced counterparts in certain situations, such as in word-final position, and when preceding a devoiced consonant in a consonant cluster, for example. Some of these voiced/devoiced pairs include В/Ф (V/F), Б/П (B/P), Г/К (G/K), Д/Т (D/T), З/С (Z/S), and Ж/Ш (Zh/Sh). Illustrated with a simple example, you might see the Russian word водка (vodka) written as Botka (votka). Similarly, Russian vowel reduction can also cause a change in orthography whereby an unstressed "– O" (O) is pronounced as an "– A" (A) and is represented as the latter in the context of the Internet. This sophisticated slang also implements a large amount of pop-culture references. As western pop-culture is well-liked in Russia, and other countries of the former Soviet Union, references to it are represented in this write-only slang. Without expert knowledge of the Russian culture and language, it would be very difficult to parse out and understand forum discussions with any precision. On the Internet the written form is not only a formal representation of thought, but is the sole vehicle of communication.

5.5 Trading Places

Many Russian forums effectively act as marketplaces (Fig. 7), where users can advertise and sell their wares. This is accomplished by setting up a board named "Commercial Area" or "For Sale" (as observed in January 2016). In this section of the forum, the first post of any thread will be written by the seller. His or her

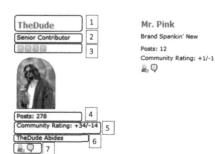

Fig. 6 Screenshot of member-id thumbnails in an English-language Darkweb-forum, taken January 2016. Key: (1) moniker; (2) community rank; (3) community rating summary; (4) number of posts; (5) positive/negative community ratings; (6) member quote; (7) contact

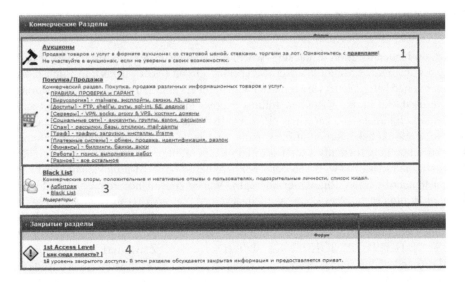

Fig. 7 Screenshot of Russian forum/marketplace, taken January 2016

post will include a description of the item for sale, the price (in one of several digital currencies), and private contact information (Jabber, ICQ or others) to invite conversation. The following posts are usually from prospective buyers asking questions, requesting demos, or even discrediting the seller and/or their products. With so many differing opinions, these discussions can get quite heated and even lead to users being banned. Many of these forum-marketplaces have wallets to deposit digital currency into, but sometimes forum administrators serve as an escrow service. Products are most often verified before any funds are released to the seller. If a seller is misleading or fails to deliver the appropriate item, they are banned from the site. Similarly, buyers can be banned for not complying with the transaction rules. This is an effective way of enforcing forum norms in that it serves as deterrent for possible future transgressions (as we will show later, banned monikers are published on "Black Lists") while at the same time reinforcing the reliability of the forum as a trading place.

Figure 7 shows a section of a forum that also acts as a marketplace. The boards within the *Purchase/ Sales*-section are broken down as follows:

1. Аукционы ("'Auctions'")—On this board threads are constituted by items which are to be auctioned off (similar to Ebay).
2. Покупка/Продажа ("Buy/Sell")—This board contains different categories of items for sale. The list includes: (Market) "Rules"; "Verification and Guarantees"; "Virusology" (malware, exploits, networking, cryptology); "Access" (FTP, Shell, rooting, SQL-injection); "Servers" (VPN, socks, proxy & VPS, hosting, domains); "Social Media" (accounts, groups, hacking and mailing); "Spam" (mailings, bases, responses, and dumps); "Traffic" (traffic, downloading,

installs, iframe); "Payment Systems" (exchange, sale, indemnification, unlock); "Finance" (billing, banks, accounts); "Work" (search, execution of work).

3. "'Black list"—This board is dedicated to resolving commercial disputes, features positive and negative reviews about members, suspicious persons, and a list of banned usernames.

4. "'Closed Forums'"—This section contains all of the forums unavailable to insufficiently vetted users. The forum administrator presides over the admission into this section.

6 Conclusion

In this chapter, we took an in-depth look at the culture of black hat communities in web-forums. In particular we were able to present the realization of social structure in an online-environment, the mechanisms of social mobility, pop-cultural references and linguistic characteristics. Social dynamics reinforce and maintain the communities and their structures. Forums are the arenas in which community norms are imposed, enforced, negotiated and altered. Understanding the social organization not only improves our knowledge, but can also aide in preventing and tackling computer crime [15]. Because forums (and marketplaces) share resources such as malicious code and knowledge (how-to), they are driving the democratization of cyber-attacks. The devaluation of expertise allows for cyber-attacks to be carried out by actors with very few skills. A key information requirement that can be addressed by monitoring online-forums is "which cyber-capabilities are available to an adversary?". The knowledge spread on these platforms through not only discussion, but provided by tutorials and free malware- and exploit kit-downloads allow for capabilities to spread quickly. New cyber tactics, techniques and procedures; identification of software vulnerabilities; and attack claims can spread rapidly in these environments. Additionally, the anonymity awarded to users of especially the darknet enables hacktivists, cyber-mercenaries, military personnel in cyber-units, and those seeking to sell malware to evade legal restrictions such as export laws. Cyber-analysts must understand the culture of these communities and their topics of discourse in order to identify emerging cyber-threats and -capabilities. As we anticipate the relevance of hacking communities to only increase in the cyber domain, understanding these specialized online communities will become of critical importance for anticipating cyber threats, intelligence analysis, understanding cyber capabilities available to adversaries who tap into the knowledge and resources of these communities.

Acknowledgements Some of the authors of this work were supported by the U.S. Department of the Navy, Office of Naval Research, NEPTUNE program as well as the Arizona State University Global Security Initiative (GSI). Any opinions, findings, and conclusions or recommendations expressed in this material are those of the author(s) and do not necessarily reflect the views of the Office of Naval Research.

References

1. Cyber criminal forum taken down - members arrested in 20 countries (2015)
2. Abbasi, A., Li, W., Benjamin, V., Hu, S., Chen, H.: Descriptive analytics: Examining expert hackers in web forums. In: IEEE – Joint Intelligence and Security Informatics Conference (JISIC), pp. 56–63. The Hague, The Netherlands (2014). DOI 10.1109/JISIC.2014.18. URL http://dx.doi.org/10.1109/JISIC.2014.18
3. Acar, G., Juarez, M., Nikiforakis, N., Diaz, C., Gürses, S., Piessens, F., Preneel, B.: Fpdetective: Dusting the web for fingerprinters. In: Proceedings of the 2013 ACM SIGSAC Conference on Computer and Communications Security, CCS '13, pp. 1129–1140. ACM, New York, NY, USA (2013). DOI 10.1145/2508859.2516674. URL http://doi.acm.org/10.1145/2508859.2516674
4. Arma: Tor security advisory: Old tor browser bundles vulnerable. The Tor Project (2013). URL https://blog.torproject.org/blog/tor-security-advisory-old-tor-browser-bundles-vulnerable
5. Bargh, J.A., McKenna, K.Y.A.: The internet and social life. pp. 573–590 (2004). DOI 10.1146/annurev.psych.55.090902.141922
6. Benjamin, V.A., Li, W., Holt, T.J., Chen, H.: Exploring threats and vulnerabilities in hacker web: Forums, irc, and carding shops. In: 2015 International Conference on Intelligence and Security Informatics (IEEE), pp. 85–90. Baltimore, MD, USA (2015). DOI 10.1109/ISI.2015.7165944. URL http://dx.doi.org/10.1109/ISI.2015.7165944
7. Boellstorff, T.: Coming of Age in Second Life: An Anthropologist Explores the Virtually Human. Princeton University Press, Princeton, NJ, USA (2008)
8. Bradbury, D.: Unveiling the dark web. pp. 14–17 (2014). DOI 10.1016/S1353-4858(14)70042-X. URL http://dx.doi.org/10.1016/S1353-4858(14)70042-X
9. Chen, H., Chung, W., Qin, J., Reid, E., Sageman, M., Weimann, G.: Uncovering the dark web: A case study of jihad on the web. pp. 1347–1359. Wiley Subscription Services, Inc., A Wiley Company (2008)
10. Chu, B., Holt, T.J., Ahn, G.J.: Examining the creation, distribution, and function of malware on line. National Institute of Justice, Washington, D.C. (2010). URL www.ncjrs.gov./pdffiles1/nij/grants/230112.pdf
11. Coleman, E.G.: Ethnographic approaches to digital media. pp. 487–505 (2010)
12. Coleman, E.G.: Coding Freedom: The Ethics and Aesthetics of Hacking. Princeton University Press (2013)
13. Dingledine, R., Mathewson, N., Syverson, P.: Tor: The second-generation onion router. In: Proceedings of the 13th Conference on USENIX Security Symposium - Volume 13, SSYM'04, pp. 21–21. USENIX Association, Berkeley, CA, USA (2004)
14. Donk, W.B.H.J.v.d.e.: Cyberprotest : New Media, Citizens, and Social Movements. Routledge London ; New York (2004)
15. Holt, T.J.: Lone hacks or group cracks: Examining the social organization of computer hackers. In: Crimes of the Internet, 1st edn., pp. 336–355. Prentice Hall Press, Upper Saddle River, NJ, USA (2009)
16. Holt, T.J., Schell, B.H.: Hackers and Hacking: A Reference Handbook. Contemporary World Issues. ABC-CLIO, LLC (2013)
17. Holt, T.J., Strumsky, D., Smirnova, O., Kilger, M.: Examining the social networks of malware writers and hackers. pp. 891–903 (2012)
18. Holt, T.J.e.: Crime On-Line: Correlates, Causes, and Context, 2 edn. Caroline Academic Press (2013)
19. Holtz, P., Kronberger, N., Wagner, W.: Analyzing internet forums: A practical guide. pp. 55–66 (2012). DOI 10.1027/1864-1105/a000062
20. Honeycutt, C., Herring, S.: Beyond microblogging: Conversation and collaboration via twitter. pp. 1–10 (2009)
21. Hutchings, A., Holt, T.J.: A crime script analysis of the online stolen data market (2014)

22. Jaishankar, K.: Space transition theory of cyber crimes. In: Crimes of the Internet, 1st edn., pp. 283–301. Prentice Hall Press, Upper Saddle River, NJ, USA (2009)
23. Java, A., Song, X., Finin, T., Tseng, B.: Why we twitter: Understanding microblogging usage and communities. In: Proceedings of the 9th WebKDD and 1st SNA-KDD 2007 Workshop on Web Mining and Social Network Analysis, WebKDD/SNA-KDD '07, pp. 56–65. ACM, New York, NY, USA (2007)
24. Jordan, T., Taylor, P.: A sociology of hackers. pp. 757–780. Blackwell Publishing Ltd (1998)
25. Juris, J.S.: Reflections on occupy everywhere: Social media, public space, and emerging logics of aggregation. American Ethnologist 39(2), 259–279 (2012)
26. Khandelwal, S.: Malware and hacking forum seized, dozens arrested (2015)
27. Khanderwal, S.: Creator of megalodonhttp ddos botnet arrested (2016)
28. Kozinets, R.V.: Netnography: Doing Ethnographic Research Online. Sage Publications Ltd. (2009)
29. Lacey, D., Salmon, P.M.: It's dark in there: Using systems analysis to investigate trust and engagement in dark web forums. In: D. Harris (ed.) Engineering Psychology and Cognitive Ergonomics, *Lecture Notes in Computer Science*, vol. 9174, pp. 117–128. Springer International Publishing (2015)
30. Levy, S.: Hackers: Heroes of the Computer Revolution. Doubleday, New York, NY, USA (1984)
31. Macdonald, M., Frank, R., Mei, J., Monk, B.: Identifying digital threats in a hacker web forum. In: Proceedings of the 2015 IEEE/ACM International Conference on Advances in Social Networks Analysis and Mining 2015, ASONAM '15, pp. 926–933. ACM, New York, NY, USA (2015)
32. McCaughey, M., Ayers, M.D. (eds.): Cyberactivism: Online Activism in Theory and Practice. Taylor and Francis, Inc., Bristol, PA, USA (2003)
33. Motoyama, M., McCoy, D., Levchenko, K., Savage, S., Voelker, G.M.: An analysis of underground forums. In: Proceedings of the 2011 ACM SIGCOMM Conference on Internet Measurement Conference, IMC '11, pp. 71–80. ACM, New York, NY, USA (2011)
34. Plesser, B.: Skilled, cheap russian hackers power american cybercrime. NBC News (2014). URL www.nbcnews.com/news/world/skilled-cheap-russian-hackers-power-american-cybercrime-n22371
35. Postill, J., Pink, S.: Social media ethnographie: The digital researcher in a messy web. Media International Australia (2012)
36. Quodling, A.: Doxxing, swatting and the new trends in online harassment. The Conversation (2015). URL theconversation.com/doxxing-swatting-and-the-new-trends-in-online-harassment-40234
37. Rausch, M., Good, N., Hoofnagle, C.J.: Searching for indicators of device fingerprinting in the javascript code of popular websites (2014)
38. Samtani, S., Chinn, R., Chen, H.: Exploring hacker assets in underground forums pp. 31–36 (2015). DOI 10.1109/ISI.2015.7165935
39. Shakarian, P., Shakarian, J., Ruef, A.: Introduction to Cyber-Warfare: A Multidisciplinary Approach, 1st edn. Syngress Publishing (2013)
40. Snodgrass, J.G.: Ethnography of online cultures. In: Handbook of Methods in Cultural Anthropology, pp. 465–496. Rowman and Littlefield, London, UK (2015)
41. Steinmetz, K.F.: Craft(y)ness: An ethnographic study of hacking. British Journal of Criminology 55(1), 125–145 (2015)
42. Steinmetz, K.F., Gerber, J.: "it doesn't have to be this way": Hacker perspectives on privacy 41(3), 29–51 (2015)
43. for Strategic, C., Studies, I.: The economic impact of cybercrime and cyber espionage (2013). URL www.mcafee.com/mx/resources/reports/rp-economic-impact-cybercrime.pdf
44. Tanenbaum, A.S., Wetherall, D.J.: Computer Networks, 5th edn. Prentice Hall Press, Upper Saddle River, NJ, USA (2010)
45. Taylor, P.A.: From hackers to hacktivists: speed bumps on the global superhighway? New Media and Society 7(5), 625–646 (2005)

46. Taylor, R.W., Fritsch, E.J., Liederbach, J.: Digital crime and digital terrorism, 3 edn. Prentice Hall Press (2014)
47. Turkle, S.: The Second Self: Computers and the Human Spirit. Simon and Schuster, Inc., New York, NY, USA (1984)
48. Wall, D.S.: Cybercrime: The Transformation of Crime in the Information Age, 1 edn. Polity (2007)
49. Wei, W.: Hunting russian malware author behind phoenix exploit kit (2013)

Anonymity in an Electronic Society: A Survey

Mauro Conti, Fabio De Gaspari, and Luigi Vincenzo Mancini

Abstract In the wake of surveillance scandals in recent years, as well of the continuous deployment of more sophisticated censorship mechanisms, concerns over anonymity and privacy on the Internet are ever growing. In the last decades, researchers have designed and proposed several algorithms and solutions that allow interested parties to maintain anonymity online, even against powerful opponents. In this chapter, we present a survey of the classical anonymity schemes that proved to be most successful, describing how they work and their main shortcomings. Finally, we discuss new directions in Anonymous Communication Networks (ACN) taking advantage of today's services, like On-Line Social Networks (OSN). OSN offer a vast pool of participants, allowing to effectively disguise traffic in the high volume of daily communications, thus offering high levels of anonymity and good resistance to analysis techniques.

1 Introduction

In an era where we are constantly connected to the Internet with a growing number of devices, consciously or unconsciously leaving traces and bits of personal information across a huge range of services, concerns over privacy and anonymity during online activities are ever growing. As a consequence, in the last years there was an increasing interest amongst the general public in cyber deception tools which allow users to anonymize and obfuscate their communications.

Anonymization is a key concept for cyber deception. Anonymity refers to one's ability to perform tasks online without leaking any information that can expose his identity, such as his IP address for instance. In an anonymized connection, if an adversary were to sniff packets he would be unable to determine both the source and destination of the connection, although in general the content of the packets could be accessed and read.

M. Conti (✉)
University of Padua, Padua, Italy
e-mail: conti@unipd.it

F. De Gaspari • L.V. Mancini
Sapienza University of Rome, Rome, Italy
e-mail: degaspari@di.uniroma1.it; mancini@di.uniroma1.it

© Springer International Publishing Switzerland 2016
S. Jajodia et al. (eds.), *Cyber Deception*, DOI 10.1007/978-3-319-32699-3_12

283

There are numerous tools that provide anonymization capabilities, but in the years there have been three in particular which have proved effective and accessible, gaining momentum and capturing a sizable amount of users [1, 2, 43]: Tor [29], Freenet [22] and I2P [5]. These tools have been to some extent effective in protecting the identity of their users, but they all suffer from several attacks that can compromise the level of anonymity provided, such as:

- Correlation Attacks.
- Sybil Attacks.
- Intersection attacks.
- Partitioning attacks.
- Protocol leaks and injection attacks.

An emerging trend which is affecting cyber communications at large and anonymous networks is the one of On-Line Social Networks (OSN). OSN are pervading every part of our daily lives, deeply altering social interactions and the way we share our interests. OSN not only provide, but encourage its users to share a variety of personal information with a select number of other users. Unfortunately, in many cases this information is directly or indirectly accessible by third parties. OSN enable sharing and communicating on a scale never seen before. However, at the same time they store a wealth of personal information about its users. This data can be aggregated and analyzed, disclosing sensitive private information, and can be used to mount sophisticated social engineering attacks against unsuspecting victims. With respect to Cyber Deception and anonymous communications, OSN become key platforms for two different aspects:

- The large use of OSN helps breaking traditional anonymization techniques (see Sect. 3).
- At the same time, the use of OSN can improve the resilience of anonymization and deception techniques.

It becomes then fundamental to understand the role of OSN in Cyber Deception. In particular, in this chapter we focus on discussing techniques regarding:

- How OSN can help re-identification of anonymized information and be an additional tool to profile/trace users (Sect. 3.1).
- How and to which extent is anonymization in OSN achievable (Sect. 3.2).
- De-anonymization in OSN (Sect. 3.3).
- Deception in OSN (Sect. 3.4).
- How OSN can be used to build tools offering improved anonymization (Sect. 3.5).

Organization The remaining part of this chapter is organized as follows. In Sect. 2, we examine traditional anonymization techniques such as Tor, I2P and Freenet, discussing their design, main vulnerabilities and shortcomings. Section 3 discusses OSN and their impact on anonymous communication networks. In particular, in Sect. 3.1 we analyze how OSN can help profile and trace users online, allowing de-anonymization of anonymous messages. In Sect. 3.2 we discuss if and to which extent is anonymization achievable in OSN, and against what type of adversary. Section 3.3 describes de-anonymization attacks in OSN and some proposed counter-measures. Section 3.4 examines deception in OSN: fake profiles and how to detect them. In Sect. 3.5 we consider how to exploit the huge number of users of OSN to build an anonymous communication network. Finally, in Sect. 4 we draw some conclusions.

2 Traditional Anonymization Techniques

This section reviews the techniques used in traditional anonymous communication networks, the level of anonymity they provide and their limitations.

2.1 Mix Networks

Mix networks [20] are a type of Anonymous Communication Network (ACN) that route packets through a series of proxies called *mixes*. Mix networks guarantee anonymity of their users by combining mixes with the use of public-key encryption. The main idea is that routing the packets through one or more mixes, adding one layer of encryption for each mix in the path, makes it hard for an adversary to link sender and receiver of the communication. Let us consider a host A that wants to communicate with another host B through a mix M. Let k_B and k_B^{-1} be the public and private keys of B, K_A and K_A^{-1} the public and private keys of A, and K_M and K_M^{-1} the public and private keys of M respectively. In this scenario, to communicate with B, host A pads the message m with a random string r and encrypts it with K_B. Then, A appends the address of host B to the encrypted message and encrypts the whole packet with K_M. Finally, A forwards the encrypted data to mix M, which will strip the outer encrypted layer with its private key K_M^{-1}. At this point, M can extract the address of host B and forward the remaining encrypted data to B. Host B can now decrypt the message using its private key K_B^{-1}, discard the random padding and process the content appropriately. A diagram of this process is illustrated in Fig. 1.

It is worth noting that host B, even though it is the recipient of the communication, does not know who the initiator of the connection is since the message is relayed by the mix M. As a consequence, a passive attacker monitoring traffic at either ends of the connection can never know both source and destination of a message:

– If the attacker is monitoring the traffic on the link between host A and mix M, then he will only know that A is communicating with the mix, but will not be able to identify host B as the recipient since the message is encrypted.

Host A   **M**   **Host B**

Fig. 1 Message delivery in Mix Networks. The message, encrypted with the public key of each mix, is decrypted and routed to destination

– If the attacker is monitoring the traffic on the link between mix M and host B, he will know only that B is receiving a communication from the mix, but the identity of the real source A will remain concealed.

Moreover, this communication scheme easily extends to multiple mixes to guarantee anonymity even in a scenario where a mix is compromised. Let us consider again the above example, but this time with the path between A and B composed of two mixes M and M'. In this case, host A first encrypts the data with the public key of M', $K_{M'}$, and successively with K_M. In this scenario, even if one of the two mixes was malicious, the anonymity of the communication would be preserved since each mix knows only its direct predecessor and successor in the path.

Mixes perform another important function: they permute their input and output in such a way that it is hard for an attacker to correlate a given input stream to the corresponding output. How this is done varies based on the specific mix network, but in general it is based on packet aggregation and introduction of random delays. This is a fundamental characteristic that gives mix networks a certain level of resistance against traffic correlation attacks [47], even against an adversary who can potentially monitor the whole network.

2.2 Onion Routing

The introduction of random delays in mix networks helps mitigating timing correlation attacks, but heavily degrades the performance of the network. Such types of networks are called high-latency anonymous networks, and are a good candidate for non-realtime online activities like exchanging emails. With the exponential growth of the World Wide Web and the advent of interactivity and low-latency communications over TCP, high-latency anonymous networks were not suitable anymore. This shift towards realtime communication created the need for a new type of ACN: low-latency anonymous networks. The most famous and widespread low-latency anonymous network is The second-generation Onion Router [29] (Tor). Given the extreme popularity of Tor, in this section we concentrate on the Tor implementation of onion routing.

Fig. 2 Message routing in Tor. Each node unwraps one layer with the negotiated symmetric key SK

Onion routing is an overlay network built on top of TCP and is based on the concept of *circuit*. A circuit is a path in the network through several nodes (which are called Onion Routers, or OR) where each OR knows only its predecessor and successor. Users learn which OR are available through a centralized system of servers, called *directory servers*, that list online OR and their public key. The packets belonging to a given communication are routed through the circuit in fixed-size units of data called *cells*, which have several layers of encryption. Every layer of the packet, one for each OR in the circuit, is encrypted with a different symmetric key. During routing, each OR in the circuit unwraps one layer of encryption (hence the name onion routing) using the pre-negotiated symmetric key. The last OR in the circuit, called the *exit node*, removes the last encrypted layer and forwards the original packet to the final destination. Figure 2 represents an illustration of this process.

In Tor, OR do not permute packets, nor add padding or delays in the retransmission. This allows the Tor network to provide its users with low latency, but—like all low-latency anonymous networks—it does not offer protection against a global adversary (i.e., an adversary who can monitor a significant portion of the network [35]).

2.2.1 Tor Design

As we said before, Tor is an overlay network composed of several interconnected OR. Each OR possesses two keys: a long term *identity key* and a short term *onion key*. The identity key is used to create and maintain TLS connections with each OR in the network and to sign *router descriptors* or *directories* (in case of a directory server). The router descriptor contains a summary of an OR's keys, address, bandwidth and exit policy, which are used to select the OR to form a given circuit. The onion key is used to decrypt requests from the users and negotiate circuit ephemeral keys. The exchanged network traffic is tokenized in fixed-size chunks of data called cells. All types of cells share a field called *circuit ID*, which is used to identify to which circuit a given cell belongs to, allowing the multiplexing of multiple circuits through a single TLS connection. In Tor, there are two types of cells: *relay* cells, which carry end-to-end data, and *control* cells, which are used to create, extend or destroy circuits.

2.2.2 Circuits

A circuit is a multiple-times encrypted connection starting at an Onion Proxy, running on the machine of the connection initiator, and extending through a set of OR. The first OR of the circuit is called the *entry guard* and is selected between a strictly chosen group of OR based on a reputation system. This entry OR is the only node in the network who knows the identity of the initiator of the connection, but does not know the destination address thanks to the layers of encryption. The last OR in the circuit is called *exit node* and must have an exit policy compatible with the type of connection requested. The exit node is the only OR who knows the destination of the connection, but does not know who the initiator is, since cells are relayed to the exit node by another OR.

In Tor, circuits are built incrementally. To build a circuit an onion proxy first chooses a trusted entry guard and sends it a special *create* cell, used to negotiate a common secret through a Diffie-Hellman handshake. Then, the negotiated shared secret is used to derive two symmetric keys; one for each direction of the communication. The circuit is then extended further through an extend cell from the onion proxy to the entry guard, specifying the next node to connect to the circuit (let us call it OR1). The entry guard then forward this cell to OR1, which will reply to the onion proxy (through the entry guard), completing a Diffie-Hellman handshake. At this point the circuit is effectively extended to two nodes, and the onion proxy shares a pair of secret symmetric keys with each OR in the circuit. The circuit can be extended further to an arbitrary length (currently Tor employs three OR [6]) by having the onion proxy tell the last OR in the circuit the next node to connect to. Once the circuit is completely established, the onion proxy can reliably and anonymously communicate with any other host, including hosts in the clearnet, by encrypting the cells once for each negotiated key, from the last to the first. The OR in the circuit will forward the cell to the next hop based on the circuit id field of the cell, which is setup in the circuit creation phase along with the shared keys. When the destination host sends back a response, each OR in the circuit encrypts the packets with the negotiated symmetric key and, upon reception, the onion proxy can unwrap all the layers and process the reply. Figure 3 illustrates the process of building a Tor circuit.

It is worth noting that connections to the clearnet in Tor are not end-to-end encrypted; therefore, if the initiator of the connection is not using a secure protocol (such as HTTPS for instance), the packets leaving the exit node are in plain text, allowing anyone monitoring the line, as well as the exit node, to see their full content. On the other hand, if the destination service of the communication is inside the Tor network (i.e., is also running a tor relay), then the connection is automatically end-to-end encrypted. A particular type of services in the Tor network are *hidden services*.

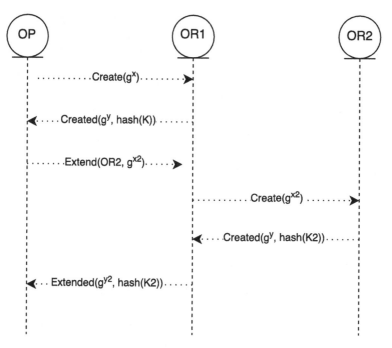

Fig. 3 Circuit creation in Tor. The onion proxy and OR1 exchange the first and second half of a Diffie-Hellman handshake, including the hash of the final key K. Then, the onion proxy asks OR1 to extend the circuit towards OR2 through the same process

2.2.3 Hidden Services

Hidden services are hosts in the Tor network that provide some kind of service to other hosts while remaining completely anonymous (i.e., while maintaining their IP address hidden). Hidden services are implemented in the Tor network through *rendezvous points* (RP) and *introduction points*. A node, let us call it *HS*, that wants to offer an hidden service needs to generate a public/private key pair, which are used to identify the service. Additionally, *HS* needs to choose some introduction points, which are OR that will wait for a connection requests to the hidden service, to build a circuit and share its public key with them. Finally, *HS* needs to create a *hidden service descriptor*, signed with its private key, containing its public key and introduction points. The descriptor is then uploaded to a Distributed Hash Table (DHT) so that clients can easily retrieve it.

When a client wants to connect to *HS*, it first fetches the descriptor of the service from the DHT to learn the public key and the identity of the introduction points. The client then selects an OR, which will act as a RP for the connection. It then builds a circuit to the RP and gives it a randomly generated cookie that can be used to later identify *HS*. Once the circuit is successfully established, the client opens another circuit to one of *HS*'s introduction points and sends it a message containing the random cookie, the identity of the RP and the first half of a Diffie-Hellman handshake. At this point, if the hidden service wants to accept the connection, it will build a circuit to the RP and send a message containing the cookie, the second half of the Diffie-Hellman handshake and the hash of the obtained shared secret. After, the RP connects the circuit of the client to the one of *HS*, and an anonymous tunnel between the two is successfully established allowing the client and the hidden service to communicate maintaining the anonymity of both intact. Using rendezvous points guarantees that no single entity in the network, not even the RP itself, knows the identity of both the client and the service at the same time.

2.3 P2P Anonymous Networks

While Tor is based on a set of OR which are distinct from its users, other ACN architectures consider each participant of the network as a possible routing node. Such peer-to-peer (P2P) anonymous networks aim to offer higher resilience and scalability, making it harder for an attacker to influence the ACN or to successfully mount DoS attacks on the network itself. Two of the most famous P2P ACN are Freenet and I2P. The former provides a distributed anonymous data storage, while the latter aims to provide anonymous communication like Tor.

2.3.1 Freenet

Freenet [21, 22] is an adaptive P2P networks that allows publication, replication and retrieval of information in an anonymous manner, effectively implementing a location-independent distributed file system. Freenet is designed to transparently manage files in the network in order to provide an efficient service which does not rely on a centralized infrastructure or expensive broadcast searches. Each node in the network maintains a local data store, which is made available to other nodes for writing/retrieving of content, and a routing table containing the addresses of other nodes and the *file keys* that they are thought to maintain. File keys are used in Freenet to uniquely identify a file in the network. There exist three different types of keys:

– Keyword-Signed Key (KSK).
– Signed-Subspace Key (SSK).
– Content-Hash Key (CHK).

KSK, the simplest type of file key, is derived from a string describing the file which is chosen by the owner of the file. This string is used to generate a public/private key pair, of which the public key is hashed to generate the KSK, while the private part is used to sign the file itself. Moreover, the file is additionally encrypted using the string description so that nodes storing the files can plausibly deny knowledge of the contents of their datastore. Since with this scheme files are identified only through the KSK, the result is a flat global namespace, which creates problems in case users choose the same descriptive string for two different files.

SSK are designed to address the flat global namespace problem. With this scheme, users can create a new namespace by randomly generating a public/private key pair used to identify it. To add a file under a namespace the user will pick a descriptive string as with KSK, hash the string and the public namespace key separately, XOR the resulting hashes and then hash the result once again. The obtained hash is used as file key and the file is subsequently signed with the private key and encrypted with the string descriptor as before. To retrieve a file a user just needs the public key identifying the private namespace and the descriptive key of the file.

The last type of keys, CHK, are used to implement updating and splitting of files. CHK are derived directly from files through hashing of the file itself, yielding a pseudo-unique file key corresponding to the content of the file. As for KSK and SSK, the user encrypts the file with a randomly generated key, which is then published together with the CHK to allow other users to access it. Additionally, combining CHK with SSK allows to obtain an updating mechanism for files. Indeed, a user can store a file using a CHK and then create an *indirect file* under his private namespace using a SSK. The indirect file will contain the CHK. When the user wants to update the original file, he will just need to store the new version in the network and update the contents of the indirect file in the private namespace. The technique of using indirect files allows to create arbitrarily complex structures, much the same way directories do. CHK and indirect files allow for files to be split too. In fact, a user who wants to store a big file in the network can split it into chunks and create a CHK for each of them. Then, it is sufficient to create an indirect file containing all the chunks' keys to be able to recombine the original file.

Retrieving Files In order to retrieve a file from the network, a user needs first to obtain the associated file key. Once the key is obtained, he will set a hops-to-live value (used to prevent infinite chains) and forward a request to the local node running on his machine. The node will first check the local storage for a correspondence and, if found, it will return the file along with a message identifying the local node as the source of the data. If the file is not found, the local node will look up in its routing table the closest key and the remote node maintaining it, and will forward the request to this remote node. If the request is successful and the file is returned, the node will cache the data in its local storage, update his routing table associating the file key with the data source and finally forward the file to the user. Since having a table of data sources for a set of files is a security risk, any node in the response path can change the data source to point to itself. In this way no node will know for certain which one is the real data source, but if a request for that data is received again they can still route it towards the correct path. If any given node fails in contacting the next step of the path, it will try with the next closest key in its routing table. If it is unable to communicate with any of the nodes in its routing table, it will report back a failure to its upstream neighbor, which will then try with the next closest key. This effectively implements a steepest-ascent hill-climbing search with backtracking. Figure 4 illustrates how requests are handled in the network.

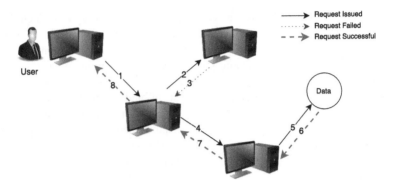

Fig. 4 Data retrieval request in Freenet. The request is routed through the nodes with the closest key. Backtracking is used when a request fails

Storing Data Storing data in the network follows a process similar to retrieving files. The first step for the user is always to calculate the file key. The user then sends a message containing the file key and a hops-to-live number to his own node. If the key is already present in the local storage of the node, the pre-existing file is returned to the user who will then know a collision has occurred. If the key is not found, the node proceeds in the same way as for files retrieval: it searches the closest key present in its routing table and forwards the store request to the associated node. This process is repeated as many times as specified by the user in the hops-to-live value. If none of the nodes in the path registers a collision, an *all clear* message is returned to the user, who will then forward the file itself to his local node. The node will store a copy of the file in the local store and then forward it to the next node, who will do the same. At the end of this process, the file will be stored in the network and replicated in as many nodes as specified in the initial hops-to-live field. Each node in the path will additionally add an entry in its routing table associating the inserter with the new key, as the data source. As for file retrieval, any node in the path can modify the store request claiming to be the source to protect the anonymity of the initiator.

Storage Management Individual nodes can configure the maximum amount of storage they will provide to the network, so no user is forced to store more than he is willing to on his local node. The local datastore of nodes is managed with a Least Recently Used (LRU) policy, with the data sorted in decreasing order based on the timestamp of the last access to the file. When a new store request is received and the storage is full, the least recently used entries are evicted until there is enough space for the new file. This expiration mechanism allows outdated or uninteresting documents to be automatically removed from the network when they are not needed anymore, but at the same time does not give any guarantees on file availability in the network.

Joining the Network Given the distributed nature of Freenet, there is no global directory listing all participating nodes. As a consequence, a distributed protocol to enter the network is required. Additionally, for security reasons it needs to be designed in such a way that no single node can heavily influence the keyspace that the new entering node will manage. To obtain this, Freenet employs a collaborative algorithm where a new node joining the network chooses a random seed and hashes it. The node then sends a *new node* message containing his identity, the hashed seed and a hops-to-live value to a random node, whose identity is learned through out-of-band means. Upon receiving a new node message, a node generates a random seed which is XORed with the received hashed seed and is then hashed again to obtain a new key called a *commitment*. The node then forwards this new hash to a randomly chosen node, which will repeat the process. The message is forwarded to as many nodes as indicated in the initial hops-to-live value. After the hops-to-live expires, each node in the chain publishes his seed, and the final value of the key for the new node is the XOR of all the seeds. This allows to collaboratively generate a key in such a way that no node has a higher influence than others on the final result. Moreover, since during the propagation of the new node message each node forwards his commitment, each node can check that the seeds published are truthful.

2.3.2 I2P

I2P [5, 46] is a low latency, fully distributed, P2P anonymous overlay network which allows users to engage in anonymous communications. Much like Tor, I2P is based on early work on mix networks [20] (see Sect. 2.1) and employs layered encryption and routing through a series of nodes (which are called routers) to guarantee anonymity. Unlike Tor, I2P is fully distributed, rather than relying on directory servers, and is based on packet-switching instead of circuit-switching. I2P employs a modified version of onion routing called *garlic routing*. While routing in Tor is based on the concept of cells (see Sect. 2.2), in I2P messages are sent into *cloves* and can be grouped in *garlic messages* [3]. Cloves are fully formed messages, padded with instructions addressed to intermediary nodes indicating how the message is to be delivered. Additionally, messages are end-to-end encrypted to preserve the confidentiality of the content [23]. Garlic messages are composed of several cloves and are encrypted multiple times, one for each node in the user's *tunnel*.

Tunnels Tunnels are the basis of the anonymity in I2P. In a similar way to how users build circuits to a destination in Tor, in I2P each user builds tunnels. A tunnel is a temporary [44] directed path through a user-selected set of routers, typically two or three. In tunnels, packets are encrypted with several layers, one for each node in the tunnel. Differently than in Tor, in I2P the user's router is also part of the tunnel. To communicate with the network, an I2P user needs to build two different sets of tunnels: outbound and inbound. Outbound tunnels are used to relay messages from the user to another node in the network, while inbound tunnels are used to receive packets from other nodes. The first node of an outbound tunnel (which corresponds to the user's node) is called *outbound gateway*, while the last node of the tunnel is

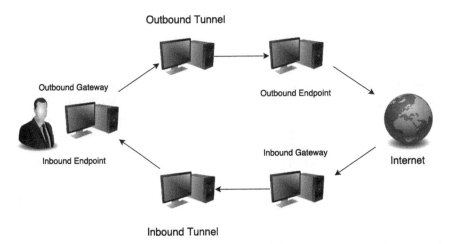

Fig. 5 Outbound and inbound tunnels in I2P

called *outbound endpoint*. For inbound tunnels the last node of the tunnel (the user's node) is called *inbound endpoint*, while the first node is called *inbound gateway*. Figure 5 illustrates inbound and outbound tunnels. When a user wants to send a message to a given destination, let us call it D, it first needs to retrieve the list of D's inbound gateways. This is accomplished by querying the *NetDB*: a global, distributed database based on the Kademlia implementation of a distributed hash table.

NetDB and Nodes A percentage of I2P nodes that satisfy a given set of criteria (generally, high bandwidth and high speed), will be appointed as *floodfill peers*. These floodfill peers compose what is known as the network database, or *NetDB*: a distributed database containing all known information about the network. Normal I2P peers publish themselves by uploading their data and lookup information in the NetDB. There are two types of information stored by each node:

– *RouterInfo*.
– *LeaseSet*.

The RouterInfo data structure contains all necessary data on a specific router, including the identity of the router (public encryption key, DSA signing key and a certificate) and its IP address, all signed with the router's private DSA key. The LeaseSet contains information, called *lease*, about how to contact a given internal I2P service, called *destination*. The lease includes the identity of the inbound gateways of the service, the tunnel ID on the gateways (since each router can be the gateway for several tunnels) and an expiration date for the tunnel. Beside the lease, the LeaseSet contains the identity of the destination, a secondary public key used for end-to-end encryption and the signature of all the LeaseSet data to guarantee its authenticity. This information allows any peer in the network to discover any other peer, once an initial set of floodfill peers is obtained by out-of-band means. The process of obtaining the first floodfill peers is called *reseeding* [33].

Establishing a Tunnel and Communicating The selection of nodes used to create tunnels is done based on tiers. While interacting with other nodes in the network, each user stores statistics on the level of the connection and the reliability of the node, creating a profile for each and organizing them into tiers. When the user needs to build new *client tunnels*, used to communicate with other peers, he will pick an ordered list of two or three nodes out of the top tier set, while for *exploratory tunnels*, which are used for NetDB communications and tunnel management, usually low-tier nodes are employed. Once the nodes are chosen, the source sends a multiple-times encrypted message (once for each selected node) to the first node in the list. This node will unwrap the exterior layer, containing a symmetric key and the next node to which the message has to be forwarded. The node will then send an (encrypted) response to its upstream node, accepting or refusing to be part of the tunnel. The response is routed back to the initiator, with each node in the path adding a layer of encryption (using the negotiated symmetric key), while the tunnel creation request is forwarded to the next node. Once at least one inbound and one outbound tunnels are established, it is possible for a user to communicate with destinations.

To start an anonymous communication with a given destination, the initiator first needs to retrieve from the NetDB the identity of an inbound endpoint of the desired service. Once it obtains the endpoint identity, the initiator will encrypt individual messages with the destination's public key and group them into a garlic message. The garlic message itself is encrypted multiple times, once for each node in the outbound tunnel, using the negotiated symmetric keys. The final encrypted message is sent to the outbound gateway, which will unwrap the first layer and forward the remainder to the next node in the path, after possibly applying delays if the message specifies so. Once the garlic message receives the outbound endpoint, the last layer of encryption is removed and the contained messages are routed to their destination inbound gateways. When the messages reach the inbound gateway of the destination, the node will encrypt them with the symmetric key negotiated in the tunnel creation phase. The encrypted garlic message is then routed through the inbound tunnel of the destination, where each node will add one layer of encryption. Finally, the destination will receive the message, decrypt and process it. If needed, it will then send back a reply to one of the inbound gateways of the initiator through one of his own outbound tunnels. It is worth noting that the destination can not learn in any way which nodes the source client chose as inbound gateways. As a consequence, to receive a reply the source client needs to specify the identity of an inbound gateway in the first message exchanged with the destination. Figure 6 illustrates the routing of a message in I2P.

The most important difference compared to Tor is in the fact that, in I2P, no node in the tunnels knows for certain who the sender and receiver are. Since all users in the network are also nodes that can be used to build tunnels, when a node receives

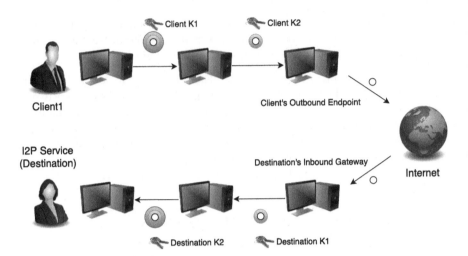

Fig. 6 Communication in I2P. The client sends an encrypted message through its outbound tunnel. Each hop unwraps one layer of encryption until the gateway, which sends the individual messages to the destination's inbound gateway. The packets are finally routed through the inbound tunnel of the destination, adding one layer of encryption at each node with the pre-negotiated symmetric key

a message it has no way of knowing if the message originated by the neighboring upstream node, or if the neighboring node is simply relaying the packet for another user [23]. On the other hand, in Tor the set of OR is well known, and users do not automatically run an OR themselves, allowing entry guards and exit nodes to know the identity of sender and receiver respectively.

2.4 Attacks and Limitations

While these classical anonymity schemes are effective and can provide good levels of anonymity for different tasks, they all suffer in different measures from a set of well known attacks aimed at exposing, or at least reducing, users' anonymity. In this section we present the most well known attacks against classical anonymity techniques. In Sect. 2.4.1 we describe timing correlation attacks, which use statistical data on packets timing to correlate network flows. Section 2.4.2 introduces the sybil attack, which allows an attacker to obtain a high influence in the ACN. In Sect. 2.4.3 we discuss intersection attacks, that can expose the identity of users through aggregation of different types of information. Section 2.4.4 analyzes partitioning attacks, which aim to segregate users into distinct groups to reduce the cardinality of their anonymity set. Finally, in Sect. 2.4.5 we describe high level protocol leaks and injection attacks, which can reveal users' information if not properly sanitized.

2.4.1 Timing Correlation Attacks

The most important threat to anonymity when using ACN comes from end-to-end attacks. As the web moves towards realtime, interactive content, ACN need to adapt and provide low latency communication. This imposes an important restriction in how ACN can operate; low latency ACN can provide high resilience against many deanonymization techniques, but the low delays in packet transmission inherently expose the network to timing correlation attacks [47]. If an adversary is strong enough to monitor traffic flows from both endpoints of a connection, it is possible to use timing correlation attacks to match two flows and confirm that there is a connection between the two hosts. These attacks are based on the statistical distribution of transmitted packets at one end and received packets at the other end, making it extremely difficult to defend against without applying long delays to packets forwarding. Extensive work shows that low latency ACN are highly vulnerable to these kind of attacks [27, 38]. Given the nature of the attack, traffic correlation is generally used as *confirmation* of a suspicion. However, when controlling entry guard and exit node simulations suggest that it is also possible to match arbitrary flows and disclose the identity of the two endpoints [35].

2.4.2 Sybil Attack

Another type of attack against ACN is the sybil attack [30]. The sybil attack aims at subverting a reputation system through forging of identities, which allows the attacker to gain more influence in the network. This type of attack can be used to increase the probability that a user will pick one or more malicious nodes to build his circuit/tunnel, allowing the attacker to harvest information about the user's traffic. In Tor, this attack can be particularly dangerous due to the fact that entry guards know the identity of the user, and exit nodes know the destination of a communication. Therefore, an attacker becoming a guard or exit node is in a good position to harvest information and mount additional attacks on the network, like in [7]. In I2P, a sybil attack can be used to gain control of all the floodfill peers in the network, since their number is fixed, and then mount additional attacks to deanonymize users [31].

2.4.3 Intersection Attack

Intersection attacks [36] aim at reducing the anonymity of ACN users through intersection of different properties of a victim. These attacks exploits all information that is possible to obtain and link to pseudo-identities. This information can be: repeated communications, posts under pseudonyms or even just the mere presence of a node in the ACN. After gathering enough information, it is possible to correlate distinct properties of the pseudonym to restrict the set of anonymous users [41]. Intersection attacks in I2P, for instance, can exploit the presence or absence of a target node in the network. At each round of monitoring, the attacker stores the total set of online nodes and the presence/absence of the target node. After several rounds, the attacker can intersect the anonymity sets of the victim and, with enough measurements, expose his identity or greatly reduce his anonymity [4].

2.4.4 Partitioning Attacks

Partitioning attacks aim at segregating users of the ACN in separate, distinct groups. Since the strength of ACN comes from the amount of traffic amongst which one can hide his communications, and therefore from the number of users in the network, an attack segregating users into separate, distinguishable groups can greatly reduce the anonymity of the ACN. In the general case, partitioning attacks exploit any tangible difference between users of the ACN; this includes for instance the set of OR a client connects to in Tor, or the number of tunnels a user employs in I2P [4]. This type of attack can be mounted, for instance, through DoS [32], which allows the attacker to group users based on the nodes employed in their circuit/tunnels, highly compromising their anonymity.

2.4.5 Information Leak

Yet another set of limitations comes from using complex high-level protocols (e.g., HTTP) through the ACN. In fact, such protocols leak information that can possibly allow an eavesdropping adversary to identify the source of the connection. A similar family of attacks on non-encrypted links of the connection (for instance, an HTTP connection between an exit node in the ACN and a host in the clearnet) are injection attacks. Indeed, if the connection exiting the ACN uses a non-encrypted protocol, it is possible for an adversary to inject malicious payload in the packets to cause the source of the connection to disclose identity-revealing information. This is particularly a problem with ACN that allow users to connect to the clearnet through exit nodes. Since the destination of the connection is outside of the ACN, the exit node must use standard protocols to communicate with it. Therefore, if this connection uses an insecure protocol, the identity of the initiator can be easily exposed through injection attacks.

3 OSN and Anonymity

On-line social networks have radically changed the way people interact and communicate online. As happened with the evolution of the web and the advent of interactivity, the changes brought forth by OSN have a strong impact on the effectiveness of anonymous communication networks and the way they will evolve. In this section, we discuss the role of social networks in cyber deception and anonymous networks, the challenges they pose and how they can be used to build the next generation of ACN. In Sect. 3.1 we discuss the potential for identity disclosure through users profiling and the analysis of social graphs. Section 3.2 analyzes techniques to provide anonymity in the context of OSN. Section 3.3 describes deanonymization attacks in OSN and introduces some defense mechanisms. In Sect. 3.4 we illustrate deception in OSN: profile cloning and countermeasures. Finally, in Sect. 3.5 we discuss how to use OSN to build an anonymous communication network.

3.1   User Profiling Through OSN

By nature, even when restrictive settings are applied, social networks leak a wealth of information about users, their habits and their connections with other users [14, 17]. This information constitutes *user profiles*, which potentially allow malicious third parties to infer sensitive information. Additionally, such user profiles can be exploited by an adversary to reduce the anonymity set of a target ACN user. For instance, let us assume that an attacker is monitoring the access patterns to a website and is interested in knowing if a specified user U visited it, but does not have access to outgoing traffic from U. Then, it would be possible for the attacker to look for possible matches between the content offered by the website and the material posted by U on an OSN. If the content matches and there is a good correspondence in the timing of the access to the web site and the posting on the OSN, then it is likely that the logged access comes from U.

Following a similar direction, in [19], the authors present a user profiling technique aimed at disclosing hidden or anonymized sensitive information through analysis of Facebook comments. In particular, the authors focus on military-related published news, where a person's name is censored for security reasons, and try to disclose his identity by analyzing comments and the social network graph of the commenters. Military information regarding officers in key positions is always censored before release to the public. This is a form of protection for military personnel, so that the officers and their families can not be identified and become potential targets for hostile forces.

In their work, the authors consider a set of 48 news with sensitive censored information and a total of over 3500 Facebook comments. Amongst these, they separate comments that leak information about the identity of the officers in the censored news article, called leakage comments. The rationale behind the approach is that, if a commenter possesses confidential information that is not disclosed in the censored article, then he is likely to have strong ties to personnel involved in the news item. Exploiting the information published in social networks, it is most likely possible to identify these ties through the use of the social graph of the commenter. Then, given a sufficient number of commenters, the intersection between their social graphs will yield with high probability the identity of the anonymized officer, or at least greatly restrict his anonymity set to a few identities.

Given the comments containing leaks, in [19] the authors use an information harvesting software called SocialSpy [17] to retrieve the social graph of the poster of each Facebook comment. The social graph contains the identities of all the poster's friends, which are likely to include the hidden officer. When all the social graphs are obtained, their intersection contains the most likely identities for the anonymous officer.

3.2 Anonymity in OSN

Given how easy it is for third parties [11, 45], or even worse for the OSN itself, to profile users and infer potentially sensitive private information, it is fundamental for users to be able to interact in an anonymous way. Unfortunately classical anonymity techniques (see Sect. 2) are powerless when applied to OSN, since users need to login in order to access the OSN platform. There is therefore the need for different approaches, that allow users to retain exclusive control of their published content and not share their sensitive data with the OSN. In fact, even if users are allowed to configure the social network's settings, their information is still stored by the OSN itself, which is subject to hacking and can potentially sell user's data for profit [42].

3.2.1 Virtual Private Social Network

To provide users with acceptable levels of privacy and to prevent users profiling, in [25, 26] the authors propose the concept of *Virtual Private Social Network* (VPSN). VPSN is a concept similar to VPN, but applied to OSN; a VPSN is an overlay network that takes advantage of the pre-existing architecture of a regular social network (called *host social network*) to allow users to interact without sharing private information with the underlying OSN. The most compelling feature of VPSN is that they do not have dedicated infrastructure, but leverage the pre-existing architecture of the host social network. Contrary to previous works on privacy-aware social networks, VPSN does not require users to migrate to new OSN, which is highly unlikely, but allows them to remain in the network used by all their friends. Also, VPSN are private, meaning that only members of the network itself can access its content and that only members know of its existence, which is otherwise hidden even from the host social network. This is because it is unlikely that an OSN would accept the use of a VPSN, since it would heavily reduce their profit. Additionally, a VPSN should be transparent to the end users, who should be able to browse his friends' profiles as if he was directly using the host social network. Therefore, a user must be able to access profiles of friends connected to the VPSN and friends not connected to it, so that there are no imposed restrictions or limitations.

In [25], the authors propose FaceVPSN, a VPSN implementation for Facebook. In their work, they acknowledge that social networks often do not allow publication of encrypted text or information, and therefore propose to use *pseudo information*.

Once a user uploads pseudo information on his profile, the real information is sent directly to the machines of his friends, where it is stored in a fashion similar to cookies. Every time a user browses the profile of a VPSN user, a plugin installed in the web browser automatically recalls the real information stored locally, transparently updates the web page and displays the real information to the user.

The process of exchanging the information required to map the pseudo profile with real user profiles is done out-of-band, meaning the two users, let us call them *A* and *B*, need to communicate their real identity information with other means than the VPSN or the OSN. This initial information required is called *FaceVPSN business card*, and includes the pseudo name of *A* in the OSN, an XML profile with the real information of *A*, a symmetric key used to encrypt future updated XML profiles and a set of *XMPP PubSub* [8] server IDs. The XMPP PubSub servers are publicly available servers implementing a publisher/subscriber system based on XMPP. Upon receiving the business card of *A*, *B* replies with his own business card and, after both users complete the subscription to the PubSub server, the system is setup. Any time a user updates his own, real profile, an update is automatically generated and is sent to the PubSub system, which forwards it to all the registered subscribers. Figure 7 illustrates this process.

3.2.2 Hiding Interactions in OSN

In the literature, there is extensive research on how to keep the content of a message confidential, but even when the content is not directly accessible it is still possible to disclose sensitive information [16, 45]. For instance, in [18] the authors show

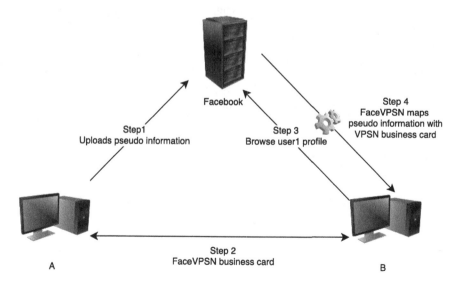

Facebook

Step 4
FaceVPSN maps
pseudo information with
VPSN business card

Step1
Uploads pseudo information

Step 3
Browse user1 profile

Step 2
FaceVPSN business card

A

B

Fig. 7 Business card exchange and translation process in FaceVPSN

how it is possible to disclose the sexual orientation of OSN users simply through the analysis of their connections (i.e., the relationships with other users in the social network). Additionally, beside establishing the existence of a relationship between users, the OSN can derive information on such relationship strength based on the number of time a user visits another user's profile, for example [12]. Therefore, OSN users need to employ techniques to prevent the social network itself from correlating the interactions between end users in order to obtain anonymity and privacy.

Building on works like [28] and [39] on using friends to secure the access to ACN, in [14] the authors propose a system to hide user interaction in social networks, both from the OSN and from other users, through a decentralized P2P system called VirtualFriendship (VF). VF exploits the existing trust between friends to relay messages for other users, creating an indirect communication that the OSN can not trace. VF distinguishes between different entities:

– The communicating users, which are the OSN users that want to interact with other OSN users. In VF, users need to run a local server that is used to relay traffic outside the OSN.
– The routing users, which are normal OSN users acting as entry/exit points from the OSN.
– An anonymous network, used to anonymize communications between users outside to the social network. The use of an ACN allows to keep the identity of the requesting user anonymous, even from the users relaying his request.
– The OSN, which is the platform storing the social information of the users and providing the infrastructure for intra-OSN communication.

When a user A wants to interact with another user B, for instance to browse B's profile, VF will automatically route the request through one of his trusted friends A_F. A will forward the request to his friend, who will in turn relay it to the local server of one of B's friends, B_F, through the anonymous network. This allows to protect the identity of A from the entry node of the ACN, while at the same time hiding the communication to the OSN, which only sees an interaction between A and his friend. Upon receiving the request, B_F checks if it should be allowed by means of an authentication token (which is possessed by all B's friends) generated by B and exchanged out-of-band. If the request is authenticated, B_F will retrieve the profile, encrypt it with the token and send it back to A_F through the ACN. Finally A_F relays the message back to A, who can decrypt it and access the profile of user B. This request process is illustrated in Fig. 8.

This protocol can be extended to posting comments by having user A sending user B a *request to post* message, containing the text to post in the profile of B. Upon successful authentication, B himself will proceed to publish the comment instead of A, hiding the interaction with A to the OSN, and allowing B to review the content before its publication. This communication protocol effectively impairs the ability of the OSN to infer the strength of friendship bonds between users, reducing the privacy leaks and additionally providing anonymity with regards to all external viewers (except for the first trusted friend A_F).

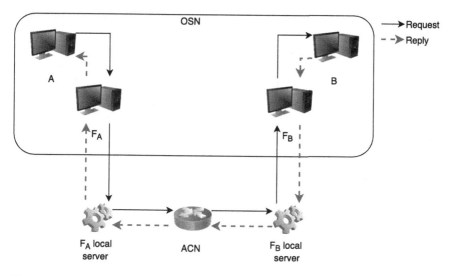

Fig. 8 Indirect user interaction through VirtualFriendship

3.3 De-Anonymization in OSN

While encryption techniques protect users' contents and communication protocols (such as VirtualFriendship) hide users' interactions, the social graph of users is still obtainable, allowing the OSN and third parties to infer sensitive information about OSN members. Indeed, in [40], Narayanan et al. propose the first demonstration of large-scale deanonymization of OSN users through passive analysis of the network structure. In their work, the authors consider an attacker with access to an anonymous, possibly sanitized network graph for an OSN and to some partial, auxiliary information about the OSN users (such as the information obtainable from another, non-anonymous social network for instance). This auxiliary information must include relationships between entities, such that it is possible to build a graph; nodes represents users and edges represent a relationship between two users. Both nodes and edges have a set of associated attributes which are weighted with a probability. For instance, an edge between two users could have the attribute "friendship" with a probability of 70 % and an attribute "contact" with a probability of 30 %. Additionally, the authors assume that the attacker possesses detailed knowledge of an extremely small set of users in the anonymized social network, and can determine if these users are also present in his auxiliary information database. This can be done by matching contextual information such as username, for example.

The identification algorithm is based on the topology of the two graphs; after obtaining an initial match between a small set of nodes in the anonymous OSN and some nodes in the auxiliary graph, it is possible to apply a propagation algorithm that extends the mapping between the two graphs based only on their topological structure. The algorithm is run in several rounds, using at each execution the output of the previous run as an input, until either all nodes are identified or the algorithm can not proceed any further. This simple but effective algorithm allowed the authors to successfully identify over 30 % of the users in their evaluation, with an error rate of 12 %. Amongst the incorrectly identified users, in over 73 % of the cases the algorithm returned either the identity of a user who was at distance 1 from the correct mapping, or the identity of someone who was in the same geographic location. Therefore, even when it was not possible to correctly identify a user, the algorithm provided useful contextual information to further reduce the anonymity set of the victims.

It is clear then that in order to preserve privacy and anonymity in OSN, it is not enough to hide the identity of users and their direct interactions, but it is necessary to deceit observers by creating indirect relationships. In [13], the authors tackle this issue of contextual privacy and propose a new approach where a common friend is used as a proxy for a relationship between two users. The main idea behind the approach is that creating an indirect relationship through the use of an extra hop will hinder deanonymization approaches based on social graph topology. The intermediary of the relationship is called *Friend in the Middle* (FiM) and needs to be connected to the two users that want to establish an untraceable relationship. In this architecture the FiM acts in a manner similar to a mixer [20] (see Sect. 2.1). Consequently, it is also possible to use multi-hop connections between users, which grant higher resilience in case some of the friend nodes used as relays are compromised. Additionally, the FiM approach requires the two users that want to establish an indirect relationship to agree on a set of FiM nodes and to exchange them out-of-band (i.e., outside the domain of the social network).

The experimental results obtained by the authors show that, when using de-anonymization attacks similar to [40], the percentage of successfully identified nodes is lower than 1 % when a user applies FiM to 5–10 % of his connections, as illustrated in Fig. 9.

3.4 Deception in OSN

While participating in an OSN exposes users to risks of profiling and invasion of privacy, not participating can lead to privacy violations too. Given the amount of information that users share in social networks and the direct connection OSN give to the friends of a possible target, OSN are a prime target for identity theft attacks. In [15], the authors propose a new identity theft attack where an attacker exploits the profile of a victim on a given social network to build a forged profile in another OSN, where the victim has not created an account yet. The authors argue and demonstrate how it is possible to automatically crawl profiles and download the complete personal information of over a million OSN users, and then use that information to clone their profiles. The cloning is done both on the same OSN where the profile has been obtained from, and in different social networks where the victim

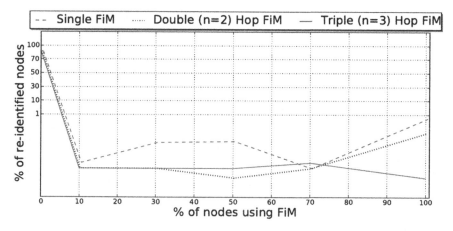

Fig. 9 Percentage of successfully identified nodes with varying rate of relationships per node using FiM. Image taken from [13]

does not possess an account yet. The rationale behind the idea of cloning an account in the same OSN as the original one is that, when receiving a friend request, most users do not pay particular attention to the requester and if it is already in their friends list. The authors show in the evaluation of their attack that, when considering profiles cloned in the same OSN as the original one, over 60 % of the requests are accepted for all the cloned accounts (reaching a peak of 90 %).

The second type of impersonation attack, the cloning of the profile to another OSN, requires that the profile to clone does not exist in the target OSN and that a reasonable number of the victim's contacts have an account in both social networks. In their evaluation, the authors clone the profile of five OSN users to the LinkedIn social network and send a request to all their friends that were also registered on LinkedIn, obtaining an acceptance rate of 56 %.

Reckoning the dangers of cloned profiles, in [24] the authors propose a mechanism for detecting fake profiles based on: the growth of their friend set over time, the real-life social interaction between users, and the evolution of the OSN graph over time. By combining these metrics obtained from the examination of real OSN data, the authors provide a framework to detect fake profiles in OSN where the victim has no real profile. This differs from previous techniques based on comparison between different profiles [34, 37], which are not effective in a scenario where there is no real profile. In their work, the authors show that in the general case the number of friends of legitimate users increases extremely rapidly in the first period of participation in the OSN, while in the long term it slows down and increases at a constant rate. Therefore, if the rate of growth of a monitored profile differs sensibly from the average, it might indicate malicious activity. Additionally, the authors note that an attacker is likely to avoid adding to their friends list any person with whom the victim interacts often in real life. This is due to the fact that close friends and relatives are likely to interact often with the victim on the social network and, when they meet in real life, there is a high chance that they might talk about posts or comments they left on the profile of the victim, therefore exposing the identity theft. Finally, the last part of the fake profile detection mechanism is based on the typical topology of the users of the social network. The authors note that the average degree of the nodes of the typical user's social graph is highly influenced by the removal of close friends (as is in the case of fake profiles, as argued earlier). Additionally, they observe that when analyzing the connected components of the graph of a typical user, on average most users have a big connected component, a few small sized ones and, finally, a significant number of isolated nodes. Therefore, monitoring the degree of the nodes of a suspect profile and the number and size of its connected components can provide valuable insight on the legitimacy of an account.

3.5   Building Anonymous Communication Networks in OSN

While the amount of information shared in social networks threatens the privacy of users, the high number of participants makes OSN a compelling platform to build distributed ACN that leverage social trust. One of the main problems with classical anonymity schemes like Tor is assessing the trustworthiness of nodes. Indeed, since anyone can run a Tor relay and become part of the network, deciding which nodes can be trusted to become, for instance, entry guards is an extremely hard problem. Leveraging the relationships of users in social networks allows to obtain ad-hoc sets of trusted nodes for each user, reducing the risks of including compromised nodes in the anonymity chain and therefore making the communication more secure.

In [28] the authors propose Drac, a distributed P2P ACN based on circuits. In particular, Drac exploits the infrastructure of an OSN and the trust relationship between connected users to increase the security of circuits. Drac assumes that there is a trust relationship between friends, meaning that friends in the OSN will not expose the identity of other friends, and that they share a pre-negotiated encryption key. Let us assume a user u_1 wants to communicate with one of its *contacts* c_1, which is a user of the OSN who is not in the friends set of u_1. Then, u_1 will build a circuit by means of a random walk, starting with one random user in his friend set. The circuit is then extended until it reaches a given length in a similar fashion to how circuits are built in Tor. The last user of the circuit is called *entry point* of user u_1, and is used to create a bridge to the entry point of the contact c_1, similar to tunnels in I2P (see Sect. 2.3.2). To ensure confidentiality of the exchanged messages, all traffic is encrypted in different layers, and at each hop the nodes in the circuit remove one layer of encryption until the message reaches the destination. Additionally, Drac requires the existence of a *presence server* (PS), that is assumed to follow the protocol but to be potentially malicious. The presence server is used to relay an initial message to a contact with whom a user wants to communicate with, containing the identity of the entry point of the user and half of a Diffie-Hellman handshake which is later used to establish a shared secret. Drac offers resilience to correlation attacks (see Sect. 2.4.1) performed by a global adversary

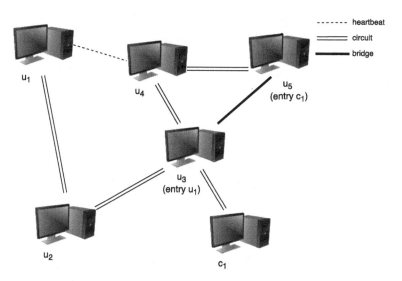

Fig. 10 Communication path between u_1 and c_1. u_1 creates a circuit to his entry point u_3, who is bridged to u_5, the entry point of c_1. Contact c_1 has a circuit to u_5 through u_3 and u_4

through *heartbeats connections*, which a users establishes with each of his friends as soon as he enters the network. Heartbeats connections are low bandwidth, bi-directional connections that are padded at a very low rate, maintaining a constant stream of data to disguise messages between users. In this way, not even the OSN can correlate incoming and outgoing packets. Figure 10 illustrates how Drac works.

While Drac is effective in preserving the anonymity of users against global adversaries, it does so by means of constant, low volume padding sent through the heartbeat connections. This might not be desirable in case of mobile connections, as usually such connections are metered and the resources of the device are limited. This is especially true for the battery, the duration of which can be severely impacted by the constant exchange of data, which prevents entering low power mode. To address this problems, in [9] the authors build on the work of Ardagna et al. [10] and propose a new low-cost protocol that guarantees (α, β)-anonymity to users of handheld devices like smartphones. In their work, the authors consider an environment with four parties:

– The smartphone users.
– The cellular network operator.
– A cloud operator, where users possess a software clone of their phone used to offload computations.
– A set of proxies, used to mediate communication with the smartphone clones in the cloud.

The authors assume that users can communicate with both their clone in the cloud, through the network operator, and with nearby phones, through ad-hoc WiFi connections. The aim of the communication protocol is to hide the sender of the communication amongst at least α devices, and the receiver amongst at least β devices, where both values are configurable by the sender. To achieve this, a sender s that wants to communicate with a destination d, randomly selects a clone c_1 in the cloud belonging to one of her friends in an OSN. Clone c_1 is selected in such a way that he has at least α connections in its social graph. S waits to be surrounded by at least α devices (to hide his identity from the network operator), then sends a multiple-times encrypted message to his proxy p_1 through probabilistic, multi-hop WiFi forwarding to devices in its proximity. Upon receiving the message, the proxy removes one layer of encryption, extracts the identity of the friend's clone c_1 and forwards the remaining message to c_1. Finally, c_1 delivers the encrypted message to α of his friends' clones, including the clone of the sender c_s, which will be the only one able to decrypt the message through a pre-shared symmetric key. At this stage, none of the actors in the system can positively identify the sender with a probability higher than $1/F_{c1}$, where F_{c1} is the number of friends that c_1 has in the social network. Since c_1 is selected such that he has at least α friends in the OSN, α-anonymity for the sender is preserved. After a fixed amount of time, all the friend clones of c_1 reply to him with a dummy message. This includes c_s, whose message is encrypted with different layers and contains instructions to forward the content to one of the friend clones of d, let us call it c_2. Node c_2 is selected such that there are at least β friends in his friends set. Upon receiving the messages, c_1 removes one layer of encryption and forwards the remaining data to c_2, which will in turn remove another encryption layer and deliver the message to all his friends' clones. Amongst these clones, only c_d, the clone of the destination, is able to remove the last layer of encryption and access the plain text message. If the message needs to be delivered to the real device d, the process is the same; all clones reply to c_2 after a fixed amount of time and c_2 will forward the message to the proxy. In turn, the proxy forwards the message to a random physical device d_1 in the proximity of d and d_1 broadcasts the message to all nearby devices. This communication process is illustrated in Fig. 11.

4 Conclusion

In this chapter, we analyzed the evolution of anonymity solutions, ranging from high-latency mix networks to realtime, low-latency onion routing and P2P overlay networks, exploiting OSN and trust relationships between users. We described how each type of ACN has its advantages and its drawbacks, making the choice of which anonymous network to use situational and subject to the strength of the adversary. Moreover, given the rising popularity of social networks and the trend of users sharing online more and more details about their lives, we have described how the analysis of large quantities of data can disclose extremely sensitive information.

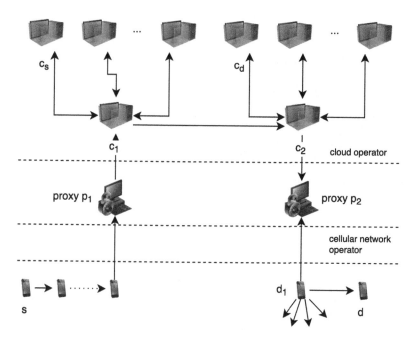

Fig. 11 Anonymous message delivery from s to d through the clones c_1 and c_2

Also, we discussed how users can exploit the infrastructure and the popularity of OSN to communicate and interact anonymously, obtaining a high level of anonymity thanks to the considerable amount of traffic present in social networks.

Future Work In the future, as we move towards the Internet of Things, the amount of data big companies will be able to obtain about their users will increase tremendously, exposing user anonymity and privacy to new, stronger threats. As a consequence, we believe that cyber deception and anonymity will remain fundamental areas of research in the future as well as in the present. Future anonymity schemes should probably exploit the unparalleled levels of connectivity offered by the sled of new, always connected devices. This allows to collaboratively relay messages in an ad-hoc, P2P manner, providing high resilience to failures and making it extremely difficult for observers to have global knowledge of the network. We believe that in a world where every device is interconnected, the benefits of decentralized solutions will outweigh their problems, making centralized (and therefore more easily observable) solutions less attractive, if not obsolete.

References

1. 2010 free software awards. http://www.fsf.org/news/2010-free-software-awards-announced.
2. Growing the network, spreading the word. https://download.i2p2.de/media/i2pcon/2015/slides/I2PCon-2015-zzz-GrowingTheNetwork.pdf.
3. I2p documentation. https://geti2p.net/en/docs/how/tech-intro.
4. I2p threat model. https://geti2p.net/en/docs/how/threat-model.
5. The invisible internet project (i2p). https://geti2p.net/en/.
6. Official Tor FAQ. http://www.torproject.org/docs/faq.html.en.
7. Tor security advisory. https://blog.torproject.org/blog/tor-security-advisory-relay-early-traffic-confirmation-attack.
8. Xmpp standard. https://xmpp.org/xmpp-software/servers/.
9. C.A. Ardagna, M. Conti, M. Leone, and J. Stefa. An anonymous end-to-end communication protocol for mobile cloud environments. *Services Computing, IEEE Transactions on*, 7(3): 373–386, 2014.
10. Claudio A. Ardagna, Sushil Jajodia, Pierangela Samarati, and Angelos Stavrou. Providing users' anonymity in mobile hybrid networks. *ACM Trans. Internet Technol.*, 12(3):7:1–7:33, 2013.
11. Marco Balduzzi, Christian Platzer, Thorsten Holz, Engin Kirda, Davide Balzarotti, and Christopher Kruegel. Abusing social networks for automated user profiling. In *Recent Advances in Intrusion Detection*, volume 6307 of *Lecture Notes in Computer Science*, pages 422–441. 2010.
12. Ero Balsa, Carmela Troncoso, and Claudia Diaz. A metric to evaluate interaction obfuscation in online social networks. *International Journal of Uncertainty, Fuzziness and Knowledge-Based Systems*, 20(06):877–892, 2012.
13. F. Beato, M. Conti, and B. Preneel. Friend in the Middle (FiM): tackling de-anonymization in social networks. In *Pervasive Computing and Communications Workshops (PERCOM Workshops), 2013 IEEE International Conference on*, pages 279–284, 2013.
14. F. Beato, M. Conti, B. Preneel, and D. Vettore. Virtualfriendship: Hiding interactions on online social networks. In *Communications and Network Security (CNS), 2014 IEEE Conference on*, pages 328–336, 2014.
15. Leyla Bilge, Thorsten Strufe, Davide Balzarotti, and Engin Kirda. All your contacts are belong to us: Automated identity theft attacks on social networks. In *Proceedings of the 18th International Conference on World Wide Web*, WWW '09, pages 551–560, 2009.
16. Joseph Bonneau, Jonathan Anderson, Ross Anderson, and Frank Stajano. Eight friends are enough: Social graph approximation via public listings. In *Proceedings of the Second ACM EuroSys Workshop on Social Network Systems*, SNS '09, pages 13–18, New York, NY, USA, 2009. ACM.
17. Andrea Burattin, Giuseppe Cascavilla, and Mauro Conti. Socialspy: Browsing (supposedly) hidden information in online social networks. In *Risks and Security of Internet and Systems*, Lecture Notes in Computer Science, pages 83–99. 2015.
18. Jernigan. C. and B. F. T. Mistree. Gaydar: Facebook friendships expose sexual orientation. First Monday Journal, 2009.
19. G. Cascavilla, M. Conti, , D. Schwartz, and I. Yahav. Revealing censored information through comments and commenters in online social networks. In *Proceedings of the 2015 IEEE/ACM International Conference on Advances in Social Networks Analysis and Mining*, ASONAM, 2015.
20. David L. Chaum. Untraceable electronic mail, return addresses, and digital pseudonyms. *Commun. ACM*, 24(2):84–90, 1981.
21. Ian Clarke, Scott G. Miller, Theodore W. Hong, Oskar Sandberg, and Brandon Wiley. Protecting free expression online with Freenet. In *IEEE Internet Computing*, volume 6, pages 40–49, 2002.

22. Ian Clarke, Oskar Sandberg, Brandon Wiley, and Theodore W. Hong. Freenet: A distributed anonymous information storage and retrieval system. In *International Workshop on Designing Privacy Enhancing Technologies: Design Issues in Anonymity and Unobservability*, pages 46–66, 2001.

23. Bernd Conrad and Fatemeh Shirazi. A survey on Tor and I2P. In *Proceedings of the 9th International Conference on Internet Monitoring and Protection (ICIMP 2014)*, 2014.

24. M. Conti, R. Poovendran, and M. Secchiero. Fakebook: Detecting fake profiles in on-line social networks. In *Advances in Social Networks Analysis and Mining (ASONAM), 2012 IEEE/ACM International Conference on*, pages 1071–1078, 2012.

25. Mauro Conti, Arbnor Hasani, and Bruno Crispo. Virtual private social networks. In *Proceedings of the First ACM Conference on Data and Application Security and Privacy*, CODASPY '11, pages 39–50, 2011.

26. Mauro Conti, Arbnor Hasani, and Bruno Crispo. Virtual private social networks and a facebook implementation. *ACM Trans. Web*, 7(3):14:1–14:31, 2013.

27. George Danezis. The traffic analysis of continuous-time mixes. In *Privacy Enhancing Technologies*, Lecture Notes in Computer Science, pages 35–50. 2005.

28. George Danezis, Claudia Diaz, Carmela Troncoso, and Ben Laurie. Drac: An architecture for anonymous low-volume communications. In *Proceedings of the 10th International Conference on Privacy Enhancing Technologies*, PETS'10, pages 202–219, 2010.

29. Roger Dingledine, Nick Mathewson, and Paul Syverson. Tor: The second-generation onion router. In *Proceedings of the 13th Conference on USENIX Security Symposium - Volume 13*, SSYM'04, pages 21–21, 2004.

30. John R. Douceur. The sybil attack. In *Revised Papers from the First International Workshop on Peer-to-Peer Systems*, IPTPS '01, pages 251–260, 2002.

31. Christoph Egger, Johannes Schlumberger, Christopher Kruegel, and Giovanni Vigna. Practical attacks against the i2p network. In *Research in Attacks, Intrusions, and Defenses*, Lecture Notes in Computer Science, pages 432–451. 2013.

32. Nathan S. Evans, Roger Dingledine, and Christian Grothoff. A practical congestion attack on Tor using long paths. In *Proceedings of the 18th Conference on USENIX Security Symposium*, SSYM'09, pages 33–50, 2009.

33. Michael Herrmann and Christian Grothoff. Privacy-implications of performance-based peer selection by onion-routers: A real-world case study using i2p. In *Privacy Enhancing Technologies*, pages 155–174. Springer Berlin Heidelberg, 2011.

34. Lei Jin, Hassan Takabi, and James B.D. Joshi. Towards active detection of identity clone attacks on online social networks. In *Proceedings of the First ACM Conference on Data and Application Security and Privacy*, CODASPY '11, pages 27–38, 2011.

35. Aaron Johnson, Chris Wacek, Rob Jansen, Micah Sherr, and Paul Syverson. Users get routed: traffic correlation on Tor by realistic adversaries. In *Proceedings of the 2013 ACM SIGSAC Conference on Computer and Communications Security*, CCS '13, pages 337–348, 2013.

36. Dogan Kedogan, Dakshi Agrawal, and Stefan Penz. Limits of anonymity in open environments. In *Revised Papers from the 5th International Workshop on Information Hiding*, IH '02, pages 53–69, 2003.

37. G. Kontaxis, I. Polakis, S. Ioannidis, and E.P. Markatos. Detecting social network profile cloning. In *Pervasive Computing and Communications Workshops (PERCOM Workshops), 2011 IEEE International Conference on*, pages 295–300, 2011.

38. BrianN. Levine, MichaelK. Reiter, Chenxi Wang, and Matthew Wright. Timing attacks in low-latency mix systems. In *Financial Cryptography*, Lecture Notes in Computer Science, pages 251–265. 2004.

39. P. Mittal, M. Wright, and N. Borisov. Pisces: Anonymous communication using social networks. NDSS, 2013.

40. A. Narayanan and V. Shmatikov. De-anonymizing social networks. In *Security and Privacy, 2009 30th IEEE Symposium on*, pages 173–187, 2009.

41. Mike Perry. Securing the Tor network.
http://www.blackhat.com/presentations/bh-usa-07/Perry/Presentation/bh-usa-07-perry.pdf.

42. Christopher Riederer, Vijay Erramilli, Augustin Chaintreau, Balachander Krishnamurthy, and Pablo Rodriguez. For sale : Your data: By : You. In *Proceedings of the 10th ACM Workshop on Hot Topics in Networks*, HotNets-X, pages 13:1–13:6, 2011.
43. Stefanie Roos, Benjamin Schiller, Stefan Hacker, and Thorsten Strufe. Measuring freenet in the wild: Censorship-resilience under observation. In *Privacy Enhancing Technologies*, Lecture Notes in Computer Science, pages 263–282. 2014.
44. Juan Pablo Timpanaro, Isabelle Chrisment, and Olivier Festor. Monitoring the i2p network. http://hal.inria.fr/inria-00632259/PDF/TMA2012-LNCS.pdf,2011.
45. Christo Wilson, Bryce Boe, Alessandra Sala, Krishna P.N. Puttaswamy, and Ben Y. Zhao. User interactions in social networks and their implications. In *Proceedings of the 4th ACM European Conference on Computer Systems*, EuroSys '09, pages 205–218, 2009.
46. Bassam Zantout and Ramzi Haraty. I2p data communication system. In *Proceedings of The Tenth International Conference on Networks*, ICN 2011.
47. Ye Zhu, Xinwen Fu, Bryan Graham, Riccardo Bettati, and Wei Zhao. On flow correlation attacks and countermeasures in mix networks. In *Proceedings of the 4th International Conference on Privacy Enhancing Technologies*, PET'04, pages 207–225, 2005.

Erratum to:

Integrating Cyber-D&D into Adversary Modeling for Active Cyber Defense

Frank J. Stech, Kristin E. Heckman, and Blake E. Strom

© Springer International Publishing Switzerland 2016
S. Jajodia et al. (eds.), *Cyber Deception*
DOI 10.1007/978-3-319-32699-3

DOI 10.1007/978-3-319-32699-3_13

The original version of this book was inadvertently published with an incorrect Table 3 in Chapter 1. The correct Table is as follows:

Table 3 MITRE ATT&CK matrix™—overview of tactics and techniques described in the ATT&CK model

Persistence	Privilege Escalation	Defense Evasion	Credential Access	Host Enumeration	Lateral Movement	Execution	C2	Exfiltration
Legitimate Credentials		Binary Padding	Credential Dumping	Account enumeration	Application deployment software	Command Line	Commonly used port	Automated or scripted exfiltration
Accessibility Features		DLL Side-Loading	Credentials in Files	File system enumeration	Exploitation	File Access	Comm through removable media	Data compressed
AddMonitor						PowerShell		
DLL Search Order Hijack		Disabling Security Tools	Network Sniffing	Group permission enumeration	Vulnerability	Process Hollowing	Custom application layer protocol	Data size limits
Edit Default File Handlers					Logon scripts			Data staged
New Service		File System Logical Offsets	User Interaction	Local network connection enumeration	Pass the hash	Registry		
Path Interception					Pass the ticket	Rundll32		
Scheduled Task						Scheduled Task	Custom encryption cipher	Exfil over C2 channel
Service File Permission Weakness		Process Hollowing			Peer connections			Exfil over alternate channel to C2 network
Shortcut Modification					Remote Desktop Protocol	Service Manipulation	Data obfuscation	
BIOS	Bypass UAC			Local networking enumeration		Third Party Software	Fallback channels	Exfil over other network medium
Hypervisor Rootkit	DLL Injection	Indicator blocking on host			Windows management instrumentation	Multiband comm	
Logon Scripts	Exploitation of Vulnerability	Indicator removal from host tools		Operating system enumeration	Windows remote management	Multilayer encryption	Exfil over physical medium
Master Boot Record		Indicator removal from host		Owner/User enumeration	Remote Services	Standard app layer protocol	
Mod. Exist'g Service		Masquerad-ing		Process enumeration	Replication through removable media		From local system
Registry Run Keys		NTFS Extended Attributes		Security software enumeration	Shared webroot	Standard non-app layer protocol	From network resource
Serv. Reg. Perm. Weakness		Obfuscated Payload		Service enumeration	Taint shared content	Standard encryption cipher	From removable media
Windows Mgmt Instr. Event Subsc.		Rootkit		Window enumeration	Windows admin shares	Uncommonly used port	Scheduled transfer
Winlogon Helper DLL		Rundll32					
		Scripting					
		Software Packing					

The online version of the original chapter can be found at
http://dx.doi.org/10.1007/978-3-319-32699-3_1

© Springer International Publishing Switzerland 2016   E1
S. Jajodia et al. (eds.), *Cyber Deception*, DOI 10.1007/978-3-319-32699-3_13

Printed in the United States
By Bookmasters